Statistical Modeling, Analysis and Management of Fuzzy Data

Studies in Fuzziness and Soft Computing

Further volumes of this series can be found at our homepage.

Vol. 65. E. Orłowska and A. Szalas (Eds.)
Relational Methods for Computer Science Applications, 2001
ISBN 3-7908-1365-6

Vol. 66. R.J. Howlett and L.C. Jain (Eds.)
Radial Basis Function Networks 1, 2001
ISBN 3-7908-1367-2

Vol. 67. R.J. Howlett and L.C. Jain (Eds.)
Radial Basis Function Networks 2, 2001
ISBN 3-7908-1368-0

Vol. 68. A. Kandel, M. Last and H. Bunke (Eds.)
Data Mining and Computational Intelligence, 2001
ISBN 3-7908-1371-0

Vol. 69. A. Piegat
Fuzzy Modeling and Control, 2001
ISBN 3-7908-1385-0

Vol. 70. W. Pedrycz (Ed.)
Granular Computing, 2001
ISBN 3-7908-1387-7

Vol. 71. K. Leiviskä (Ed.)
Industrial Applications of Soft Computing, 2001
ISBN 3-7908-1388-5

Vol. 72. M. Mareš
Fuzzy Cooperative Games, 2001
ISBN 3-7908-1392-3

Vol. 73. Y. Yoshida (Ed.)
Dynamical Aspects in Fuzzy Decision, 2001
ISBN 3-7908-1397-4

Vol. 74. H.-N. Teodorescu, L.C. Jain and A. Kandel (Eds.)
Hardware Implementation of Intelligent Systems, 2001
ISBN 3-7908-1399-0

ol. 75. V. Loia and S. Sessa (Eds.)
Soft Computing Agents, 2001
ISBN 3-7908-1404-0

Vol. 76. D. Ruan, J. Kacprzyk and M. Fedrizzi (Eds.)
Soft Computing for Risk Evaluation and Management, 2001
ISBN 3-7908-1406-7

Vol. 77. W. Liu
Propositional, Probabilistic and Evidential Reasoning, 2001
ISBN 3-7908-1414-8

Vol. 78. U. Seiffert and L.C. Jain (Eds.)
Self-Organizing Neural Networks, 2002
ISBN 3-7908-1417-2

Vol. 79. A. Osyczka
Evolutionary Algorithms for Single and Multicriteria Design Optimization, 2002
ISBN 3-7908-1418-0

Vol. 80. P. Wong, F. Aminzadeh and M. Nikravesh (Eds.)
Soft Computing for Reservoir Characterization and Modeling, 2002
ISBN 3-7908-1421-0

Vol. 81. V. Dimitrov and V. Korotkich (Eds.)
Fuzzy Logic, 2002
ISBN 3-7908-1425-3

Vol. 82. Ch. Carlsson and R. Fullér
Fuzzy Reasoning in Decision Making and Optimization, 2002
ISBN 3-7908-1428-8

Vol. 83. S. Barro and R. Marín (Eds.)
Fuzzy Logic in Medicine, 2002
ISBN 3-7908-1429-6

Vol. 84. L.C. Jain and J. Kacprzyk (Eds.)
New Learning Paradigms in Soft Computing, 2002
ISBN 3-7908-1436-9

Vol. 85. D. Rutkowska
Neuro-Fuzzy Architectures and Hybrid Learning, 2002
ISBN 3-7908-1438-5

Vol. 86. Marian B. Gorzałczany
Computational Intelligence Systems and Applications, 2002
ISBN 3-7908-1439-3

Carlo Bertoluzza
María Á. Gil
Dan A. Ralescu
Editors

Statistical Modeling, Analysis and Management of Fuzzy Data

With 29 Figures
and 9 Tables

Springer-Verlag Berlin Heidelberg GmbH

Professor Carlo Bertoluzza
Università degli Studi di Pavia
Dipartimento di Informatica e Sistemistica
Via Ferrata 1
27100 Pavia
Italy
carlo.bertoluzza@unipv.it

Professor María-Ángeles Gil
Universidad de Oviedo
Departamento de Estadística e I.O. y D.M.
C/Calvo Sotelo, s/n
33007 Oviedo
Spain
angeles@pinon.ccu.uniovi.es

Professor Dan A. Ralescu
University of Cincinnati
Department of Mathematical Sciences
Cincinnati, OH 45221-0025
USA
Dan.Ralescu@math.uc.edu

DOI 10.1007/978-3-7908-1800-0

Cataloging-in-Publication Data applied for
Die Deutsche Bibliothek – CIP-Einheitsaufnahme
Statistical modeling, analysis and management of fuzzy data: with 9 tables / Carlo Bertoluzza ... (ed.). – Heidelberg; New York: Physica-Verl., 2002
(Studies in fuzziness and soft computing; Vol. 87)

Originally published by Physica-Verlag Heidelberg New York in 2002
MyCopy version of the original edition 2002

Hardcover Design: Erich Kirchner, Heidelberg
www.springer.com/mycopy

Foreword

"Statistical Modeling, Analysis and Management of Fuzzy Data," or SMFD for short, is an important contribution to a better understanding of a basic issue –an issue which has been controversial, and still is though to a lesser degree. In substance, the issue is: are fuzziness and randomness distinct or coextensive facets of uncertainty? Are the theories of fuzziness and randomness competitive or complementary? In SMFD, these and related issues are addressed with rigor, authority and insight by prominent contributors drawn, in the main, from probability theory, fuzzy set theory and data analysis communities.

First, a historical perspective. The almost simultaneous births –close to half a century ago– of statistically-based information theory and cybernetics were two major events which marked the beginning of the steep ascent of probability theory and statistics in visibility, influence and importance. I was a student when information theory and cybernetics were born, and what is etched in my memory are the fascinating lectures by Shannon and Wiener in which they sketched their visions of the coming era of machine intelligence and automation of reasoning and decision processes. What I heard in those lectures inspired one of my first papers (1950) "An Extension of Wiener's Theory of Prediction," and led to my life-long interest in probability theory and its applications to information processing, decision analysis and control. At Columbia University, where I taught before moving to Berkeley, I had a close relationship with the Department of Mathematical Statistics, and especially with the late Herbert Robbins, a brilliant mathematician who was a dominant figure in probability theory and its applications.

With the passage of time, I began to realize that there is a need for differentiation between fuzziness, randomness and vagueness. This realization was the genesis of my 1965 paper on fuzzy sets –a paper which was followed in 1968 by a paper in which fuzzy set theory was linked to probability theory through the concept of a fuzzy event.

Shortly after the publication of my first paper on fuzzy sets, it was noted by Loginov, and subsequently by Orlov, Goodman, Wang and others, that a fuzzy set may be generated by a random set. This is a constructive observation which is dicussed in detail in SMDF. It does not imply, however, that fuzziness and randomness are merely different labels for the same phenomenon or that fuzzy set theory (FST) is subsumed by probability theory (PT) or vice-versa. Many of the misconceptions about the relationships between fuzzy set theory and probability theory center on these issues.

Among those who take a skeptical view of fuzzy set theory there are some who claim that any problem that can be solved through the use of fuzzy set theory can be solved equally well or better through the use of probabilty theory. To me this contention is a manifestation of a lack of familiarity with

FST. Two simple problems whose solutions lie beyond the reach of PT will suffice to refute this contention.

The first is what I call the Robert example. A simple version of this example is the following. Suppose that I know that (*a*) usually Robert leaves his office at about 5:30 pm; and (*b*) usually it takes him about thirty minutes to get home by car. What is the probability that Robert is home at 6:30 pm? What is the earliest time at which the probability that Robert is home is high?

The second example is the Box problem. A box contains about ten balls. Most are large and a few are small. Most large balls are heavy and most small balls are light. What is the total weight of the balls in the box? What is the probability that a ball drawn at random is neither large nor small? What is the probability that a ball drawn at random is neither heavy nor light?

A basic reason why PT cannot deal with problems of this kind is that PT does not have the capability to operate on perceptions expressed in a natural language. This incapability is one of the most serious limitations of PT, since much of the information on which decisions are based on real-world settings is a mixture of measurements and perceptions.

In a series of recent papers, I have outlined how this capability can be added to PT through the use of fuzzy-logic-based computational theory of perceptions (CTP). This involves three stages of generalization of PT.

The first stage, labeled f-generalization, adds to PT the capability to compute with fuzzy probabilities, fuzzy events and fuzzy relations. The f-generalized version of PT is denoted as PT+.

The second stage, labeled f.g-generalization, adds to PT+ the capability to deal with granulated variables, distributions and relations, with the understanding that a granule is a clump of values drawn together by indistinguishability, similarity, proximity or functionality. In this sense, granular computing (GrC) is a fuzzy-logic-based collection of concepts and techniques in which the objects of computation are granules defined by so-called generalized constraints. The f.g-generalization of PT+ transforms PT+ into PT++.

The third stage of generalization, denoted as nl-generalization, adds to PT++ the capability to operate on propositions drawn from a natural language. Assuming that perceptions are described in a natural language, this added capability transforms PT++ into what is called perception-based probability theory, denoted as PTp.

The capability of PTp to deal with real-world problems is far greater than that of PT. This is the reason why contentions to the effect that anything that can be done with FST can be done equally well or better with PT, are manifestations of a lack of understanding of the tools that are needed to operate on perception-based information.

Viewed in the perspective of generalization of PT, SMFD addresses the basic concepts, issues and problems which relate to PT+ and, to a lesser extent, to PT++. In an important way, SMFD lays the groundwork for PTp but stops short of entering the still largely unexplored domain of representation of perception-based information in a natural language. This is a major task that entails a far-reaching paradigm shift in probability theory –a shift from manipulation of measurements to manipulation of perceptions.

SMFD presents a wealth of information and deep insights into the basic issues which arise in dealing with uncertainty and imprecision on a high level of rigor and mathematical sophistication. In addition, SMFD describes techniques, especially in the realm of fuzzy data analysis, which are of high importance in practical applications.

The editors of SMFD, Professors Bertoluzza, Ralescu and Gil have produced a serious work which commands attention and respect. The editors, the contributors, the Series editor, Professor Kacprzyk, and the publisher, Physica-Verlag, deserve our thanks and congratulations.

Computer Science Division
University of California
Berkeley, California

Lotfi A. Zadeh

May 2001

Preface

In 1968, three years after his seminal work introducing Fuzzy Sets, Professor Zadeh published a paper in which the probability of fuzzy events associated with a random experiment was presented. He wanted then to point out, for the first time, that (as he often has asserted) Probability Theory and Fuzzy Logic are complementary rather than competitive ways to deal formally with uncertainty.

Fuzziness and randomness are present in many aspects of real-life. As Karl Popper indicated "... Both, precision and certainty, are false ideals. They are impossibe to attain, ..., it is always undesirabe to make an effort to increase precision for its own sake -specially linguistic precision- since this usually leads to lack of clarity, ...: one should never try to be more precise than the problem situation demands."

Fuzziness and randomness often arise "combined" in practical situations. As a "real-life" example, we can refer to a visit we made to Professor Zadeh a long time ago in his office at the University of California at Berkeley. In this visit we found two messages on Zadeh's office door revealing that he is a real practitioner of his theory. One of the messages said "I am in today; if there is no answer when you knock on the door, I am out *temporarily*", and the other one "Will be back *later* in the afternoon". As we did not get an answer after knocking at the office door, probabilistic uncertainty (due to randomness) appears because we ignored what the last message on his door was, whereas fuzzy impecision (fuzziness) was a consequence of the ill-definition of terms used in each message.

This book presents some views and approaches on the connection between Fuzzy Set Theory and Statistics/Probability Theory, in which several models and solutions to some problems are included. Furthermore, it also tries to trigger debates on a subject for which only some answers have been given, although in the last years the topic is receiving more research oriented attention.

The book is divided into four parts covering different aspects, namely: fuzziness and randomness connections; fuzzy-valued random elements; possibility, probability and fuzzy measures; Statistics and fuzzy data analysis.

Part 1 presents formal connections between fuzziness and randomness, by discussing on the random representation of fuzzy concepts and some of the implications of this representation.

Part 2 contains six contributions concerning fuzzy-valued random elements. The two first papers refer to some generalized families of variation measures for these random elements, one of them extending classical variance of a random variable, and the other one extending inequality indices of positive random variables. Although analyses of these measures are mainly theoretical or descriptive, some inferential statistical conclusions are also obtained. The other four contributions in Part 2 regard probabilistic studies of fuzzy-valued random elements, like limit theorems for fuzzy random variables (in the supremum metric sense) and for fuzzy-valued martingales, submartingales and supermartingales (in the graph convergence sense), a Korovkin-type approximation theorem for fuzzy random variables, and a differentiability notion for fuzzy-valued mappings where connections with previous ones are examined.

Part 3 includes five contributions. The first one refers to the introduction and analysis of an index associated with a fuzzy set, which is its average level defined in terms of the Kudo-Aumann integral of a random set. The other four contributions concern different types of measures. Thus, the concept of probability induced by a random variable is extended to the one related to set- and fuzzy set-valued mappings, the extension being connected with second order possibility measures. On the other hand, a theorem extending measures from a so-called meet-system to the σ-field generated by this system is given as a basis to develop falling measures representation theorems. Furthermore, convex families of measures are studied and their elements are represented as sums of primitive measures, allowing us to get conclusions on the algebraic structure of fuzzy measures and to extend some classical probabilistic results. Finally, a method to classify statistical classes based on level sets is considered, and these classes are generalized and their connection with possibility theory is discussed.

In Part 4 several approaches on statistical problems involving fuzzy elements are gathered. The first contribution in this part refers to a theory modeling basic notions in one-dimensional Statistics concerning fuzzy data. The problem of testing fuzzy hypotheses from fuzzy data is also examined in two papers in which a method based on the necessity index of strict dominance is introduced, the concept of the p-value is generalized to the case of fuzzy data, and indices of the possibility and necessity of dominance are used to test hypotheses. Finally, the three last papers are focussed on different models and solutions of regression analysis problems with fuzzy data/parameters, and they provide us with an overview on most of the common ways to deal with this analysis in a fuzzy setting.

We want to express our sincere thanks to Professor Janusz Kacprzyk for his kind invitation to prepare this book, and for his encouragement and support at every moment. We wish also to thank all who have contributed with their papers to the book (especially, those sending their contributions a long time ago and being very patient in waiting for the book to be completed and edited). We are very grateful to Springer-Verlag, and in particular to Dr. Martina Bihn, because of the help and understanding we have received in different respects in editing the book. Last, but not least, to Professor Lotfi A. Zadeh for being our permanent source of inspiration.

University of Pavia, Italy — Carlo Bertoluzza
University of Oviedo, Spain — María Ángeles Gil
University of Cincinnati, Ohio — Dan A. Ralescu

May 2001

Contents

Part 1

FUZZINESS AND RANDOMNESS

Fuzziness and randomness

Irwin R. Goodman[1] and Hung T. Nguyen[2]

[1] Code D4223, SPAWARSYSCEN, San Diego, CA 92152-7446-USA
[2] Department of Mathematical Sciences, New Mexico State University, Las Cruces, NM 88003-8001-USA

Abstract. This paper presents a survey and some new results of the mathematical investigation into formal connections between two types of uncertainty: fuzziness and randomness.

1 Introduction

In his pioneering work on random elements in metric spaces, Fréchet (1948) pointed out that besides standard random objects (such as points, vectors, functions), nature, science and technology offer other random elements, some of which *cannot be described mathematically*. For example, to each population of humans, chosen at random, one might be interested in its "morality," its "political spirit;" to each town chosen at random, one might be interested in its "form," its "beauty,".... It is clear that when we use natural language to describe properties of things, we often run into such situations. A property p on a collection of objects Ω defines a subset A of Ω, namely those elements of Ω which possess the property p, *provided that p is crisp*, in the sense that each element of Ω has the property or does not have it. If a property p stands for "tall" in a human population, it is not clear how p determines a subset of Ω. Concepts such as "tall" are called *fuzzy concepts.* Thus, examples of random elements mentioned by Fréchet are *fuzzy concepts chosen at random.* In 1965, Zadeh (1965) proposed a mathematical theory for modeling of fuzzy concepts, in which fuzziness is a matter of degree and is described by membership functions.

While it appears that fuzziness and randomness are two distinct types of uncertainty, it is of interest to find out whether or not there exist formal relations between them. This is similar to Potential Theory and Markov Processes where a formal connection between them is beneficial : Potential Theory provides powerful tools for studying Markov processes, Markov processes provide probability interpretations for various concepts in Potential Theory. This dual aspect is also reminiscent in many other areas of mathematics : Fourier or Laplace tranforms are considered as appropriate according to the need for a time domain analysis or a frequency domain analysis; or, as an analogue with complex analyis : using complex variable $z = x + iy$ such as in residue representation theorems and complex Taylor series expansions, but alternatively, at times, considering the real part (x) and the imaginary part (y) such as in the Cauchy-Riemann criterion and in harmonic analysis.

In subsequent sections we will investigate formal relationships between fuzziness and randomness and explore their consequences.

2 Generalities on fuzzy set theory

A fuzzy (sub) set A of a set U is a map $A : U \to [0,1]$. Note that we use the same symbol A to denote the fuzzy concept and its mathematical modeling by the membership function. For $u \in U$, $A(u)$ stands for the degree to which u is compatible with the meaning of A. As generalized sets (where membership functions generalize indicator functions of ordinary (crisp) sets), operations on them form the so-called *fuzzy logic*. Basically these operations are defined via operations called t-norms (as copulas and co-copulas), t-conorms, negations, implications. When the set of membership values is the unit interval $[0,1]$, the associated fuzzy logic is termed *first order fuzzy logic.* If $[0,1]$ is replaced by a complete lattice L, then we talk about L- fuzzy sets. For example, L could be the lattice of all $[0,1]$-valued functions on U (i.e. first order fuzzy sets) with the natural partial order between functions, and in this case, the associated logic is called *second order fuzzy logic.* As an intermediate level, *interval-valued fuzzy logic* could be considered. For background on mathematical foundations of fuzzy logic, see e.g. Nguye and Walker (1996). In view of the popular *interval computations*, sometimes the study of membership functions can be carried out by considering their *α-level sets*, i.e.,

$$A_\alpha = \{u \in U : A(u) \geq \alpha\} \quad for \quad \alpha \in [0,1]$$

Perhaps, for philosophical or application reasons, the question of how to obtain membership functions for fuzzy concepts is essential. Although membership functions are assigned, in general, subjectively, say, by experts in specific domains, it is useful to understand some factors used by experts in this process. We mention here one explanation due to Orlowski (1994) in the context of Decision Theory. In modeling a given fuzzy concept W, a set-valued map S, defined on some space Ω with values in the power set $\mathcal{P}(U)$ of U is available. Also, a set-function μ defined on $\mathcal{P}(\Omega)$ with values in $[0,1]$ is specified by experts. This set-function μ satisfies simply the conditions : $\mu(\varnothing) = 0$ and the monotonicity : if A, B are (crisp) subsets of Ω such that $A \subseteq B$ then $\mu(A) \leq \mu(B)$. Set-functions such as these are called *fuzzy measures* by Sugeno (1974), see also Nguyen and Prasad (1998). Then the membership function W is obtained as $W(u) = \mu\{\omega : u \in S(\omega)\}$. Early in 1976, Goodman showed that any membership function W can be represented as $W(u) = P\{\omega : u \in S(\omega)\}$, where P is a *probability measure* on some σ-field of Ω, and S is a *random set* in U.

3 Generalities on random sets

Roughly speaking, a random set is a set chosen at random. Formally, by a random set we mean a set-valued random variable. Specifically, let $\mathbb{E}$ be a subset of $\mathcal{P}(U)$ and $\mathcal{E}$ be a σ-field of $\mathbb{E}$, and $(\Omega, \mathcal{A})$ a measurable space. Then a *random set* S in U is a map from $\Omega \to \mathbb{E}$ such that $S^{-1}(\mathcal{E}) \subseteq \mathcal{A}$. If P is a probability on $(\Omega, \mathcal{A})$, then the probability law of S is the probability measure PS^{-1} on $(\mathbb{E}, \mathcal{E})$. In applications, we are mainly concerned with locally compact (Hausdorff) topological spaces such as euclidean spaces $U = \mathbb{R}^k$. When $\mathbb{E}$ is the class of all closed set $\mathcal{F}(\mathbb{R}^k)$ of $\mathbb{R}^k$ and $\mathcal{E}$ is the Borel σ-field on $\mathcal{F}(\mathbb{R}^k)$ where $\mathcal{F}(\mathbb{R}^k)$ is topologized according to the so-called *hit-and-miss topology* (see, e.g. Matheron, 1975), we talk about *random closed sets* (see also Goutsias *et al.*, 1997). Probability laws of random closed sets can be characterized by their *capacity functionals* (counter-parts of cumulative distribution functions of random vectors), namely by set-functions T defined on the class of compact sets $\mathcal{K}(\mathbb{R}^k)$ with values in $[0,1]$ such that:

(i) If K_n is a decreasing sequence in $\mathcal{K}$ then $T(K_n) \searrow T(\cap K_n)$,

(ii) T is *alternating of infinite order* on $\mathcal{K}$, i.e. T is monotone increasing and for any $K_1, K_2, \ldots, K_n$ in $\mathcal{K}$, $n \geq 2$,

$$T(\bigcap_{i=1,\ldots,n} K_i) \leq \sum_{\varnothing \neq I \subseteq \{1,2,\ldots,n\}} (-1)^{|I|+1} T(\bigcup_{i \in I} K_i)$$

where $|I|$ denotes the cardinality of the set I. This important result is referred to as *Choquet Theorem*, see,e.g. Matheron (1975).

The capacity functional T of the random closed set S is defined as

$$T(K) = P\{\omega : S(\omega) \cap K \neq \emptyset \}.$$

T induces the *covering function* (or *the one-point-coverage function*) of S :

$$\pi : U \to [0,1], \qquad \pi(u) = T(\{u\}) = P\{\omega : u \in S(\omega)\}$$

Obviously, the knowledge of the covering function π is not enough to specify the capacity functional T. However, various quantities related to S can be computed in terms of π alone. For example, let λ denote the Lebesgue measure on $\mathbb{R}^k$, or a σ-finite measure. Then, $\lambda(S)$ is a non-negative random variable whose expected value is given by

$$E\lambda(S) = \int_{\mathbb{R}^k} \pi(x) d\lambda(x).$$

This formula is refered to as *Robbins' Formula* (1944).

4 Random set representation of membership functions

Let A be a fuzzy set of U. The α-level sets of A can be obtained by randomizing the level α. Suppose we do that by choosing α uniformly in $[0,1]$. That is, we view α as a random variable, defined on some probability space $(\Omega, \mathcal{A}, P)$ with values in $[0,1]$ such that

$$P(\omega : \alpha(\omega) \leq x) = x \quad for \quad x \in [0,1].$$

Then α induces a random set S in U as $S(\omega) = \{u \in U \colon A(u) \geq \alpha(\omega)\}$ which is simply the randomized level set $A_{\alpha(\omega)}$. Now the covering function of this *canonical* random set, denoted as S_A, is

$$\pi(u) = P(\omega : u \in S_A) = P(\omega : A(u) \geq \alpha(\omega)) = A(u).$$

In other words, S_A is a random set in U whose covering function coincides with the membership function of the fuzzy concept A. It is important to note that we start with the function $A(.)$ and obtain a random set S_A such that $A(u) = P(\omega : u \in S_A(\omega))$. Thus the class $\mathcal{S}(A)$ of random sets in U whose covering functions are identical to $A(.)$ is not empty. Elements in $\mathcal{S}(A)$ are considered as equivalent in the sense that they have the same covering function. Of course, the covering function of any random set in U defines formally a fuzzy subset of U.

In the case where $U = \mathbb{R}$ and $A(.)$ is upper semi-continuous (usc), i.e. all α-level sets A_α are closed (for example, when A is a *fuzzy number*), S_A is a random closed set in $\mathbb{R}$ whose capacity functional is precisely

$$T(K) = \sup\{A(u) : u \in K\} \quad for \quad K \in \mathcal{K}(\mathbb{R}) \tag{1}$$

Indeed,

$$\begin{aligned} P(\omega \ : \ S_A(\omega) \cap K \neq \varnothing) &= P(\omega : \alpha(\omega) \leq A(x) \text{ for some } x \in K\ ) \\ &= P(\omega : \alpha(\omega) \leq \sup\{A(x) : x \in K\}) = \sup\{A(x) : x \in K\} \end{aligned}$$

This is a probabilistic proof of the fact that (1) is capacity functional of some random set. A direct proof of this fact is interesting and of independent interest.

(i) Let $K_n \in \mathcal{K}(\mathbb{R}^k)$ with $K_n \searrow K \in \mathcal{K}$. We have $T(K) \leq \inf T(K_n)$ $= a$, say. Let $\varepsilon > 0$ and $B_n = A_{a-\varepsilon} \cap K_n$, $n \geq 1$. Since $A(.)$ is u.s.c., the B_n ' s are closed. Also the B_n 's are all contained in the compact K_1, and hence (since each $B_n \neq \emptyset$) $\cap B_n \neq \varnothing$. Now, $B_n \subseteq K_n$, we have $T(K) \geq \sup$ $\{A(x) : x \in \cap B_n\ \} \geq a - \epsilon$.

(ii) The fact that a set-function like $T(K) = \sup\ \{A(x) : x \in K\}$ is alternating of infinite order is a consequence of the following general result.

Let $\mathcal{C}$ be a class of subsets of some space Ω, such that $\mathcal{C}$ contains the empty set $\varnothing$ and is stable under finite intersections and unions. Let $T : \mathcal{C} \rightarrow$ $[0, \infty)$ be *maxitive*, i.e.

$$\text{for any } B, C \text{ in } \mathcal{C}, \quad T(B \cup C) = \max\{T(B), T(C)\}$$

Then, T is alternating of infinite order. The proof goes as follows.

Let $B_i \in \mathcal{C}, i = 1, 2, \ldots, n$. Then $T(\cup_1^n B_i) = \max \{T(B_i), i = 1, 2, \ldots, n\}$. Without loss of generality, assume

$$0 \leq b_n = T(B_n) \leq \ldots \leq b_2 = T(B_2) \leq T(B_1) = b_1.$$

For $k \in \{1, 2, \ldots, n\}$, let $\mathcal{T}(k) = \{I \subseteq \{1, 2, \ldots, n\} : |I| = k\}$, and for $I \in \mathcal{T}(k)$, and $i = 1, 2, \ldots, n-k+1$, let $\mathcal{T}_i(k) = \{I \in \mathcal{T}(k) : \min I = i\}$. Then $T(\cup_{j \in I} B_j) = b_j$ for every $I \in \mathcal{T}_i(k)$, and $i = 1, 2, \ldots, n-k+1$. Next

$$\begin{aligned} \sum_{I \in \mathcal{T}(k)} T(\cup_{i \in I}) &= \sum_{i=1}^{n-k+1} \sum_{I \in \mathcal{T}(k)} T(\cup_{j \in I} B_j) \\ &= \sum_{i=1}^{n-k+1} \sum_{I \in \mathcal{T}(k)} b_i = \sum_{i=1}^{n-k+1} \tbinom{n-i}{k-1} b_i. \end{aligned}$$

Thus,

$$\begin{aligned} &\sum_{\{I : \varnothing \neq I \subseteq \{1,2,\ldots,n\}\}} (-1)^{|I|+1} T(\cup_{i \in I} B_i) \\ &= \sum_{k=1}^{n} (-1)^{k+1} \sum_{I \in \mathcal{T}(k)} T(\cup_{i \in I} B_i) \end{aligned}$$

$$= \sum_{k=1}^{n} (-1)^{k+1} \sum_{i=1}^{n-k+1} \tbinom{n-i}{k-1} b_i = \sum_{i=1}^{n} \left[\sum_{k=0}^{n-i} \tbinom{n-i}{k} (-1)^k \right] b_i.$$

But

$$\sum_{k=0}^{n-i} \tbinom{n-i}{k} (-1)^k = 0 \text{ for every } i = 1, 2, \ldots, n-1$$

Hence

$$\sum_{\{I : \varnothing \neq I \subseteq \{1,2,\ldots,n\}\}} (-1)^{|I|+1} T(\cup_{i \in I}) = b_n = T(B_n) \geq T(\cap_{k=1}^n B_k).$$

The *range* $\mathcal{R}(S_A)$ of the *canonical random set* S_A associated with the fuzzy set A is *nested* in the sense that $\mathcal{R}(S_A) = \{A_t : 0 \leq t \leq 1\}$ is totally ordered (by set-inclusion). It is clear that $P(\omega : S_A(\omega) \in \mathcal{R}(S_A)) = 1$. Moreover, for any $A_t \in \mathcal{R}(S_A)$,

$$\{\omega : A_t \subseteq S_A(\omega)\} = \{\omega : A(A_t) \subseteq [\alpha(\omega), 1]\} = \{\omega : \alpha(\omega) \leq \inf A(A_t)\} \in \mathcal{A}$$

where, as usual, $A(A_t)$ is the image set of the set A_t by the function $A(.)$, i.e. $\{A(x) : x \in A_t\}$. In fact,

$$\forall B \subseteq U, \{\omega : B \subseteq S_A(\omega)\} = \{\omega : \alpha(\omega) \leq inf A(B)\} \in \mathcal{A}.$$

This leads to the following formal definition:

Definition 1. Let $S : (\Omega, \mathcal{A}, P) \to (\mathbb{E}, \mathcal{E})$ be a random set in U. We say that S is *nested* if

(*i*) $\mathcal{R}(S)$ is a totally ordered (by set-inclusion) family of subsets of U,
(*ii*) $\forall B \in \mathcal{R}(S)$, $\{\omega : B \subseteq S(\omega)\} \in \mathcal{A}$.

It turns out that S_A is the only nested random set representing A.

Theorem 1. *Let $S : (\Omega, \mathcal{A}, P) \to (\mathbb{E}, \mathcal{E})$ be a random set in U, with covering function π. Is S is nested, then $S = S_\pi$ in distribution where S_π is the canonical random set representing the fuzzy set π.*

Proof. In view of Theorems 3 and 5 in Goodman and Nguyen (1985) (pp. 341-343), it suffices to show that

$$\forall B \in \mathcal{R}(S),\ P(\omega : B \subseteq S(\omega)) = P(\omega : B \subseteq S_\pi) \tag{2}$$

Let $B \in \mathcal{R}(S)$. If $B = \varnothing$, then (2) is clearly satisfied since both sides are equal to 1. Thus, consider the case where $B \neq \varnothing$. Let $\mathcal{W}(B) = \{D \in \mathcal{R}(S) : \varnothing \neq D \subset B\}$.

If $\mathcal{W}(B) = \varnothing$, then for any $x \in B$, we have, since $\mathcal{R}(S)$ is totally ordered

$$\{\omega : x \in S(\omega)\} = \{\omega : B \subseteq S(\omega)\}. \tag{3}$$

If $\mathcal{W}(B) \neq \varnothing$, then there exists some $x \in B \cap D'$ for any $D \in \mathcal{W}(B)$. For that x, (3) is clearly satisfied. Thus, whenever $B \neq \varnothing$, there is an $x \in B$ such that (3) holds.
Now

$$\begin{aligned} P(\omega\ :\ B \subseteq S_\pi(\omega)) &= P(\omega : \pi(B) \subseteq [\alpha(\omega), 1]\} \\ &= P(\omega : \inf \pi(B) \geq \alpha(\omega)\} = \inf \pi(B). \end{aligned}$$

Next,

$$\begin{aligned} P(\omega\ :\ B \subseteq S(\omega)) &= P(\omega : y \in S(\omega), \forall y \in B) \\ &\leq P(\omega : z \in S(\omega) \text{ for some } z \in B). \end{aligned}$$

Thus,

$$P(\omega : B \subseteq S(\omega)) \leq \inf\{P(\omega : y \in S(\omega)), y \in B\} = \inf \pi(B)$$

But, in view of (3), we have

$$\begin{aligned} \inf \pi(B) &= \inf\{P(\omega : y \in S(\omega)) : y \in B\} \\ &\leq P(\omega : x \in S(\omega)) = P(\omega : B \subseteq S(\omega)). \end{aligned}$$

Therefore (2) is proved. ■

Remark 1. Nested random set representations might not be appropriate for *multimodal membership functions.* For example, the fuzzy concept "happy" can have a multimodal membership function due to happiness periods in human childhood, not adolescence, again in middle age, but not in old age, etc... On the other hand, membership functions of fuzzy concepts can arise durectly from covering functions of arbitrary random sets, and as such, they might have quite irregular shapes. As an extreme case, consider a discrete random variable X, sat taking values in the set of integers $\mathbb{N}$ with probability density function f. X is a special case of a random set $S = \{X\}$, a singleton random set. Its covering function $\pi(n) = P\{\omega : n \in S(\omega)\} = P\{\omega : X(\omega) = n\} = f(n)$. Obviously, the random set S which represents the fuzzy set π is not nested- quite the opposite, with all values (here, points in $\mathbb{N}$) being mutually disjoint.

Remark 2. If A is a fuzzy subset of U, then according to Zadeh, A induces a possibility measure on the power set $\mathcal{P}(U) : \mu(B) = \sup\{ A(u) : u \in B\}$. This set-function μ is a fuzzy measure and can be represented as a Choquet integral with respect to another fuzzy measure, namely $\nu(B) = 0$ or 1 according to B empty or not. Indeed, we have

$$\mu(B) = \int_o^\infty \nu\{u : (A(u) \geq t) \cap B\}dt \text{ for any } B \subseteq U$$

By analogy with standard Measure Theory, we can write $d\mu/d\nu = f$, and call f the Radon-Nikodym derivative of μ with respect to ν.

Remark 3. The problem of finding all possible random sets S which represent a given fuzzy set A of a (finite) set U will be studied in Section 6. In the next section, we are going to establish isomorphic and homomorphic-like relations between Zadeh fuzzy logic operators and ordinary set operators over canonical random sets.

5 Connections with fuzzy logic

Since canonical nested random sets representing fuzzy sets via covering functions are randomized level sets, the connections between operations on fuzzy sets and operations on (random) sets are essentially based upon results concerning α-level sets.

Let J be an arbitrary index set, and $A^{(j)}$, $j \in J$, be fuzzy subsets of U. Then $\bigwedge_{j\in J} A^{(j)}$, $\bigvee_{j\in J} A^{(j)}$ stand for fuzzy sets of U with membership functions $\inf\{ A^{(j)}(.) : j \in J \}$, $\sup\{ A^{(j)}(.) : j \in J \}$, respectively. For any $\alpha \in [0,1]$, we have:

$$\left(\bigwedge_{j\in J} A^{(j)}\right)_\alpha = \bigcap_{j\in J} A^{(j)}_\alpha$$

but in general,

$$\bigcup_{j\in J} A_{\alpha}^{(j)} \subseteq \left(\bigvee_{j\in J} A^{(j)}\right)_{\alpha}$$

with equality if $\forall x \in U$, $\sup\ \{A^{(j)}(x) : j \in J\}$ is attained on J. This fact is somewhat related to *Nguyen's Theorem* (1978) on level-sets of fuzzy sets defined via Zadeh's extension principle. See also Fuller and Keresztfalvi (1990), Bertoluzza and Bodini (1998). When J is finite, the above condition is satisfied and hence equality holds for $\vee$. Recall that the (fuzzy/as well as the ordinary) negation of A, denoted as A', has membership function $1-A(.)$. Obviously, $(A')_{\alpha} \neq (A_{\alpha})'$.

The probabilistic versions of the above are as follows. Let A and B be fuzzy sets of U. Viewing now α as a random variable uniformly distributed on $[0,1]$. We have, $\forall x \in U$:

$$\begin{aligned} P(\omega\ :\ x \in A_{\alpha(\omega)} \cap B_{\alpha(\omega)}) &= A(x) \wedge B(x) \\ &= P(\omega : x \in A_{\alpha(\omega)}) \wedge P(\omega : x \in B_{\alpha(\omega)}). \\ P(\omega\ :\ x \in A_{\alpha(\omega)} \cup B_{\alpha(\omega)}) &= A(x) \vee B(x) \\ &= P(\omega : x \in A_{\alpha(\omega)}) \vee P(\omega : x \in B_{\alpha(\omega)}). \end{aligned}$$

More generally, let J be an arbitrary index set and $A^{(j)}, j \in J$, be fuzzy subsets of U,then

$$\begin{aligned} \forall x \in U,\quad & P(\omega : x \in \cap_{j\in J} A^{(j)}_{\alpha(\omega)}) = \inf\{A^{(j)}(x) : j \in J\} \\ &= \inf\ \{P(\omega : x \in A^{(j)}_{\alpha(\omega)}) : j \in J\} \end{aligned}$$

Although only inclusion holds for $\cup$ in general, we always have

$$P(\omega : x \in \cup_{j\in J} A^{(j)}_{\alpha(\omega)}) = \sup\{A^{(j)}(x) : j \in J\} :$$

Indeed, the above equation holds since

$$\begin{aligned} \{\omega\ :\ x \in A^{(i)}_{\alpha(\omega)}\} &= \{\omega : A^{(i)}(x) \geq \alpha(\omega)\} \\ &\subseteq \{\omega : \alpha(\omega) \leq \sup(A^{(j)}(x) : j \in J)\}, \end{aligned}$$

it follows that

$$\begin{aligned} P(\omega\ :\ x \in \cup_{j\in J} A^{(j)}_{\alpha(\omega)}) &= P(\cup_{j\in J}\{\omega : x \in A^{(j)}_{\alpha(\omega)}\}) \\ &\leq P(\omega : x \in (\vee_{j\in J} A^{(j)})_{\alpha(\omega)}) = \sup\{A^{(j)}(x) : j \in J\}. \end{aligned}$$

On the other hand, for each j,

$$\{\omega : x \in A^{(j)}_{\alpha(\omega)}\} \subseteq \cup_{j\in J}\{\omega : x \in A^{(j)}_{\alpha(\omega)}\},$$

thus

$$\sup\{A^{(j)}(x) \ : \ j \in J\} = \sup\{P(\omega : x \in A^{(j)}_{\alpha(\omega)}) : j \in J\}$$
$$\leq P(\cup_{j\in J}\{\omega : x \in A^{(j)}_{\alpha(\omega)}\}).$$

As for negation, we note that, for each ω,

$$(A')_{\alpha(\omega)} = \{x \in U : 1 - A(x) \geq \alpha(\omega)\}$$
$$= \{x \in U : A(x) > 1 - \alpha(\omega)\}'$$

and

$$A'_{\alpha(\omega)} = \{x \in U : A'(x) > 1 - \alpha(\omega)\}.$$

Thus,

$$P\{\omega \ : \ x \in (A')_{\alpha(\omega)}\} = 1 - A(x)$$
$$= P\{\omega : x \in (A_{\alpha(\omega)})'\}, \quad \forall x \in U$$

despite the fact that $(A')_{\alpha(.)} \neq (A_{\alpha(.)})'$, although they are identically distributed. In summary, the map $A \to S_A$,between fuzzy sets of U and their nested canonical random sets, is not, in general, a "full" isomorphism with respect to logical operations "and," "or," and "not" but it is an "and"-"or" isomorphism. This is expected since the class of all fuzzy subsets of U equiped with Zadeh's logical operations (max, min, 1-(.)) is not boolean, but rather it is a browerian lattice or, equivalently, a (non-complemented) Stone lattice.

Other operations on fuzzy sets also have similar corresponding relations in terms of covering functions of canonical random sets. For example, the *cartesian product* of fuzzy sets $A^{(j)}$ of U_j, for $j \in J$, denoted as $\prod_{j\in J} A^{(j)}$, has membership function on $\prod_{j\in J} U_j$ given by $\inf\{A^{(j)}(u_j) : j \in J\}$ which is equal to

$$P\left\{\omega : (u_j, j \in J) \in \prod_{j\in J} A^{(j)}_{\alpha(\omega)}\right\} = P\left\{\omega : (u_j, j \in J) \in \left(\prod_{j\in J} A^{(j)}\right)_{\alpha(\omega)}\right\}.$$

The *cartesian sum* $\bigoplus_{j\in J} A^{(j)} = \left(\prod_{j\in J} A^{(j)\prime}\right)'$ has the following dual relations :

$$P\left\{\omega : (u_j, j \in J) \in \bigoplus_{j\in J} A^{(j)}_{\alpha(\omega)}\right\}$$

$$= P\left\{\omega : (u_j, j \in J) \in \left(\bigoplus_{j\in J} A^{(j)}\right)_{\alpha(\omega)}\right\} = \sup\{A^{(j)}(u_j) : j \in J\}.$$

Also, multivariate versions of covering functions can be considered. Replacing the uniformly distributed random variable $\alpha(.)$ on $[0,1]$ by a random vector, still denoted as $\alpha = (\alpha_1, \alpha_2, \ldots, \alpha_n) : \Omega \to [0,1]^n$ where each marginal component α_i is uniformly distributed on $[0,1]$. The (joint) cumulative distribution function of such a random vector is called an *n-copula* (see e.g., Schweizer and Sklar, 1983), that is a function $C : [0,1]^n \to [0,1]$ such that:

(*i*) C is *n-increasing*, i.e. for any box B in $[0,1]^n$ where $B = [a_1, e_1] \times \ldots \times [a_n, e_n]$, the quantity $\sum \operatorname{sgn}_B(\mathbf{x})C(\mathbf{x}) \geq 0$, where the summation is taken over all vertices $\mathbf{x} = (x_1, \ldots, x_n)$ of B (i.e. $\mathbf{x} \in [0,1]^n$ such that each x_i is equal to either a_i or e_i), and $\operatorname{sgn}_B(\mathbf{x}) = 1$ or -1 according to $x_i = a_i$ for an even number or for an odd number of i's.

(*ii*) C is *grounded*, i.e. $C(x_1, \ldots, x_n) = 0$ for all $(x_1, \ldots, x_n)$ such that $x_i = 0$ for at least one i.

(*iii*) Any (one-dimensional) margin C_i of C satisfies $C_i(x) = x$ for $x \in [0,1]$, where $C_i : [0,1] \to [0,1]$ is defined as $C_i(x) = C(1, \ldots, 1, x, 1, \ldots, 1)$, x being located at the ith position.

Copulas are closely related to *t-norms*: associative 2-copulas are t-norms, 2-increasing t-norms are 2-copulas. The (DeMorgan) *dual* (or *co-*) *copula* of an n-copula C is defined as $C^*(x_1, \ldots, x_n) = 1 - C(1 - x_1, \ldots, 1 - x_n)$. The basic *Sklar Theorem* is this. If F is the joint cumulative distribution function of the random vector $(X_1, \ldots, X_n)$ with marginal $F_j, j = 1, \ldots, n$, then there exists a unique n-copula C such that $F(x_1, \ldots, x_n) = C(F_1(x_1), \ldots, F_n(x_n))$ for all $x_j \in \mathbb{R}$. Conversely, any n-copula C together the one-dimensional F_j's define a legitimate joint distribution function this way. See Schweizer and Sklar (1983).

With the above notations, it can be verified that : Let $A^{(j)}$ be fuzzy sets of U_j, then

$$P\left\{\omega : (u_1, \ldots, u_n) \in (\prod_{j \in J} (A^{(j)})_{\alpha_j(\omega)})\right\}$$

$$= P(\cap_{i=1}^n \{\omega : u_i \in (A^{(i)})_{\alpha_i(\omega)}\}) = C(A^{(1)}(u_1), \ldots, A^{(n)}(u_n))$$

$$= P\left\{\omega : (u_1, \ldots, u_n) \in \bigoplus_{j \in J} (A^{(j)})\alpha_j(\omega)\right\}$$

$$= 1 - P\left\{\omega : (u_1, \ldots, u_n) \in \left(\prod_{j \in J} ((A^{(j)})\alpha_j(\omega)\right)\right\}$$

$$= P(\cup_{j \in J} \{\omega : u_j \in (A^{(j)})_{\alpha(\omega)}\})$$

$$= \sum_{\varnothing \neq I \subseteq \{1, \ldots, n\}} ((-1)^{|I|} C(A^{(j)}(u_j), j \in I)).$$

6 A solution for random set representation

Let A be a fuzzy subset of U. Let $\mathcal{S}(A)$ denote the class of all random sets on U having A as their *common* covering function. We are going to specify $\mathcal{S}(A)$ in the case where U is *finite.*

Without lost of generality, assume $U = \{1, 2, \ldots, n\}$. A random set S on U is characterized by its probability function $f : \mathcal{P}(U) \to [0, 1]$ where

$$f(B) = P(\omega : S(\omega) = B),$$

$\mathcal{P}(U)$ denoting the power set of U. If we let $V_i(\omega) = 1_{S(\omega)}(i)$ (indicator function of the set $S(\omega)$), $i \in U$, then V_i is a random variable taking values in $\{0, 1\}$ with

$$P(\omega : V_i(\omega) = 1) = P(\omega : i \in S(\omega)) = 1 - P(\omega : V_i(\omega) = 0)$$

The distribution of the random vector $V_S = (V_1, V_2, \ldots, V_n)$ is completely determined by that of S and vice versa. Indeed, for any $\mathbf{x} = (x_1, \ldots, x_n) \in [0, 1]^n$,

$$\begin{aligned} P(\omega \ : \ V_S(\omega) = \mathbf{x}) &= P\{\omega : (V_1(\omega), \ldots, V_n(\omega)) = \mathbf{x}\} \\ &= P(\omega : S(\omega) = B) \text{ where } B = \{i \in U : x_i = 1\} \end{aligned}$$

For any $S \in \mathcal{S}(A)$, the cumulative distribution function (cdf) of each V_i is

$$G_{A(i)}(x) = \begin{cases} 0, & \text{if } \ x < 0 \\ 1 - A(i), & \text{if } 0 \leq x < 1 \\ 1, & \text{if } \ 1 \leq x \end{cases}$$

If the fuzzy set A is given, then the marginal cdfs $G_{A(i)}, i = 1, \ldots, n$

$$F(x_1, \ldots, x_n) = C(G_{A(1)}(x_1), \ldots, G_{A(n)}(x_n))$$

where C is an *n-copula.*

Thus,

Theorem 2. *Let A be a fuzzy subset of a finite set U. Then all possible distributions for random sets S in $\mathcal{S}(A)$ are of the form $C(G_{A(u)}, u \in U)$ with C being any $|U|$-copula. In particular, the canonical nested random set $A_{\alpha(.)}$ corresponds to the choice of the copula $C(x_1, \ldots, x_{|U|}) = \min\{x_i : i = 1, \ldots, |U|\}$, $x_i \in [0, 1]$.*

Remark 4. The above result can be readily extended to the case of a finite number of fuzzy sets on finite domains. Specifically, let $A^{(j)}$ be fuzzy sets of U_j, $j \in J$ (a finite index set). Then the joint distribution of $(S_j, j \in J)$, where each $S_j \in \mathcal{S}(A^{(j)})$, is of the form $C(G_{A^{(j)}(x)}, x \in U_j, j \in J)$ where C is a $|\bigcup_{j \in J} (U_j \times \{j\})|$-copula

Moreover, we have the following homomorphic-like relations :

Theorem 3. *For any $u_j \in U_j$, $j \in J$,*

(i)

$$P(\cap_{j\in J}\{\omega : u_j \in S_j(\omega)\}) = \sum_{\emptyset \neq K \subseteq J} ((-1)^{|K|+1} C^*(A^{(j)}(u_j), j \in K))$$

where each S_j has distribution $C(G_{A^{(j)}(u)}, u \in U_j)$.

(ii)

$$P(\cup_{j\in J}\{\omega : u_j \in S_j(\omega\}) = C^*(A^{(j)}(u_j) : j \in J)$$

Proof. (*i*) Let V_j denote the indicator function of $\{\omega : u_j \in S_j(\omega)\}$. We have

$$P(\cap_{j\in J}\{\omega : V_j(\omega) = 1\}) = \sum_{K \subseteq J} ((-1)^{|K|} P(\cap_{j\in K}\{\omega : V_j = 0\}))$$

with the convention that when $K = \emptyset$, the corresponding term is 1. But for $K \neq \emptyset$,

$$\begin{aligned} P(\cap_{j\in K}\{\omega \ : \ V_j(\omega) = 0\}) &= C(P(\omega : V_j(\omega) \leq 0) : j \in K) \\ &= C(P(\omega : V_j(\omega) = 0) : j \in K) = C((1 - A^{(j)}(u_j)) : j \in K) \\ &= 1 - C^*(A^{(j)}(u_j) : j \in K). \end{aligned}$$

Thus,

$$\begin{aligned} P(\cap_{j\in J}\{\omega \ : \ u_j \in S_j(\omega)\}) &= P(\cap_{j\in J}\{\omega : V_j(\omega) = 1\}) \\ &= 1 + \sum_{\emptyset \neq K \subseteq J} ((-1)^{|K|}(1 - C^*(A^{(j)}(u)) : j \in K) \\ &= \sum_{\emptyset \neq K \subseteq J} ((-1)^{|K|+1} C^*(A^{(j)}(u_j) : j \in K). \end{aligned}$$

(*ii*) By DeMorgan dual:

$$\begin{aligned} P(\cup_{j\in J}\{\omega : u_j \in S_j(\omega)\}) &= 1 - P(\cap_{j\in J}\{\omega : u_j \in (S_j(\omega))'\}) \\ &= 1 - P(\cap_{j\in J}\{\omega : V_j(\omega) = o\}) = C^*(A^{(j)}(u_j) : j \in J). \end{aligned}$$ ■

Remark 5. The expression on the right hand side of Theorem 3((*i*)) in general does not coincide with $C(A^{(j)}(u_j), j \in J)$. But, by choosing a restricted family for C (and thus C^*) it does coincide. This includes C = min or product. This is also related to the issue of characterizing homomorphic-like relations for arbitrary finite combinations of C and C^* acting upon the $A^{(j)\prime}$s. See Goodman (1994).

7 Some applications

Random set representations of fuzzy concepts can be useful in interpretating fuzzy concepts or in manipulating them using random set operations. On the other hand, random sets can be used to define more appropriate fuzzy concepts**a)** First, here are some formal random set representations of other fuzzy concepts. A discrete *fuzzy partition* of a set U is a countable collection of fuzzy subsets A_n of U. As such we have, for any $u \in U$, $\sum_n A_n(u) = 1$. A clear connection with probability is this. For each $u \in U$, we have a discrete probability density function on $\mathbb{N}$, namely

$$f_u : \mathbb{N} \to [0,1], f_u(n) = A_n(u).$$

Extending the concept of projection of ordinary sets, the *fuzzy projection* of a fuzzy subset A of a cartesian product space $U \times V$ onto the factor space U has membership function given by

$$proj_U(A)(u) = \sup\{A(u,v) : v \in V\}, u \in U$$

In view of results in Section 5, we have

$$proj_U(A)(u) = P(\cup_{v \in V}\{\omega : u \in A^v_{\alpha(\omega)}\}) = P\{\omega : u \in proj_U(A_{\alpha(\omega)})\}$$

where $A^v(u) = A(u,v)$.

Operations of fuzzy numbers are defined via the well known extension principle. For example addition of two fuzzy numbers A and B (fuzzy subsets of the real line $\mathbb{R}$ with appropriate properties) is defined as

$$(A+B)(x) = \sup\{\min(A(u), B(v)) : u+v = x\}$$

From Section 5, we get

$$(A+B)(x) = P\{\omega : x \in A_{\alpha(\omega)} + B_{\alpha(\omega)}\}.$$

Next let us discuss some interpretations of relative attributes of populations involving fuzzy cardinality. Consider, for example, two attributes $A =$ "blond," and $B =$ "tall," fuzzy subset of a (finite) population U. Recall the the fuzzy cardinality of A, say, is $\sum_{x \in U} A(x)$. Let C be a 2-copula, then a fuzzy conjunction can be defined as $H(t,s) = s + t - C^*(t,s)$, and the total fuzzy cardinality of A conjoined with B is given by $\sum_{x \in U} H(A(x), B(x))$. The average degree to which attribute B is implied by A can be defined as

$$av(A|B) = \sum_{x \in U} H(A(x), B(x)) / \sum_{x \in U} B(x).$$

It turns out that *Robbins' formula* can provide a probabilistic interpretation for this quantity. Indeed, let S and W be random subsets of U corresponding to A and B for some choice of copula C. Expressing Robbins' formula for the case of U being finite and the measure μ being the counting measure on U, using the fact that

$$\forall x \in U, A(x) = P(\omega : x \in S(\omega)) = PS^{-1}(\mathcal{F}_x)$$

where $\mathcal{F}_x = \{B \subseteq U : x \in B\}$, we get

$$\sum_{x \in U} A(x) = \int_{a \in \mathcal{P}(U)} \mu(a) d(PS^{-1})(a) = E(|S|).$$

Applying this to the two components of $av(A|B)$ yields

$$av(A|B) = E(|S \cap W|)/E(|W|).$$

which can be interpreted as a sort of conditional expectation of cardinalities of the two random set representations.

Note that for $U \subseteq \mathbb{R}^n$ and λ being the Lebesgue measure, a continuous version of the above interpretation is also valid since $\int_U A(x) d\lambda(x) = E(|S|)$.

Another application of Robbins'formula is to a new random set interpretation of the well-established concept of center of area (COA) or center of gravity of a membership function which is widely used in fuzzy inference, e.g. in fuzzy control designs, see e.g. Nguyen and Sugeno, 1998. Specifically, consider the situation where control laws of complex dynamical systems are derived from rules of the form

$$\mathcal{R}_j = \text{"If } x_1 \text{ is } A_{j1}, \ldots, x_n \text{ is } A_{jn}, \text{ then } y \text{ is } B_j,\text{"} \; j = 1, 2, \ldots, r$$

where $x = (x_1, \ldots, x_n) \in \mathbb{R}^n$ is the input vector, the A_{ji}'s and B_j's are linguistic labels (fuzzy concepts in a natural language), y is the output scalar, say. The approach to modeling and deriving control laws using Fuzzy Sets Theory is called *Fuzzy Control* in the literature. First, it consists of modeling the above linguistic labels by fuzzy sets of appropriate spaces. Next, decide upon some inferential method to translate and combine information. For example, the strength of each above rule $\mathcal{R}_j$ is defined as

$$w_j(x) = \left[\prod_{i=1}^{n} A_{ji}(x_i)\right] \cdot \left[\sum_{j=1}^{r} \prod_{i=1}^{n} A_{ji}(x_i)\right]^{-1}$$

then the (typical) fuzzy output is obtained as a weighted average of the fuzzy sets B_j's, via the extension principle, generalizing Minkowski's operations on ordinary sets :

$$B_x(.) = \sum_{j=1}^{r} w_j(x) B_j(.).$$

For a given input x, the typical value B_x is precisely the " expected value" of the *random fuzzy set* which takes values B_j with corresponding probabilities $w_j(x)$, $j = 1, 2, \ldots, r$. The numerical output y, when the input is x, is obtained by "defuzzifying " the fuzzy set B_x :

$$y(x) = \left[\int_{\mathbb{R}} yB_x(y)dy\right] \cdot \left[\int_{\mathbb{R}} B_x(y)dy\right]^{-1}.$$

This value is called the center-of-area (or centroid) of the function $B_x(.)$, which is in fact the expected value of a random variable with conditional density

$$f(y|x) = [B_x(y)] \cdot \left[\int_{\mathbb{R}} B_x(y)dy\right]^{-1}.$$

Now, let D be a Borel set of $\mathbb{R}$. Then, the *centroid* of D is

$$\mathcal{C}(D) = [1/\lambda(D)] \int_D x d\lambda(x),$$

where λ denotes Lebesgue measure. Using again the fact that, if S is a random set in U ($\subseteq \mathbb{R}$)having the fuzzy set (on U) A as its covering function, then

$$\forall u \in U, \quad A(u) = PS^{-1}(\mathcal{F}_u)$$

we get

$$COA(A) = [E(\mathcal{C}(S)\lambda(S))] \cdot [E(\lambda(S))]^{-1}$$

which is a sort of conditional expected centroid of S with respect to $\lambda(S)$. A multivariate version (i.e. for $D \subseteq \mathbb{R}^n$) of the above exists using vector means. Note that in many situations, such as when $A(.)$ is strictly unimodal, symmetric, then, with S being the canonical random set, COA(A) can be taken as $E(\mathcal{C}(S))$ in some approximate fashion.

Modeling and manipulating fuzzy rules are essential in applying fuzzy technology to real world problems. A *fuzzy rule* is an uncertain conditional of the form : "If X is A then Y is B" where X, Y are variables and A, B are fuzzy subsets. Since such rules are in general uncertain, it is necessary to attach to them their degrees of applicability, indicating their "strength". Two issues arise. What is the logic behind using some form of implication modeling? How the empirical "weights of evidence" of rules are obtained? The two issues are related. In fact, from a practical point of view, the (empirically obtained) strengths of the rules should dictate how the rules are going to be modelled. This is typical in probabilistic systems in which conditional probabilities are empirical strengths for "If... Then...." rules. While in fuzzy systems, i.e. systems involving fuzzy sets, there are several ways to model rules and to extract their strengths consistently, e.g. in standard fuzzy control, it seems that a satisfactory mathematical concept for " fuzzy conditionals" is lacking. In the following, we will indicate a solution to this problem, based upon our work on *conditional events* (see. e.g. Goodman *et al.*, 1991, Goodman *et al.*, 1997) and on random set representations of fuzzy sets.

A rule can be written as a conditional $A \Longrightarrow B$, where A and B are crisp sets, elements of a (σ)-algebra $\mathcal{A}$ of subsets of Ω. In the context of boolean logic, it is natural to interprete such a rule as material implication, i.e. $A' \cup B$. However, if the strength of $A \Longrightarrow B$ is quantified as conditional probability $P(B|A)$ $(= P(A \cap B)/P(A))$, then obviously the material implication is not appropriate, since

$$P(A' \cup B) = P(B|A) + P(A')P(B'|A) > P(B|A)$$

in general, unless $P(A) = 1$ or $P(B' \cap A) = 0$. Thus, a basic question is : how to model $A \Longrightarrow B$ mathematically so that the assignment of $P(B|A)$ to it is consistent ? One thing is clear : $(B|A)$ cannot be an ordinary event, i.e. an element of $\mathcal{A}$, in view of the so-called *Lewis'Triviality Result* (Goodman *et al.*, 1991). An approach to this problem is given in Goodman *et al.* (1991), see also Hailperin (1996) and Milne (1997), where "intervals" in the boolean ring $\mathcal{A}$ of the form

$$[AB, A \to B] = \{C \in \mathcal{A} : A \cap B \subseteq C \subseteq A' \cup B\}$$

are taken to be *conditional objects*, denoted as $(B|A)$. It is shown that, with interval operations, the space of such conditional objects forms a Stone algebra containing the sub Stone algebra of *rough sets*. See Pawlak (1992), Nguyen (1992). Here, we present another approach which is boolean, called the *product space approach.*

Theorem 4. *Let $(\Omega, \mathcal{A})$ be a measurable space. There exist a measurable space $(\Omega^*, \mathcal{A}^*)$ and a map $\Psi : \mathcal{A} \times \mathcal{A} \to \mathcal{A}^*$ which embeds $(\Omega, \mathcal{A})$ into $(\Omega^*, \mathcal{A}^*)$, such that for any probability measure P on $(\Omega, \mathcal{A})$, there is a probability measure P^* on $(\Omega^*, \mathcal{A}^*)$ such that $P^*(\Psi(A, B)) = P(B|A)$.*

Proof. Ω^* is taken to be the infinite countable cartesian product of Ω's, and $\mathcal{A}^*$ is the associated product σ-field. Define

$$\Psi : \mathcal{A} \times \mathcal{A} \to \mathcal{A}^*$$

$$\Psi(A, B) = (AB) \cup (A' \times AB) \cup (A' \times A' \times AB) \cup \ldots$$

where AB stands for $A \cap B$, and terms like $AB, A' \times AB$ are short hand notation for $AB \times \Omega \times \Omega \times \ldots$. i.e. representing the cylinder in Ω^* with base AB, and thus $\cup$ among these terms is the union of subsets in Ω^*. Next, for any P on $(\Omega, \mathcal{A})$, let P^* denote the infinite product measure on $(\Omega^*, \mathcal{A}^*)$ whose one- dimensional marginals are all identical to P. Note that the terms in the definition of $\Psi(A, B)$ are disjoint elements of $\mathcal{A}^*$. We have, by construction of P^*

$$
\begin{aligned}
P^*(\Psi(A, B)) &= \sum_{n=0}^{\infty} P^*[(A')^n \times AB] \\
&= \sum_{n=0}^{\infty} P(AB)[P(A')]^n = P(AB) \sum_{n=0}^{\infty} [P(A')]^n \\
&= P(AB)/[1 - P(A')] = P(AB)/P(A) = P(B|A). \qquad \blacksquare
\end{aligned}
$$

The object $\Psi(A, B)$ constructed above is called a *conditional event*, denoted as $(B|A)$ (by itself, without the operator P), which is an event, not in $\mathcal{A}$, but in a bigger space $\mathcal{A}^*$. It is this type of mathematical objects that we use to model conditionals $A \Longrightarrow B$ consistently with conditional probabilities. When A and B are fuzzy subsets of U, we can use their random set representations to define *fuzzy conditionals.* If U is finite, then such random set representations can be chosen by $|U|$-copulas, otherwise, canonical nested random sets (uniformily randomized level-sets) can be used. So let S_A, S_B denote any random set representations of A, B, respectively. Then $(S_B|S_A)$ is a *random conditional event.* The *conditional fuzzy set*, denoted as $(B|A)$, is a fuzzy subset of U^* (countable product of U's) with membership function given by

$$(B|A)(u^*) = P^*(u^* \in (S_B|S_A))$$

which can be explicitly evaluated. For details see Goodman *et al.* (1997).

Another situation where covering functions of random sets can be related to fuzzy concepts is in the usual way fuzzy logic employs modifiers. In general, the modifier of a fuzzy concept A of U is obtained by composition with a fuzzy subset M of the unit interval $[0, 1]$: $M \circ A$, for example, "very A" is $(A)^2$, i.e. take $M(x) = x^2$. Now the canonical random set of such a modifier is clearly

$$(M \circ A)_{\alpha(\omega)} = (M \circ A)^{-1}[\alpha(\omega), 1] = A^{-1}(M^{-1}[\alpha(\omega), 1]) = A^{-1}(M_{\alpha(\omega)}).$$

Thus, for all $u \in U$

$$M(A(u)) = P\{\omega : u \in (M \circ A)_{\alpha(\omega)}\} = P\{\omega : A(u) \in M_{\alpha(\omega)}\}.$$

A more interesting way in representing fuzzy set modifiers is as follows. Suppose $g : [0, 1] \to [0, 1]$ is such that for any probability space $(\Omega, \mathcal{A}, P)$ there always exists a larger probability space $(\Omega^*, \mathcal{A}^*, P^*)$ and a corresponding relational operator $g^* : \mathcal{A} \to \mathcal{A}^*$, not depending on P, such that $g(P(A)) = P^*(g^*(A))$, for any $A \in \mathcal{A}$ (g commutes with P). Here is a simple example. Let m be a positive integer. Define $g(s) = s^m$ for $s \in [0, 1]$, and $g^*(A) = A \times A \times \ldots \times A$ (m times). In the same flavor, conditional events correspond to arithmetic division of numbers. It is thus natural to look for probabilistic counter-parts of other operations such as exponentiation, linear functions, polynomials, analytic functions, etc. Such probabilistic counterparts are called *relational events*, extending the concept of conditional events. Here is an interesting example of *constant-probability events*. Let m, n be non negative integer with $m \leq n$. Then there exists an event $\Xi(m.n)$ in some appropriate space such that the probability of it is equal to m/n . Specifically, let $C \in \mathcal{A}$ such that $P(C) \in (0,1)$ for P on $(\Omega, \mathcal{A})$. Form

$$\Gamma(m,n,C) = C^{m-1} \times C' \times C^{n-m}$$

where $C^k = C \times C \times \ldots \times C$ (k times), with the convention that $C^o \times C' = C'$, $C' \times C^o = C'$. Consider the corresponding n-fold product probability space $(\Omega^n, \mathcal{A}^n, P^n)$. Each $\Gamma(m,n,C)$ is an element of $\mathcal{A}^n$, so that we can consider conditional events in the corresponding space $((\Omega^n)^*, (\mathcal{A}^n)^*, (P^n)^*)$ built from $(\Omega^n, \mathcal{A}^n, P^n)$ as in the construction of conditional events presented above. Let

$$\Xi(m,n,C) = (\cup_{j=1}^{m} \Gamma(j,n,C) | \cup_{j=1}^{n} \Gamma(j,n,C)).$$

Since the $\Gamma(j,n,C)$'s are mutually disjoint, we have

$$(P^n)^*(\Xi(m,n,C)) = [m(P(C))^{n-1}P(C')]/[n(P(C))^{n-1}P(C')] = m/n.$$

Note that we can in fact suppress C in the above because the construction of $\Xi(n,m,C)$ is valid for any such C. Applications of relational event algebra, including constant-probability events, arise mainly in combination, comparisons and testing for similarity of models. See again Goodman *et al.* (1997) for details.

References

1. Bertoluzza, C. and Bodini, A. (1998). A new proof of Nguyen's compatibility theorem in a more general context, *Fuzzy Sets and Systems* **95**, 99-102.
2. Fréchet, M. (1948). Les elements aleatoires de nature quelconque dans un espace distancié, *Ann. Inst. Henri Poincaré* **X-IV**, 215-310.
3. Fuller, R. and Keresztfalvi, T. (1990). On generalization of Nguyen's theorem, *Fuzzy Sets and Systems* **41**, 371-374.
4. Goodman, I.R. (1976). Some relations between fuzzy sets and random sets. Preprint (cited in Nguyen, H.T. (1979) Some mathematical tools for linguistic probabilities, *Fuzzy Sets and Systems* **2**, 53-65.)
5. Goodman, I.R. (1994). A new characterization of fuzzy logic operators producing homomorphic-like relations with one-point coverages of random sets. In *Advances in Fuzzy Theory and Technology*. vol **II** (P.P.Wang, Ed.). Duke Univ., Durham, NC, 133-159.
6. Goodman, I.R. and Nguyen, H.T. (1985). *Uncertainty Models for Knowledge-Based Systems*. North-Holland, Amsterdam.
7. Goodman, I.R., Nguyen, H.T. and Walker, E.A. (1991). *Conditional Inference and Logic for Intelligent Systems*. North-Holland, Amsterdam.

8. Goodman, I.R., Mahler, R. and Nguyen, H.T. (1997). *Mathematics of Data Fusion.* Kluwer Academic.
9. Goutsias, J., Mahler, R. and Nguyen, H.T. (1997). *Random Sets : Theory and Applications.* Springer-Verlag, Heidelberg.
10. Hailperin, T.(1996). *Sentential Probability Logic.* Lehigh Univ. Press.
11. Matheron, G. (1975). *Random Sets and Integral Geometry.* J. Wiley & Sons, New York.
12. Milne, P. (1997). Bruno de Finetti and the logic of conditional events, *British J. Philos. Sci.* **48**, 195-232.
13. Nguyen, H.T. (1978). A note on the extension principle for fuzzy sets, *J. Math. Anal. Appl.* **64**, 369-380.
14. Nguyen, H.T. (1992). Interval in boolean rings: approximation and logic,. *J. Foundations of Computing and Decision Sciences* **17**, 131-138.
15. Nguyen, H.T. and Sugeno, M.(1998). *Fuzzy Systems : Modeling and Control.* Kluwer Academic.
16. Nguyen, H.T. and Prasad, N.R.(1998). *Fuzzy Modeling and Control : Selected Works of M.Sugeno.* CRC Press, Boca Raton.
17. Nguyen, H.T. and Walker, E.A. (1996). *A First Course in Fuzzy Logic.* CRC Press, Boca Raton.
18. Orlowski, S.A. (1994), *Calculus of Decomposable Properties, Fuzzy Sets and Decisions.* Allerton Press.
19. Pawlak, Z. (1992). *Rough Sets.* Kluwer Academic.
20. Robbins, H.E. (1944). On the measure of a random set. *Ann.Math. Statist.* **15**, 70-74.
21. Schweizer, B. and Sklar, A. (1983). *Probabilistic Metric Spaces.* North-Holland, Amsterdam.
22. Sugeno, M. (1974). Theory of fuzzy integrals and its applications. PhD thesis. Tokyo Institute of Technology, Japan.
23. Zadeh, L.A. (1965). Fuzzy sets, *Inform. Control* **8**, 338-353.

Part 2

FUZZY-VALUED RANDOM ELEMENTS

On the variance of random fuzzy variables

Ralf Körner and Wolfgang Näther

Faculty of Mathematics and Computer Sciences, Freiberg University of Mining and Technology, 09596 Freiberg, Germany

Abstract. In this paper a generalized definition of the variance of a random fuzzy variable is introduced on the basis of a suitable generalized metric defined on the wide class of the variable values.

1 Introduction and overview

In many real situations uncertainty of data comes from two sources: from randomness and from fuzziness. Randomness models the stochastic variability of all possible outcomes of an experiment and fuzziness describes the vagueness of the given or just realized outcome. Randomness is more an instrument of a normative analysis which thinks about the question "What will happen in future?"; fuzziness is more an instrument of a descriptive analysis reflecting questions like "What has happened?" or "What is meant by the data?".

Therefore, in modeling realistic situations, fuzziness is often tied to randomness since possible random outcomes have to be described by fuzzy sets, especially in the case of linguistically expressed outcomes. This leads to the concept of a fuzzy random variable Y. Following Kwakernaak (1978), a fuzzy random variable Y is considered as a vague perception of a crisp but unobservable random variable X. A conceptional different approach is given by the so called probabilistic fuzzy sets introduced by Hirota (1981). Here, Y is considered, first of all, as a fuzzy set with random membership values. The most successful approach to rfv's, however, was presented by Puri and Ralescu (1986) where Y is considered as a fuzzification of a random set (therefore, sometimes Y is called random fuzzy set, too). An exact definition is the following:

Definition 1. Denote by $\mathcal{F}_c(\mathbb{R}^d)$ the set of all normal compact convex fuzzy subsets of $\mathbb{R}^d$, i.e. any $A \subset \mathcal{F}_c(\mathbb{R}^d)$ with membership function m_A satisfies

i) $\forall \alpha \in (0,1]$ the α-level sets $A_\alpha = \{x \in \mathbb{R}^d : m_A(x) \geq \alpha\}$ are convex and compact

ii) A is normal, i.e. $A_1 = \{x \in \mathbb{R}^d : m_A(x) = 1\} \neq \emptyset$.

Let $(\Omega, \mathcal{B}, P)$ be a probability space. Then, $Y : \Omega \to \mathcal{F}_c(\mathbb{R}^d)$ is called a *random fuzzy variable* (rfv) on $\mathbb{R}^d$ if for any $\alpha \in (0,1]$ the α-cut Y_α is a convex compact random set (e.g. in the sense of Matheron, 1975).

The concept of rfv's in the sense of Definition 1 has been studied successfully, e.g. for limit theorems (starting with Klement *et al.*, 1986) and has been applied to asymptotical statistics with vague data by Kruse and Meyer (1987).

In the following we will concentrate us on moments of rfv's. Asking for laws of large numbers for rfv's, Puri and Ralescu (1986) have proposed to use the so called Aumann-expectation as a suitable expectation of rfv's. The Aumann-expectation goes back to the paper Aumann (1965) on integrals of set valued functions and is defined as follows:

Definition 2. Let Y be a rfv on $\mathbb{R}^d$. The *Aumann-expectation* of Y is defined as the fuzzy set $\mathbb{E}^{(A)}Y \in \mathcal{F}_c(\mathbb{R}^d)$ with

$$\forall \alpha \in (0,1] : (\mathbb{E}^{(A)}Y)_\alpha = \mathbb{E}^{(A)}Y_\alpha$$

where $\mathbb{E}^{(A)}Y_\alpha$ is the Aumann-expectation of the random set Y_α defined by

$$\mathbb{E}^{(A)}Y_\alpha = \{\mathbb{E}\eta : \eta(\omega) \in Y_\alpha(\omega)\ P-\text{a.e. and } \eta \in L^1(\Omega, \mathcal{B}, P)\}\,.$$

Note that there are also another proposals to define an expectation of rfv's. E.g. for a rfv in form of a probabilistic fuzzy set Y an expectation $\mathbb{E}^{(m)}Y$ can be defined via the expected membership function, i.e.

$$m_{\mathbb{E}^{(m)}Y}(x) = \mathbb{E}m_Y(x);\ x \in \mathbb{R}^d,$$

which in general does not coincide with Definition 2. There are further definitions of expectations for random sets (see e.g. Molchanov, 1993, Stoyan and Stoyan, 1994) which can be used for further alternative α-cut-wise definitions of expectations for rfv's.

In Section 2, we will discuss the reasons why the Aumann-expectation is preferable.

Only few investigations are known w.r.t. a suitable definition of a variance for rfv's. Kruse (1987) follows the selector-principle, i.e. he defines an "Aumann-like" fuzzy-set-valued variance by

$$(\mathrm{Var}^{(A)}Y)_\alpha = \{\mathrm{Var}\,\eta : \eta(\omega) \in Y_\alpha(\omega)\ P-\text{a.e. and } \eta \in L^2(\Omega, \mathcal{B}, P)\}.$$

This variance, however, measures not only the variability generated by randomness, but also, in some sense, the size of the random fuzzy set Y. For instance, a deterministic set B has a non-zero set-valued variance, e.g. for the deterministic interval $B = [-a, a]$ we have $\mathrm{Var}^{(A)}(B) = [0, a^2]$.

In this paper, we will discuss a real valued variance for rfv's. We will follow the approach by Körner (1997b) and Näther (1997) and we will use the well known Fréchet principle where the variance of an random variable Z is defined as the (real-valued) expectation of the squared distance d of Z from its expectation, i.e.

$$\mathrm{Var}(Z) = \mathbb{E}d(Z, \mathbb{E}Z)^2. \tag{1}$$

The paper is organized as follows: In Section 2 the Fréchet principle is formulated in more detail and suitable distances between fuzzy sets are introduced. We justify the Aumann-expectation as a Fréchet-expectation w.r.t. the L^2-distance δ_2 in the associated space of support functions and define $\operatorname{Var} Y$ via (1) w.r.t. this δ_2. Using results from Banach-space-valued random variables we generalize this approach which now includes several other definitions of a variance, e.g. the definition by Lubiano *et al.* (2000) which uses distances introduced by Bertoluzza *et al.* (1995). Moreover, Section 2 presents the essential properties and an unbiased estimator of $\operatorname{Var} Y$.

Section 3 is devoted to applications: At first, supposing the existence of $\operatorname{Var} Y$, a strong law of large numbers for rfv's can be proved totally similar as in the classical way, and a central limit theorem can be formulated which uses known results on random variables in Hilbert spaces. Second, an approximative test w.r.t. hypothesis on the expectation of a rfv Y is presented.

2 A suitable definition of the variance

2.1 The Fréchet-principle

For defining expectation and variance of a rfv Y, as a methodological principle, the Fréchet-principle is used. Fréchet (1948) has defined the expectation $\mathbb{E}^{(d)} Z$ for a random variable Z with values in a metric space (M, d) as a (not necessary unique) solution of the problem

$$\mathbb{E}d(Z, \mathbb{E}^{(d)} Z)^2 = \inf_{a \in M} \mathbb{E}d(Z, a)^2 \,. \tag{2}$$

Note that $\mathbb{E}d(Z, a)^2$ is the usual expectation of the real-valued variable $d(Y, a)^2$. The variance of Z, denoted by $\operatorname{Var}^{(d)} Z$, is then defined by

$$\operatorname{Var}^{(d)} Z = \mathbb{E}d(Z, \mathbb{E}^{(d)} Z)^2. \tag{3}$$

This is a generalization of the known fact that for a real valued random variable X the expectation $\mathbb{E}X$ minimizes $\mathbb{E}|X-a|^2$ and $\operatorname{Var} X$ equals $\mathbb{E}|X - \mathbb{E}X|^2$.

In the following, $\mathbb{E}^{(d)} Z$ satisfying (2) is called Fréchet-expectation w.r.t. d. For rfv Y, the Fréchet approach opens the way for defining several expectations and their (via (3)) associated variances, each induced by a given metric between fuzzy sets. Therefore, first of all, we have to discuss on suitable distances between fuzzy sets.

2.2 Distances between fuzzy sets

It seems to be natural to start with the Hausdorff-metric between crisp sets $A, B \subset \mathbb{R}^d$, given by

$$d_H(A, B) = \max\left\{ \sup_{b \in B} \inf_{a \in A} \|a - b\|, \ \sup_{a \in A} \inf_{b \in B} \|a - b\| \right\}.$$

For two fuzzy sets A, B this can be generalized to

$$d_p(A,B) = \begin{cases} \left(\int_0^1 (d_H(A_\alpha, B_\alpha))^p d\alpha\right)^{\frac{1}{p}}, & p \in [1,\infty) \\ \sup_{\alpha\in(0,1]} d_H(A_\alpha, B_\alpha) & p = \infty \end{cases}$$

where especially d_1 and d_∞ are investigated in the literature.

For example, $(\mathcal{F}_c(\mathbb{R}^d), d_1)$ appears as a complete and separable metric space, $(\mathcal{F}_c(\mathbb{R}^d), d_\infty)$, however, is a complete but non-separable metric space (see Puri and Ralescu, 1986).

Another type of distances can be defined via so called support functions. For any compact convex set $A \subset \mathbb{R}^d$ the support function s_A is defined as

$$s_A(u) = \sup_{y\in A} < u, y >; \qquad u \in S^{d-1},$$

where $< \cdot, \cdot >$ is the scalar product in $\mathbb{R}^d$ and S^{d-1} the $(d-1)$-dimensional unit sphere in $\mathbb{R}^d$. Note that for convex and compact $A \subset \mathbb{R}^d$ the support function s_A is uniquely determined. A fuzzy set $A \in \mathcal{F}_c(\mathbb{R}^d)$ can be characterized α-cut-wise by its support function:

$$s_A(u,\alpha) := s_{A_\alpha}(u); \qquad \alpha \in (0,1],\ u \in S^{d-1}\,. \tag{4}$$

Thus, via support functions (4), $\mathcal{F}_c(\mathbb{R}^d)$ can be embedded in a space of functions on $S^{d-1} \times (0,1]$ and we can define a metric in $\mathcal{F}_c(\mathbb{R}^d)$ using e.g. the L_2-metric in $L_2(S^{d-1} \times (0,1])$, i.e.

$$\delta_2(A,B) = \sqrt{d \int_0^1 \int_{S^{d-1}} (s_A(u,\alpha) - s_B(u,\alpha))^2 \nu(du)d\alpha}, \tag{5}$$

where ν is the Lebesgue measure on S^{d-1}. Note that $(\mathcal{F}_c(\mathbb{R}^d), \delta_2)$ is separable (see Diamond and Kloeden, 1994).

As an example, consider socalled LR-fuzzy numbers $A := \langle m, l, r\rangle_{LR}$ with modal value $m \in \mathbb{R}^1$, left and right spreads $l, r \in \mathbb{R}^+$, decreasing left and right shape functions $L, R : \mathbb{R}^+ \to [0,1]$ with $L(0) = R(0) = 1$ and finite support, i.e. a fuzzy set A with

$$m_A(x) = \begin{cases} L(\frac{m-x}{l}) & \text{if } x \le m \\ R(\frac{x-m}{r}) & \text{if } x > m. \end{cases} \tag{6}$$

Note that the α-cuts are given by the intervals

$$A_\alpha = [m - L^{-1}(\alpha)l,\ m + R^{-1}(\alpha)r]; \quad \alpha \in (0,1]$$

and that the support function of an interval is defined on $S^0 = \{-1, 1\}$ by

$$s_{[a,b]}(u) = \begin{cases} -a & \text{if } u = -1 \\ b & \text{if } u = 1. \end{cases}$$

Then, we have for two totally symmetric fuzzy numbers ($L = R$ and $\Delta := l = r$) $A = (m_A, \Delta_A)_L$, $B = (m_B, \Delta_B)_L$:

$$d_H(A_\alpha, B_\alpha) = |m_A - m_B| + L^{-1}(\alpha)|\Delta_A - \Delta_B|$$

$$d_1(A, B) = |m_A - m_B| + 2L_1|\Delta_A - \Delta_B|; \qquad L_1 = \frac{1}{2}\int_0^1 L^{-1}(\alpha)d\alpha$$

$$d_\infty(A, B) = |m_A - m_B| + L_\infty|\Delta_A - \Delta_B|; \qquad L_\infty = \sup_{\alpha\in(0,1]} L^{-1}(\alpha)$$

$$\delta_2(A, B) = \sqrt{(m_A - m_B)^2 + 2L_2(\Delta_A - \Delta_B)^2} \qquad L_2 = \frac{1}{2}\int_0^1 (L^{-1}(\alpha))^2 d\alpha.$$

There are more complicated formula for LR-fuzzy numbers.

2.3 Definition of the variance

If we use the generalized Hausdorff-metric d_p for defining expectation and the associated variance of a rfv Y we have a lot of disadvantages:

i) The Aumann-expectation is not Fréchet w.r.t. d_p (see for an example in Näther, 1997).
ii) The Fréchet-expectation w.r.t. to d_p is a nonlinear operator, i.e.,

$$\mathbb{E}^{(d_p)}(Y_1 \oplus Y_2) \neq \mathbb{E}^{(d_p)}Y_1 \oplus \mathbb{E}^{(d_p)}Y_2,$$

where $\oplus$ denotes the addition extended via extension principle (see Näther, 1997).
iii) The variance w.r.t. d_p is not additive for the sum of two independent rfv's (see Körner, 1997a), i.e. $\mathrm{Var}^{(d_p)}(Y_1 \oplus Y_2) \neq \mathrm{Var}^{(d_p)}Y_1 + \mathrm{Var}^{(d_p)}Y_2$.

Since the Aumann-expectation $\mathbb{E}^{(A)}$ fulfills the desired properties of linearity, i.e.

$$\mathbb{E}^{(A)}(\lambda_1 Y_1 \oplus \lambda_2 Y_2) = \lambda_1 \mathbb{E}^{(A)} Y_1 \oplus \lambda_2 \mathbb{E}^{(A)} Y_2,$$

the question is: Can $\mathbb{E}^{(A)}X$ be interpreted as a Fréchet-expectation w.r.t. a certain metric? If this would be true, this metric could be used by (3) for a well defined variance. The answer presents the following theorem.

Theorem 1. *The Aumann-expectation $\mathbb{E}^{(A)}$ is a Fréchet-expectation w.r.t. δ_2 from (5).*

For the proof see Näther (1997). Thus, we can define:

Definition 3. Let Y be a rfv on $\mathbb{R}^d$. Then, the variance of Y (if it exists) is defined by

$$\begin{aligned} \operatorname{Var} Y &= \mathbb{E}\delta_2(Y, \mathbb{E}^{(A)}Y)^2 \qquad (7) \\ &= d \int_0^1 \int_{S^{d-1}} \mathbb{E}(s_Y(u,\alpha) - s_{\mathbb{E}^{(A)}Y}(u,\alpha))^2 \nu(du)d\alpha \,. \end{aligned}$$

Following standard arguments from random sets (see e.g. Stoyan and Stoyan, 1994), it holds $s_{\mathbb{E}^{(A)}Y} = \mathbb{E}s_Y$ and we have

$$\operatorname{Var} Y = d \int_0^1 \int_{S^{d-1}} \operatorname{Var} s_Y(u,\alpha)\ \nu(du)d\alpha,$$

which presents the connection between $\operatorname{Var} Y$ and the classical defined $\operatorname{Var} s_Y(u,\alpha)$. We use for both the same notion, because a random number is a special random fuzzy set.

For random fuzzy numbers $Y = \langle m,l,r\rangle_{LR}$ with uncorrelated random m,l,r the variance reduces to a very simple form:

$$\operatorname{Var} \langle m,l,r\rangle_{LR} = \operatorname{Var} m + L_2 \operatorname{Var} l + R_2 \operatorname{Var} r; \quad R_2 = \frac{1}{2}\int_0^1 R^{-1}(\alpha))^2 d\alpha.$$

2.4 The general approach

Using support functions, the class $\mathcal{F}_c(\mathbb{R}^d)$ of all non-empty compact convex fuzzy sets on $\mathbb{R}^d$ is embedded in the Banach space $L(S^{d-1} \times (0,1])$ (cf. Radström, 1952). By this embedding the semi-linear structure is preserved, i.e.

$$s_{A\oplus B} = s_A + s_B \quad \text{and} \quad s_{\lambda A} = \lambda s_A$$

for $A, B \in \mathcal{F}_c(\mathbb{R}^d)$ and $\lambda \geq 0$. Furthermore, the mapping is isometric, i.e.

$$d_p(A,B) = \begin{cases} (\int_0^1 \sup_{u\in S^{d-1}} |s_A(u,\alpha) - s_B(u,\alpha)|^p d\alpha)^{\frac{1}{p}}, & p \in [1,\infty) \\ \sup_{\alpha\in(0,1]} \sup_{u\in S^{d-1}} |s_A(u,\alpha) - s_B(u,\alpha)|, & p = \infty. \end{cases}$$

Hence, results of Banach space valued random variables can be used to define an expectation, a variance and to prove limit theorems.

First let us state some facts about random variables in Banach space and deduce that only an L_2-distance leads to an appropriate variance.

Let $(\Omega, \mathcal{B}, P)$ be a probability space, and $(B, \|\cdot\|)$ be a separable Banach space equipped with the Borel σ-algebra generated by $\|\cdot\|$. Then a B-valued random variable Z is a measurable mapping from Ω into B (cf. Araujo and Giné, 1980).

Definition 4. Let Z be a B-valued random variable with $\mathbb{E} \| Z \| < \infty$. The *Pettis expectation* $\mathbb{E}^{(P)} X$ is that element of B with

$$f(\mathbb{E}^{(P)} Z) = \mathbb{E} f(Z)$$

for each $f \in B'$ (space of continuous linear mappings from B into $\mathbb{R}$).
If $\mathbb{E} f(Z)^2 < \infty$ for any $f \in B'$ then the non-negative symmetric bilinear form on B' defined by

$$\Big(\operatorname{Cov}(Z)\Big)(f, g) = \mathbb{E}\Big(f(Z) - \mathbb{E} f(Z)\Big)\Big(g(Z) - \mathbb{E} g(Z)\Big)$$

is called the *covariance of* Z.

Theorem 2. *Let $(\Omega, \mathcal{B}, P)$ be a non-atomic probability space. If for any B-valued random variable Z with $\mathbb{E} \| Z \|^2 < \infty$*

$$\inf_{a \in B} \mathbb{E} \| Z - a \|^2 = \mathbb{E} \| Z - \mathbb{E}^{(P)} Z \|^2$$

then $\|\cdot\|$ is a Hilbert norm, i.e. for $a, b \in B$ the distance $d(a, b) = \| a - b \|$ is an L_2-distance in B. Furthermore, the Frechét-variance (2.2), i.e. $\operatorname{Var}(Z) = \mathbb{E} \| Z - \mathbb{E}^{(P)} Z \|^2$ *is additive, i.e. for independent random variables Z_1 and Z_2 we obtain*

$$\operatorname{Var}(Z_1 + Z_2) = \operatorname{Var}(Z_1) + \operatorname{Var}(Z_2)\,.$$

In our Banach space $L(S^{d-1} \times (0, 1])$ any L_2-distance has the structure

$$\rho_2(f', g') = \int_{[0,1]^2 \times (S^{d-1})^2} \Big(f'(u, \alpha) - g'(u, \alpha)\Big)\Big(f'(v, \beta) - g'(v, \beta)\Big) dK(u, \alpha, v, \beta)\,,$$

with a symmetric and positive definite kernel K.

Using the Riesz-Fischer theorem, each linear functional $f \in B'$ of the separable Hilbert space B can be written as $f(\cdot) = \langle f', \cdot \rangle$ with $f' \in B$ and the induced inner product $\langle \cdot, \cdot \rangle$. Hence for a rfv Y we obtain

$$f(Y) = \int_0^1 \int_{S^{d-1}} f'(u, \alpha) s_Y(u, \alpha)\, dK(u, \alpha) = \langle f', s_Y \rangle_K$$

and

$$\mathbb{E}\Big(f(Y) - \mathbb{E} f(Y)\Big)\Big(g(Y) - \mathbb{E} g(Y)\Big)$$

$$= \int_{[0,1]^2 \times (S^{d-1})^2} f'(u,\alpha) C_Y(u,\alpha,v,\beta) g'(v,\beta)\, dK(u,\alpha,v,\beta)\,,$$

where $C_Y(u,\alpha,v,\beta) = \mathbb{E}\Big(s_Y(u,\alpha) - \mathbb{E}s_Y(u,\alpha)\Big)\Big(s_Y(v,\beta) - \mathbb{E}s_Y(v,\beta)\Big)$ is the covariance of s_Y.

Theorem 3. *Let Y be a rfv with $\mathbb{E} \parallel Y \parallel < \infty$.*
Then the Aumann expectation is equal to the Pettis expectation.

Therefore, the Aumann expectation $\mathbb{E}^{(A)}Y$ is the unique Frechét expectation with respect to any ρ_2-distance

$$\rho_2(A,B)^2 = \int_{[0,1]^2 \times (S^{d-1})^2} \Big(s_A(u,\alpha) - s_B(u,\alpha)\Big)\Big(s_A(v,\beta) - s_B(v,\beta)\Big) dK(u,\alpha,v,\beta)\,. \quad (8)$$

and the general definition of $\mathrm{Var}\, Y$ is given by

$$\mathrm{Var}\, Y = \mathbb{E}\rho_2(Y, \mathbb{E}^{(A)}Y)^2 = \int_{[0,1] \times (S^{d-1})^2} C_Y(u,\alpha,v,\beta) dK(u,\alpha,v,\beta) \quad (9)$$

In the following we drop the upper index (A) and write $\mathbb{E}Y$ for the Aumann expectation.

2.5 Comparison with other approaches

Especially, the variance defined in Definition 3 is obtained from (8) and (9) by the kernel

$$dK(u,\alpha,v,\beta) = d \cdot \delta_u(v)\delta_\alpha(\beta)\nu(du)d\alpha.$$

Then $\mathrm{Var}(Y) = d \cdot \mathbf{tr}(C_Y)$, where $\mathbf{tr}$ is the trace of the covariance C_Y.

The class of distances introduced by Bertoluzza *et al.* (1995) are special cases of (8). They defined a distance between two normal convex fuzzy sets A and B of the real line $\mathbb{R}^1$ by

$$D(A,B)^2 = \int_0^1 \int_0^1 [t(\inf A_\alpha - \inf B_\alpha) + (1-t)(\sup A_\alpha - \sup B_\alpha)]^2 dg(t)\, d\varphi(\alpha)\,,$$

where g and φ are normalized weight measures on $[0,1]$. Straightforward calculations show that

$$\begin{aligned} D(A,B)^2 = c_2 \int_0^1 (\inf A_\alpha - \inf B_\alpha)^2 d\varphi(\alpha) \\ +(1 - 2c_1 + c_2) \int_0^1 (\sup A_\alpha - \sup B_\alpha)^2 d\varphi(\alpha) \\ +2(c_1 - c_2) \int_0^1 (\inf A_\alpha - \inf B_\alpha)(\sup A_\alpha - \sup B_\alpha)\, d\varphi(\alpha) \end{aligned}$$

with

$$c_1 = \int_0^1 t\, dg(t) \qquad \text{and} \qquad c_2 = \int_0^1 t^2 dg(t)\,.$$

Because $S^{d-1} = S^0 = \{+1, -1\}$, the integral with respect to S^{d-1} is a sum of two terms $u = +1$ and $u = -1$. Hence, with the kernel

$$dK(u,\alpha,v,\beta) = \delta_\alpha(\beta) d\varphi(\alpha) \begin{cases} 1 - 2c_1 + c_2 & \text{for } u = v = +1 \\ c_2 & \text{for } u = v = -1 \\ c_1 - c_2 & \text{for } u = -v \end{cases}$$

$D(A,B)^2$ is a special case of (8). Especially, the $\boldsymbol{\lambda}$-mean squared dispersion defined by Lubiano *et al.* (2000) is based on a special D-distance and is included in the variance concept above. Lubiano defines the $\boldsymbol{\lambda}$-mean squared dispersion of a rfv Y by

$$\Delta_{\boldsymbol{\lambda}}^2(Y) = \mathbb{E} D_{\boldsymbol{\lambda}}(Y, \mathbb{E}Y)^2$$

with

$$D_{\boldsymbol{\lambda}}(A,B)^2 = \int_0^1 \Big[\lambda_1(\sup A_\alpha - \sup B_\alpha)^2 + \lambda_2(\text{mid } A_\alpha - \text{mid } B_\alpha)^2 + + \lambda_3(\inf A_\alpha - \inf B_\alpha)^2\Big] d\alpha\,,$$

where $\boldsymbol{\lambda} = (\lambda_1, \lambda_2, \lambda_3)$ with $\lambda_i \in [0,1)$, $\lambda_1 + \lambda_2 + \lambda_3 = 1$ and $\text{mid } A_\alpha = 1/2(\sup A_\alpha + \inf A_\alpha)$ denotes the middle point of the interval. Here

$$dK(u,\alpha,v,\beta) = \delta_\alpha(\beta) d\alpha \begin{cases} \lambda_1 + \lambda_2/4 & \text{for } u = v = +1 \\ \lambda_3 + \lambda_2/4 & \text{for } u = v = -1 \\ -\lambda_2/4 & \text{for } u = -v\,. \end{cases}$$

At least, the Hagaman distance (cf. Bardossy *et al.*, 1992) between two different LR-fuzzy numbers $A = \langle m_A, l_A, r_A \rangle_{L_A R_A}$ and $B = \langle m_B, l_B, r_B \rangle_{L_B R_B}$ (see (4))

$$D_f(A,B)^2 := \int_0^1 \Big\{(m_A - l_A L_A^{-1}(\alpha) - m_B + l_B L_B^{-1}(\alpha))^2 + (m_A + r_A R_A^{-1}(\alpha) - m_B - r_B R_B^{-1}(\alpha))^2\Big\} f(\alpha)\, d\alpha$$

with density function f is obtained by

$$dK(u,\alpha,v,\beta) = \frac{1}{2}\delta_\alpha(\beta) f(\alpha) d\alpha\,.$$

In this way, also Diamonds distance (Diamond, 1988) in the space of triangular fuzzy numbers $A = \langle m_A, l_A, r_A \rangle_\Delta$ and $B = \langle m_B, l_B, r_B \rangle_\Delta$

$$d(A,B)^2 := (m_A - l_A - m_B + l_B)^2 + (m_A + r_A - m_B - r_B)^2 + (m_A - m_B)^2$$

is included.

Using Hilbert space properties, it is straightforward to prove that:

Theorem 4. *All topologies generated by (8) are equivalent.*

2.6 Properties and estimation

We will summarize some properties of the variance, and we will show that the variance can be estimated in a similar way as in the classical case.

Theorem 5. *Let Y_1, Y_2 and Y be rfv's and define the variance with respect to any L_2 distance ρ_2 in (8) and let $\| \cdot \|_\rho = \rho_2(\cdot, \{o\})$ be the induced norm. Then for any positive squared integrable random variable ξ, for any $A \in \mathcal{F}_c(\mathbb{R}^d)$ and for any real number λ we obtain*

1. $\operatorname{Var}\{\xi\} = \mathbb{E}(\xi - \mathbb{E}\xi)^2$,
2. $\operatorname{Var} Y = \mathbb{E} \| Y \|_\rho^2 - \| \mathbb{E}Y \|_\rho^2$,
3. $\operatorname{Var}(\lambda Y) = \lambda^2 \operatorname{Var} Y$,
4. $\operatorname{Var}(\xi A) = \| A \|_\rho^2 \cdot \operatorname{Var} \xi$,
5. $\operatorname{Var}(A \oplus Y) = \operatorname{Var} Y$,
6. $\operatorname{Var}(\xi Y) = \mathbb{E} \| Y \|_\rho^2 \operatorname{Var} \xi + \mathbb{E}\xi^2 \operatorname{Var} Y$ *if ξ and Y are independent.*
7. $\operatorname{Var}(Y_1 \oplus Y_2) = \operatorname{Var} Y_1 + \operatorname{Var} Y_2$ *if Y_1 and Y_2 are independent.*

For the proof see Körner (1997b)

Theorem 6. *Let $Y_1, Y_2, \ldots$ be independent and identically distributed rfv's with $\mathbb{E} \| Y \|_\rho < \infty$. Then $\overline{Y}_n = (Y_1 + \cdots Y_n)/n$ is an unbiased and consistent estimator of the Aumann expectation, i.e.*

$$\mathbb{E}\overline{Y}_n = \mathbb{E}Y_1 \qquad \text{and} \qquad \overline{Y}_n \xrightarrow[n\to\infty]{P-\text{a.s.}} \mathbb{E}Y_1 \,.$$

Proof. Because of the linearity of the Aumann expectation we obtain that $\overline{Y}_n$ is unbiased. The consistency follows by the law of large numbers of Klement *et al.* (1986). ∎

Theorem 7. *Let $Y_1, Y_2, \ldots$ be a sequence of independent and identically distributed rfv's with $\mathbb{E} \| Y_1 \|_\rho^2 < \infty$.*
Then $S_n^2 = 1/(n-1) \sum_{k=1}^n \rho_2(Y_k, \overline{Y}_n)^2$ is an unbiased and consistent estimator of the variance $\operatorname{Var}(Y_1)$, i.e.

$$\mathbb{E}S_n^2 = \operatorname{Var}(Y_1) \qquad \text{and} \qquad S_n^2 \xrightarrow[n\to\infty]{P-\text{a.s.}} \operatorname{Var}(Y_1) \,.$$

Proof. Note that

$$\frac{n-1}{n} S_n^2 = \frac{1}{n} \sum_{k=1}^n \| Y_k \|_\rho^2 - \| \overline{Y}_n \|_\rho^2 \,.$$

Now, $\overline{Y}_n \xrightarrow[n\to\infty]{P-\text{a.s.}} \mathbb{E}Y_1$ and $\| \cdot \|_\rho^2$ is continuous, therefore,

$$\| \overline{Y}_n \|_\rho^2 \xrightarrow[n\to\infty]{P-\text{a.s.}} \| \mathbb{E}Y_1 \|_\rho^2 \qquad\qquad (\| \mathbb{E}Y_1 \|_\rho^2 \le \mathbb{E} \| Y_1 \|_\rho^2 < \infty) \,.$$

Furthermore, $d_k = \| X_k \|_\rho^2$; $k = 1, \ldots, n$; is a sequence of independent identically distributed variables, such that

$$\overline{d}_n := \frac{1}{n}\sum_{k=1}^{n} d_k \underset{n\to\infty}{\overset{P-\text{a.s.}}{\longrightarrow}} \mathbb{E}d_1 = \mathbb{E} \| Y_1 \|_\rho^2 \ .$$

Hence,

$$S_n^2 = \underset{n\to\infty}{\overset{P-\text{a.s.}}{\longrightarrow}} \mathbb{E} \| Y_1 \|_\rho^2 - \| \mathbb{E}Y_1 \|_\rho^2 = \mathrm{Var}(Y_1) \ .$$

Furthermore S_n^2 is unbiased, because

$$\begin{aligned}
\mathbb{E}\frac{n-1}{n}S_n^2 &= \mathbb{E}\frac{1}{n}\sum_{k=1}^{n} \| Y_k \|_\rho^2 - \mathbb{E} \| \overline{Y}_n \|_\rho^2 \\
&= \mathbb{E} \| Y_1 \|_\rho^2 - \mathbb{E} \| \overline{Y}_n \|_\rho^2 \\
&= \mathbb{E} \| Y_1 \|_\rho^2 - \Big(\frac{1}{n}\mathbb{E} \| Y_1 \|_\rho^2 + \frac{n-1}{n} \| \mathbb{E}Y_1 \|_\rho^2\Big) \\
&= \frac{n-1}{n}\Big(\mathbb{E} \| Y_1 \|_\rho^2 - \| \mathbb{E}Y_1 \|_\rho^2\Big)
\end{aligned}$$

and with Theorem 5: $\mathbb{E}S_n^2 = \mathbb{E} \| Y_1 \|_\rho^2 - \| \mathbb{E}Y_1 \|_\rho^2 = \mathrm{Var}\, Y_1$. ∎

Boswell and Taylor (1987) establish that for independent rfv's Y_1 and Y_2 the relation $\mathbb{E}(Y_1 \odot Y_2) = \mathbb{E}Y_1 \odot \mathbb{E}Y_2$ is valid, with its own concept of independence. But a indepenence assumption is not really necessary. The product of two fuzzy numbers A and B can be defined α-cut-wise by

$$(A \odot B)_\alpha = \{a \cdot b \ : \ a \in A_\alpha,\ b \in B_\alpha\} \ .$$

Theorem 8. *Let Y_1 and Y_2 be independent integrable rfv's of $\mathbb{R}^1$. Then $\mathbb{E}(Y_1 \odot Y_2) = \mathbb{E}Y_1 \odot \mathbb{E}Y_2$.*

Proof. The operation $A \odot B$ is linear in both arguments, i.e. for all fuzzy sets A, B_1, B_2 and all real numbers λ

$$A \odot (\lambda B_1) = \lambda(A \odot B_1) \quad \text{and} \quad A \odot (B_1 \oplus B_2) = (A \odot B_1) \oplus (A \odot B_2)$$

(for the first argument use the symmetry $A \odot B = B \odot A$). By the independence of Y_1 and Y_2 and by the linearity of the expectation we obtain

$$\mathbb{E}(Y_1 \odot Y_2) = \mathbb{E}_{Y_1}\mathbb{E}_{Y_2}(Y_1 \odot Y_2 \mid Y_1) = \mathbb{E}_{Y_1}(Y_1 \odot \mathbb{E}Y_2 \mid Y_1) = \mathbb{E}Y_1 \odot \mathbb{E}Y_2. \quad ∎$$

3 Applications

3.1 Limit theorems with respect to ρ_2

Klement, Puri and Ralescu (1986) gave a first law of large numbers for rfv's, using results from random variables with values in a Banach space. Lyashenko

(1983) proved a strong law of large numbers for sums of compact convex random sets, based on the three-series theorem of Kolmogorov, by use of an L_2-based variance. Our definition of variance makes it possible to prove a strong law of large numbers by a direct application of classical methods (e.g. used in Rényi, 1970). For the proof of a strong law of large numbers the Kolmogorov inequality for rfv's is used:

Theorem 9. (cf. Körner, 1997b) *Let $Y_1, \ldots, X_n$ be independent rfv's with $\mathbb{E} \parallel Y_k \parallel_\rho^2 < \infty$. For $\varepsilon > 0$*

$$P\left(\max_{1 \le k \le n} \rho_2 \left(\sum_{j=1}^{k} Y_j, \sum_{j=1}^{k} \mathbb{E}Y_j\right) \ge \varepsilon\right) \le \frac{\sum_{k=1}^{n} \operatorname{Var}(Y_k)}{\varepsilon^2}.$$

Theorem 10. (cf. Körner, 1997b) *Let $Y_1, Y_2, \ldots$ be a sequence of independent rfv's, with $\mathbb{E} \parallel Y_k \parallel_\rho^2 < \infty$. If the series $\sum_{k=1}^{\infty} \operatorname{Var}(X_k)/k^2$ converges, then*

$$P\left(\lim_{n \to \infty} \rho_2 \left(\frac{1}{n} \sum_{k=1}^{n} X_k, \frac{1}{n} \sum_{k=1}^{n} \mathbb{E}X_k\right) = 0\right) = 1.$$

A first central limit theorem for rfv's was given by Klement *e al.* (1986). But this central limit theorem is too stringent, because of the Lipschitz condition on the membership function. Fuzzy numbers with membership functions $m_A(x) = \max\{1 - x^2, 0\}$ are not Lipschitzean.

In the central limit theorem of Boswell and Taylor (1987) the concept of selectors is used. For this, the existence of the moment generating function of each selector is assumed, i.e. each selector has moments of all orders. Again the conditions are too restrictive, because the existence of the moment generating function implies that the rfv's have moments of all orders, i.e. $\mathbb{E} \parallel X \parallel_\rho^k < \infty$ for any $k \in N$.

Since fuzzy sets can be characterized by their sequence of α-level sets results of the theory of compact convex sets can be used (Weil, 1982, Lyashenko, 1983).

Here, by the isomorphic embedding of the class of convex fuzzy sets in the function space $L_2(S^{d-1} \times (0,1])$ results in the Hilbert space $L_2(S^{d-1} \times (0,1])$ are transferred into the space of rfv's.

Remember that a random element Z in a linear normed space is called Gaussian, if, for each linear continuous functional f, the real-valued random variable $f(Z)$ is Gaussian.

Again, a Gaussian random element Z is characterized by the mean $\mathbb{E}Z$ and the covariance C_Z.

Theorem 11. (Application of Vakhania, 1981, and Sazonov, 1958) *Let $Y_1, Y_2, \ldots$ be a sequence of independent, identically distributed rfv's with*

$\mathbb{E} \| Y_1 \|_\rho^2 < \infty$ and covariance C. Then the normalized sum

$$S_n := \frac{1}{\sqrt{n}} \sum_{k=1}^{n} (s_{Y_k} - s_{\mathbb{E}Y_k})$$

converges, in ρ_2, to a Gaussian variable Z in $L_2(S^{d-1} \times (0,1])$ with expectation zero and covariance C.

Furthermore, the random variable

$$\sqrt{n}\rho_2\left(\overline{Y}_n, \mathbb{E}Y_1\right)$$

converges in distribution to the random variable $\| Z \|_\rho$.

Furthermore, the distribution of $\| Z \|_\rho$ can be characterized. This is important for the construction of asymptotical tests and for asymptotical confidence ellipsoids for the mean of rfv's.

Theorem 12. (Application of Vakhania, 1981, Dunford and Schwartz, 1988) *Let Z be a Gaussian random variable on the separable Hilbert space with covariance C_Z and let $\lambda_1 \geq \lambda_2 \geq \ldots$ be the eigenvalues of C_Z in decreasing order (each written as often as its multiplicity). Then, the distribution of the norm can be expressed by*

$$\| Z - \mathbb{E}Z \|_\rho^2 \overset{\mathrm{D}}{\sim} \sum_{k=1}^{\infty} \lambda_k \xi_k^2 , \tag{10}$$

where $\xi_1, \xi_2, \ldots$ are independent normalized Gaussian random numbers.

The random variable $\omega^2 := \sum_{k=1}^{\infty} \lambda_k \xi_k^2$ is called ω^2-distributed (omega square-distributed, cf. Martynov, 1978).

The ω^2-distribution is known only for special cases, i.e. for special sequences of eigenvalues $\lambda_1, \lambda_2, \ldots$.

3.2 An asymptotic α-test for the expectation

In test theory, significance testing (α-test) is a decision about whether the distribution of a stochastic quantity ξ belongs to a class of distribution $H_0 : \{P_\theta \,|\, \theta \in \Theta_0\}$ or not. This decision is based on a test variable

$$t(\xi_1, \ldots, \xi_n) \overset{\mathrm{D}}{\sim} P_t ,$$

whereby the distribution P_t of $t(\xi_1, \ldots, \xi_n)$ is known under H_0. For example for an iid. normally distributed sample $\xi_1, \ldots, \xi_n \overset{\mathrm{D}}{\sim} \mathcal{N}(\mu, \sigma^2)$ for the hypothesis $H_0 : \mu = \mu_0$ the test variable

$$t(\xi_1, \ldots, \xi_n) = \frac{\overline{\xi}_n - \mu_0}{S_n / \sqrt{n}}$$

is t-distributed under H_0 and the hypothesis H_0 is rejected if

$$|t(\xi_1, \dots, \xi_n)| = \left| \frac{\overline{\xi}_n - \mu_0}{S_n/\sqrt{n}} \right| > t_{n-1,1-\alpha/2} \, ,$$

where $t_{n-1,1-\alpha/2}$ is the $(1-\alpha/2)$-quantile of the t-distribution with $(n-1)$ degrees of freedom.

Note that $|t(\xi_1, \dots, \xi_n)| = \sqrt{n} \left| \overline{\xi}_n - \mu_0 \right| / S_n$ is a distance between $\overline{\xi}_n$ and μ_0.

One way to construct tests for fuzzy samples is given by the extension of classical α-tests to fuzzy tests by the extension principle (cf. Casals and Gil, 1986, Kruse and Meyer, 1987, Watanabe and Imaizumi, 1993, Viertl, 1996). But the extensions are not α-tests.

Gebhardt *et al.* (1998) is recommended for a summary.

Since there are only trivial normally distributed rfv's, there is not a test theory with normally distributed fuzzy samples. We propose a method to test hypotheses about the expectation with respect to the limit distribution. By the central limit theorem (cf. Theorem 11) the asymptotical distribution of the distance between the sample mean $\overline{Y}_n$ and the expectation $\mathbb{E}Y$ can be calculated by an ω^2-distribution (cf. Theorem 12). Using the asymptotical distribution, an asymptotic α-test for fuzzy data is constructed, i.e. a hypotheses $H_0 : \mathbb{E}Y = \mu_0$ will be rejected if the distance between the sample mean $\overline{Y}_n$ and the hypothetical value μ_0 is greater than the $(1-\alpha)$-quantile of the asymptotical ω^2-distribution.

We restrict ourselves to the distance δ_2 from (5).

Theorem 13. *Let $Y_1, \dots, Y_n$ be independent and identically distributed rfv's with $\mathbb{E} \parallel Y_1 \parallel_2^2 < \infty$. Then, the test variable*

$$n \, \delta_2 \left(\overline{Y}_n, \, \mu_0 \right)^2$$

is under $H_0 : \mathbb{E}Y = \mu_0$ asymptotically ω^2-distributed and the test:

$$\textit{reject } H_0 \qquad \textit{if} \qquad n \, \delta_2 \left(\overline{Y}_n, \, \mu_0 \right)^2 > t_{1-\alpha}$$

is an asymptotical α-test, where t_q is the q-quantile of the ω^2-distribution with respect to the eigenvalues $\lambda_1, \lambda_2, \dots$ of the covariance of Y_1.

In particular, with three real-valued random variables m, l, r, with $P\{l \geq 0\} = P\{r \geq 0\} = 1$ a random LR-fuzzy number is defined by $Y = \langle m, l, r \rangle_{LR}$. The covariance C_Y of a random LR-fuzzy number Y has only three eigenvalues.

Theorem 14. *Let $Y_1, \dots, Y_n$ be a sample of LR-fuzzy numbers with given covariance structure $\{C_{mm}, C_{ll}, C_{rr}, C_{lm}, C_{rm}, C_{lr}\}$. Then*

$$n \, \delta_2(\overline{Y}_n, \mathbb{E}Y)^2 \xrightarrow[n \to \infty]{\mathcal{L}} \lambda_1 \xi_1^2 + \lambda_2 \xi_2^2 + \lambda_3 \xi_3^2 \, ,$$

where ξ_1, ξ_2, ξ_3 are independent $\mathcal{N}(0,1)$-distributed random variables and $\lambda_1, \lambda_2, \lambda_3$ are the eigenvalues of the matrix K_Y

$$K_Y = \begin{pmatrix} C_{mm} - L_1 C_{lm} + R_1 C_{rm} & L_2 C_{lm} - L_1 C_{mm} & R_1 C_{mm} + R_2 C_{rm} \\ C_{lm} - L_1 C_{ll} + R_1 C_{rl} & L_2 C_{ll} - L_1 C_{lm} & R_1 C_{lm} + R_2 C_{rl} \\ C_{rm} - L_1 C_{rl} + R_1 C_{rr} & L_2 C_{rl} - L_1 C_{rm} & R_1 C_{rm} + R_2 C_{rr} \end{pmatrix},$$

where $R_1 = 1/2 \int_0^1 R^{-1}(\alpha) d\alpha$ and for $z, y \in \{m, l, r\}$:

$$C_{zy} = \mathbb{E}(z - \mathbb{E}z)(y - \mathbb{E}y) .$$

Moreover, an asymptotical α-test of the hypothesis

$$H_0 : \mathbb{E}Y = \mu_0 = \langle m_\mu, l_\mu, r_\mu \rangle_{LR} \qquad \textit{against} \qquad H_1 : \mathbb{E}Y \neq \mu_0$$

is formulated by:

$$\textit{reject } H_0 \qquad \textit{if} \qquad n\,\delta_2\left(\overline{Y}_n,\, \mu_0\right)^2 > t_{1-\alpha} ,$$

where t_q is the q-quantile of the ω^2-distribution with respect to the eigenvalues λ_1, λ_2 and λ_3.

If m, l, r are independent, i.e. $C_{lm} = C_{lr} = C_{mr} = 0$, then the eigenvalues of covariance C_Y are equal to the eigenvalues of

$$\begin{pmatrix} C_{mm} & -L_1 C_{mm} & R_1 C_{mm} \\ -L_1 C_{ll} & L_2 C_{ll} & 0 \\ R_1 C_{rr} & 0 & R_2 C_{rr} \,. \end{pmatrix}$$

For totally symmetric fuzzy numbers ($L = R$, $l = r$) the structure of covariance is simple

$$K_Y = \begin{pmatrix} C_{mm} & 2L_2\, C_{lm} \\ C_{lm} & 2L_2\, C_{ll} \end{pmatrix}$$

and the eigenvalues are

$$\lambda_{1,2} = \frac{1}{2}\left(2L_2\, C_{ll} + C_{mm} \pm \sqrt{(2L_2\, C_{ll} + C_{mm})^2 - 8L_2(C_{ll}C_{mm} - C_{lm}^2)}\right) .$$

The density of $\lambda_1\xi_1^2 + \lambda_2\xi_2^2 + \lambda_3\xi_3^2$ can be expressed in terms of density of χ^2-distributed variables (cf. Körner, 2000). Of course, a hypothesis will be accept if $n\,\delta_2(\overline{Y}_n, \mathbb{E}Y)^2$ is less than the expectation of $\lambda_1\xi_1^2 + \lambda_2\xi_2^2 + \lambda_3\xi_3^2$, where $\mathbb{E}(\lambda_1\xi_1^2 + \lambda_2\xi_2^2 + \lambda_3\xi_3^2) = \lambda_1 + \lambda_2 + \lambda_3 = \mathbf{tr}(K_Y) = \operatorname{Var} Y_1$. Moreover, as a fast decision for accepting hypothesis, we can use the Gaussian distribution. Let be $\lambda_1 \geq \lambda_2 \geq \lambda_3 > 0$, then

$$\lambda_1\xi_1^2 + \lambda_2\xi_2^2 + \lambda_3\xi_3^2 \geq \lambda_1\xi_1^2 \quad \text{i.e.} \quad \lambda_1\chi_{1,1-\alpha}^2 \leq t_{1-\alpha}$$

where $\chi^2_{1,1-\alpha}$ is the $(1-\alpha)$-quantile of the χ^2_1-distribution. Furthermore, $\chi^2_{1,1-\alpha} = \left(z_{1-\alpha/2}\right)^2$, where $z_{1-\alpha/2}$ is the $(1-\alpha/2)$-quantile of the Gaussian distribution. Hence the hypothesis will not be rejected if

$$n\,\delta_2(\overline{Y}_n, \mathbb{E}Y)^2 < \lambda_1 \left(z_{1-\alpha/2}\right)^2 .$$

On the other hand, we obtain

$$\lambda_1\xi_1^2 + \lambda_2\xi_2^2 + \lambda_3\xi_3^2 \le \lambda_1(\xi_1^2 + \xi_2^2 + \xi_3^2) \quad \text{i.e.} \quad t_{1-\alpha} \le \lambda_1\chi^2_{3,1-\alpha} ,$$

where $\chi^2_{3,1-\alpha}$ is the $(1-\alpha)$-quantile of the χ^2_3-distribution. Hence we obtain a simply decision for rejecting the hypothesis. The hypothesis will be rejected if

$$n\,\delta_2(\overline{Y}_n, \mathbb{E}Y)^2 > \lambda_1\chi^2_{3,1-\alpha} .$$

Example 1. Test with LR-fuzzy numbers
Let m be Gaussian distributed $m \overset{D}{\sim} \mathcal{N}(3, \frac{1}{2})$ and l and r be independent uniformly distributed on $[0, 1]$. Then, $C_{ml} = C_{mr} = C_{lr} = 0$ and

$$\mathbb{E}m = 3 , \quad \mathbb{E}l = \mathbb{E}r = \frac{1}{2} , \quad C_{mm} = \frac{1}{2} \quad \text{and} \quad C_{ll} = C_{rr} = \frac{1}{12} .$$

The matrix corresponding to the covariance (11)

$$K_Y = \begin{pmatrix} \frac{1}{2} & -\frac{1}{8} & \frac{1}{8} \\ -\frac{1}{48} & \frac{1}{72} & 0 \\ \frac{1}{48} & 0 & \frac{1}{72} \end{pmatrix}$$

has eigenvalues

$$\lambda_1 = 0.5105, \ \lambda_2 = 0.0139, \ \lambda_3 = 0.0034 .$$

Let $\{Y_1, \ldots, Y_{10}\}$ be a sample of the triangular random fuzzy quantity $Y = \langle m, l, r \rangle_{LR}$

$$\begin{array}{ll} Y_1 = \langle 3.1852, 0.4682, 0.1370 \rangle_{LR} & Y_2 = \langle 2.6886, 0.1936, 0.4585 \rangle_{LR} \\ Y_3 = \langle 1.7157, 0.4016, 0.6891 \rangle_{LR} & Y_4 = \langle 3.3739, 0.9307, 0.2790 \rangle_{LR} \\ Y_5 = \langle 2.7512, 0.4672, 0.4093 \rangle_{LR} & Y_6 = \langle 3.1022, 0.3784, 0.7399 \rangle_{LR} \\ Y_7 = \langle 3.5732, 0.5176, 0.0313 \rangle_{LR} & Y_8 = \langle 2.5636, 0.2278, 0.3039 \rangle_{LR} \\ Y_9 = \langle 2.1573, 0.8993, 0.1926 \rangle_{LR} & Y_{10} = \langle 1.5278, 0.9858, 0.8132 \rangle_{LR} \end{array}$$

with $\overline{Y}_n = \langle 2.6639, 0.5470, 0.4054 \rangle_{LR}$.
By Theorem 14 we obtain

$$n\,\delta_2(\overline{Y}_n, \mathbb{E}Y)^2 \xrightarrow[n\to\infty]{\mathcal{L}} \lambda_1\xi_1^2 + \lambda_2\xi_2^2 + \lambda_3\xi_3^2,$$

where ξ_1, ξ_2, ξ_3 are independent $\mathcal{N}(0, 1)$-distributed. Now, the hypothesis

$$H_0 : \mathbb{E}Y = \mu_0 = \langle 3, 1/2, 1/2 \rangle_{LR} \qquad \text{against} \qquad H_1 : \mathbb{E}Y \ne \langle 3, 1/2, 1/2 \rangle_{LR}$$

will not be rejected, because the distance between the average and the hypothetical value $n\,\delta_2(\overline{Y}_n, \mu_0)^2 = 1.1485$ is less than $\lambda_1 \left(z_{1-\alpha/2}\right)^2 = 1.96106$. The critical value for $\alpha = 0.05$ is $t_{0.95} = 1.984$.
Moreover, the hypothesis

$$H_0: \ \mathbb{E}Y = \langle 3,0,0\rangle_{LR} \qquad \text{against} \qquad H_1: \mathbb{E}Y \neq \langle 3,0,0\rangle_{LR} \ ,$$

i.e. there is no fuzziness, is rejected, because the distance between the average mean and the hypothesis $n\,\delta_2(\overline{Y}_n, \langle 3,0,0\rangle_{LR})^2 = 2.1406$ is larger than the critical value $t_{0,95} = 1.984$. (The simply decision for rejecting the hypothesis is not applicable, because $\lambda_1 \chi^2_{3,1-\alpha} = 3.98942$.)

3.3 Concluding remark

Having linearity properties of $\mathbb{E}Y$ and $\operatorname{Var} Y$, we can ask for linear unbiased estimation with minimal variance. For this application we refer to our second paper in this volume (see Näther and Körner, 2001).

References

1. Araujo, A., Giné, E. (1980). *The central limit theorem for real and Banach valued random variables.* John Wiley and Sons.
2. Aumann, R.J. (1965). Integrals of set-valued functions, *J. Math. Anal. Appl.* **12**, 1–12.
3. Bardossy, A., Hagaman, R., Duckstein, L., Bogardi, I. (1992). Fuzzy least squares regression: Theory and application. In *Fuzzy Regression Analysis* (J. Kacprzyk and M. Fedrizzi, Eds.). Physica-Verlag, Heidelberg, 183–193.
4. Bertoluzza, C., Salas A., Corral, N. (1995). On a new class of distances between fuzzy numbers, *Mathware & Soft Computing* **2**, 71–84.
5. Boswell, S.B., Taylor, M.S. (1987). A central limit theorem for fuzzy random variables, *Fuzzy Sets and Systems* **24**, 331–344.
6. Casals, M.R., Gil, M.A., Gil, P. (1986). The fuzzy decision problem: an approach of testing statistical. hypotheses with fuzzy information, *J. Oper. Res.* **27**, 371–382.
7. Diamond, P. (1988). Fuzzy least squares, *Inform. Sci.* **46**, 141–157.
8. Diamond, P., Kloeden, P. (1990). Metric spaces of fuzzy sets, *Fuzzy Sets and Systems* **35**, 241–249.
9. Dunford, N., Schwartz, J.T. (1988). *Linear operators.* Part I. John Wiley & Sons Ltd.
10. Fréchet, M. (1948). Les éléments aléatoires de natures quelconque dans un éspace distancié, *Ann. Inst. H. Poincaré* **10**, 215–310.
11. Gebhardt, J., Gil, M.A., Kruse, R. (1998). Fuzzy set-theoretic methods in Statistics. In *Fuzzy Sets in Decision Analysis, Operations Research and Statistics* Ch. 10 (R. Słowiński, Eds.), 311-347. Kluwer Acad. Pub., Norwell.
12. Hirota, K. (1981). Concepts of probabilistic sets, *Fuzzy Sets and Systems* **5**, 31–46.

13. Klement, E.P., Puri, M.L., Ralescu, D.A. (1986). Limit theorems for fuzzy random variables, *Proc Royal Soc. London - Series A* **19**, 171–182.
14. Körner, R. (1997a). *Linear Models with Random Fuzzy Variables*, PhD Thesis, Faculty of Mathematics and Computer Sciences, Freiberg University of Mining and Technology.
15. Körner, R. (1997b). On the variance of fuzzy random variables, *Fuzzy Sets and Systems* **92**, 83–93.
16. Körner, R. (2000). An asymptotic α-test for the expectation of random fuzzy variables, *J. Statist. Plan. Infer.* **83**, 331-346.
17. Kruse, R. (1987). On the variance of random sets, *J. Math. Anal. Appl.* **122**, 469–473.
18. Kruse, R., Meyer, K.D. (1987). *Statistics with Vague Data.* D. Reidel Publ. Comp. Dordrecht, Boston.
19. Kwakernaak, H. (1978). Fuzzy random variables - I. Definitions and theorems, *Inform. Sci.* **15**, 1–29.
20. Lubiano, M.A., Gil, M.A., López, A.T., López-Diaz, M. (2000). The $\boldsymbol{\lambda}$-mean squared dispersion associated with a fuzzy random variable, *Fuzzy Sets and Systems* **111**, 307–317
21. Lyashenko, N.N. (1983). Statistics of random compacts in Euclidean space, *J. Sovet. Math.* **21**, 76–92.
22. Martynov, G.V. (1978). *Omega-square criteria.* Nauka, Moscow (in Russian).
23. Matheron, G. (1975). *Random Sets and Integral Geometry.* J. Wiley & Sons, New York.
24. Molchanov, I.S. (1993). *Limit Theorems for Unions of Random Closed Sets.* Lecture Notes in Mathematics **1561**, Springer-Verlag, Berlin-Heidelberg-New York.
25. Näther, W. (1997). Linear statistical inference for random fuzzy data, *Statistics* **29**, 221–240.
26. Näther, W. and Körner, R. (2001). Linear Regression with Random Fuzzy Observations. (In Part 4 of this volume).
27. Puri, M.L., Ralescu, D.A. (1986). Fuzzy random variables, *J. Math. Anal. Appl.* **114**, 409–422.
28. Radström, H. (1952). An embedding theorem for spaces of convex sets, *Proc. Amer. Math. Soc.* **3**, 165–169.
29. Rényi, A. (1979). *Probability Theory.* Akadémiai Kiadó, Budapest.
30. Sazonov, V.V. (1958). A remark on characteristic functionals, *Theory of Probability and Its Applications* **3**.
31. Stoyan, D., Stoyan, H. (1994). *Fractals, Random Shapes and Point Fields.* J. Wiley & Sons, Chichester.
32. Vakhania, N.N. (1981). *Probability Distribution on Linear Spaces.* Elsevier Science Publishers B.V., North Holland.
33. Viertl, R. (1996). *Statistical Methods for Non-Precise Data.* CRC Press, Boca Raton, New York, London, Tokyo.
34. Watanabe, N., Imaizumi, T. (1993). A fuzzy statistical test of fuzzy hypothesis, *Fuzzy Sets and Systems* **53**, 167–178.
35. Weil, W. (1982). An application of the central limit theorem for Banach-space-valued random variables to the theory of random sets, *Z. Wahrscheinlichkeitsth. verw. Geb.* **14**, 582–599.

f-inequality indices for fuzzy random variables

María Asunción Lubiano and María Ángeles Gil

Departamento de Estadística e I.O. y D.M., Universidad de Oviedo, 33071
Oviedo, Spain

Abstract. This paper presents a generalized family of real-valued inequality indices associated with fuzzy-valued random elements. This family is first defined and later several general and particular desirable properties of the indices are examined. The unbiased estimation of an index of the above family is stated. Examples are considered to illustrate the studies developed in this paper.

1 Introduction

One of the fundamental aims in Statistics is describing a set of observations in terms of a few measures summarizing this set. Among the main summary measures one can observe the location ones (and especially the expected value), and the variation ones. When there is no variation of observations, the statistical methodology has no real interest. Furthermore, the quantification of the variation makes greater sense when we wish to compare populations, samples, variables, estimators, etc.

The variation associated with the random magnitude providing us with the observations, can be quantified by means of either absolute or relative measures. The first ones usually measure the variation in the units (or the squared units) of the magnitude, and they try to achieve an idea of the extent to which the location measures (or more generally, certain reference values) represent the values of the magnitude. The measures of the relative variation (more precisely, the inequality measures) are usually dimensionless indices that try to achieve an idea of the extent to which the location measures (or referential values) are above or below the values of the considered magnitude.

The most common measures of the relative variation of real-valued random variables are defined in terms of the ratios between certain values, so that they are scale invariant. Consequently, they are especially suitable to deal with random variables which are measured in a ratio scale.

In this work we will present some relative variation measures for random elements taking on fuzzy values. The model employed to characterize random elements taking on fuzzy values is that of fuzzy random variables (also referred to often as random fuzzy sets) in the sense formalized by Puri and Ralescu (1986).

The measures of the relative variation defined in this paper are assumed to take on real values. Nevertheless, some authors have emphasized that the inequality is an intrinsically imprecise characteristic, even when it is quantified for real-valued magnitudes. However, since the main aim of inequality indices is serving as the basis to compare populations, magnitudes, etc., we have chosen to define real-valued indices allowing a direct comparison.

In Section 2 the preliminaries for the study developed in this paper are recalled. In Section 3 a generalized measure of the (relative) inequality of a fuzzy random variable is defined and properties of this measure are examined. In Section 4 the estimation of a particular measure in random samplings from finite populations is established. Finally, illustrative examples are considered.

2 Preliminaries

In this section some basic concepts along with some supporting results are recalled.

Throughout this work, the involved experimental data are assumed to be imprecise. Models dealing with this imprecision will be certain fuzzy sets of the space of real numbers $\mathbb{R}$. The fuzzy subsets of $\mathbb{R}$ we will handle in the present work satisfy the conditions indicated in the following definition:

Definition 1. $\mathcal{F}_c(\mathbb{R})$ denotes the class of fuzzy subsets of $\mathbb{R}$, $\widetilde{V} : \mathbb{R} \to [0,1]$, satisfying that

i) the α-level set of $\widetilde{V}$, $\widetilde{V}_\alpha = \{x \in \mathbb{R} \mid \widetilde{V}(x) \geq \alpha\}$, is compact for all $\alpha \in (0,1]$,

ii) $\widetilde{V}_1 = \{x \in \mathbb{R} \mid \widetilde{V}(x) = 1\} \neq \emptyset$ (i.e., $\widetilde{V}$ is normal),

iii) $\widetilde{V}$ is a convex fuzzy subset, that is, for any $\alpha \in (0,1]$ the α-level set $\widetilde{V}_\alpha$ is a convex subset of $\mathbb{R}$,

iv) the closed convex hull of the support of $\widetilde{V}$ (where $\operatorname{supp} \widetilde{V} = \{x \in \mathbb{R} \mid \widetilde{V}(x) > 0\}$), which in this case coincides with the closure of $\operatorname{supp} \widetilde{V}$ and is denoted by $\widetilde{V}_0$, is compact.

From now on, we will refer generically to the elements of $\mathcal{F}_c(\mathbb{R})$ as *fuzzy numbers*.

The statistical management of fuzzy numbers usually requires considering elementary operations between them. The *arithmetic of fuzzy numbers* is stated on the basis of *Zadeh's extension principle* (Zadeh, 1975), and on the basis of some results from Nguyen (1978), and because of working with compact α-levels, one can prove (see, for instance, López Díaz, 1996) for the algebraic operations that the fuzzy sum $\oplus$ (or, alternatively, $\textstyle\bigoplus\!\!\sum$), fuzzy substraction $\ominus$, fuzzy product by a real number $\odot$, and fuzzy quotient $\oslash$, satisfy for $\widetilde{V}$ and $\widetilde{W} \in \mathcal{F}_c(\mathbb{R})$ that

$$\left(\widetilde{V} \oplus \widetilde{W}\right)_\alpha = \left[\inf \widetilde{V}_\alpha + \inf \widetilde{W}_\alpha, \sup \widetilde{V}_\alpha + \sup \widetilde{W}_\alpha\right],$$

$$\left(\widetilde{V} \ominus \widetilde{W}\right)_\alpha = \left[\inf \widetilde{V}_\alpha - \sup \widetilde{W}_\alpha, \sup \widetilde{V}_\alpha - \inf \widetilde{W}_\alpha\right],$$

$$\left(\lambda \odot \widetilde{W}\right)_\alpha = \left[\lambda \cdot \inf \widetilde{W}_\alpha, \lambda \cdot \sup \widetilde{W}_\alpha\right], \text{ if } \lambda \geq 0,$$

$$\left(\lambda \odot \widetilde{W}\right)_\alpha = \left[\lambda \cdot \sup \widetilde{W}_\alpha, \lambda \cdot \inf \widetilde{W}_\alpha\right], \text{ if } \lambda < 0,$$

and if $\widetilde{V}$ and $\widetilde{W}$ are in $\mathcal{F}_c((0,+\infty))$ (i.e, $\widetilde{V}$, $\widetilde{W} \in \mathcal{F}_c(\mathbb{R})$ and $\widetilde{V}_0$, $\widetilde{W}_0 \subset (0,+\infty)$) it holds for all $\alpha \in [0,1]$ that

$$\left(\widetilde{V} \oslash \widetilde{W}\right)_\alpha = \left[\inf \widetilde{V}_\alpha / \sup \widetilde{W}_\alpha, \sup \widetilde{V}_\alpha / \inf \widetilde{W}_\alpha\right].$$

On the basis of the fuzzy sum and product by a real number above, it is possible to establish in an analogous way to the real-valued case, the matricial operations, when the elements involved in the matrices are fuzzy numbers (see Dubois and Prade, 1980).

When we develop a statistical study with fuzzy numbers, and especially whenever our aim is to compare populations, variables, etc., we have often to order them. There are several criteria to *ranking fuzzy numbers* (see, for instance, Adamo, 1980, Yager, 1981, Bortolan and Degani, 1985, Ramík and Římanek, 1985, Kołodziejczyk, 1986, Nakamura 1986, Delgado *et al.* 1988, Campos and González 1989, Tseng and Klein, 1989, González and Vila, 1992, and Sánchez de Posada Martínez, 1998).

Some of these criteria are based on crisp relations, that is, they determine a categorical ordering (or preordering) on the class $\mathcal{F}_c(\mathbb{R})$. Among them, we can distinguish the group of the criteria leading to a total ordering (or preordering), and the group of the criteria leading to a partial one.

In Section 3 we will make use of two different criteria, which we now present. The application of the criterion stated by Ramík and Římanek (1985) leads to a partial ordering on $\mathcal{F}_c(\mathbb{R})$ which is universally accepted, and is formalized as follows:

Definition 2. If $\widetilde{V}$, $\widetilde{W} \in \mathcal{F}_c(\mathbb{R})$, then $\widetilde{V} \succeq_S \widetilde{W}$ (or, $\widetilde{W}$ is said to be strongly dominated by $\widetilde{V}$ in accordance with *Ramík-Římanek's criterion*) if, and only if,

$$\inf \widetilde{V}_\alpha \geq \inf \widetilde{W}_\alpha \text{ and } \sup \widetilde{V}_\alpha \geq \sup \widetilde{W}_\alpha,$$

for all $\alpha \in [0,1]$.

The inconveniences of employing this criterion in practice are due to the fact that there are many pairs of fuzzy numbers this criterion cannot compare. However, it leads to an ordering since $\widetilde{V} \succeq_S \widetilde{W}$ and $\widetilde{W} \succeq_S \widetilde{V}$ (that will be denoted by $\widetilde{V} \sim_S \widetilde{W}$) is equivalent to $\widetilde{V} = \widetilde{W}$.

The following criterion avoids the situations which arise when fuzzy numbers cannot be compared, and it agrees with Ramík and Římánek's criterion for the pairs it is applicable to. This criterion is based on one of the ranking functions introduced by Yager (1981) which is stated as follows:

Definition 3. If $\widetilde{V}$, $\widetilde{W} \in \mathcal{F}_c(\mathbb{R})$, $\widetilde{V}$ is said to be greater than or equal to $\widetilde{W}$ in accordance with *Yager's ranking criterion*, and it will be denoted by $\widetilde{V} \geq_Y \widetilde{W}$, if, and only if, $F(\widetilde{V}) \geq F(\widetilde{W})$, where F is the ranking function defined for any $\widetilde{V} \in \mathcal{F}_c(\mathbb{R})$ by

$$F(\widetilde{V}) = \frac{1}{2} \int_{[0,1]} \left[\sup \widetilde{V}_\alpha + \inf \widetilde{V}_\alpha\right] \, d\alpha.$$

Yager's criterion determines a total preordering on $\mathcal{F}_c(\mathbb{R})$ and it is a special case of the parameterized criterion introduced by Campos and González (1989).

On the other hand, there are several *metrics* which can be defined on the class $\mathcal{F}_c(\mathbb{R})$. In Diamond and Kloeden (1994) we can find a broad review on many of these metrics. In this section we only recall the one we will use throughout the paper, which is sometimes referred to as the sometimes called "generalized Hausdorff metric" (or supremum metric on $\mathcal{F}_c(\mathbb{R})$) and was introduced by Puri and Ralescu (1981, 1983).

Definition 4. If $\widetilde{V}$, $\widetilde{W} \in \mathcal{F}_c(\mathbb{R})$, the *generalized Hausdorff metric* , between $\widetilde{V}$ and $\widetilde{W}$, $d_\infty(\widetilde{V}, \widetilde{W})$, is given by

$$d_\infty(\widetilde{V}, \widetilde{W}) = \sup_{\alpha \in [0,1]} d_H(\widetilde{V}_\alpha, \widetilde{W}_\alpha),$$

where d_H denotes the well-known Hausdorff metric on the class of the nonempty convex compact subsets of $\mathbb{R}$, $\mathcal{K}_c(\mathbb{R})$, which for K, $K' \in \mathcal{K}_c(\mathbb{R})$ is given by

$$d_H(K, K') = \max\left\{ \sup_{a \in K} \inf_{b \in K'} |a - b|, \sup_{b \in K'} \inf_{a \in K} |a - b| \right\},$$

and, because of the convexity of elements in $\mathcal{K}_c(\mathbb{R})$, it can be alternatively defined by

$$d_H(K, K') = \max\{\, |\inf K - \inf K'| \, , |\sup K - \sup K'| \,\}.$$

In this paper we will consider random experiments involving quantification processes which cannot be identified with a real-valued random variable but rather with *fuzzy-valued quantification processes* (more precisely, taking on values on $\mathcal{F}_c(\mathbb{R})$). It should be emphasized that the quantification processes will be assumed to be intrinsically imprecise, so that imprecision does not come from an imprecise perception or report of an existing real-valued quantification.

Assume that the original random experiment is mathematically modeled by means of a probability space $(\Omega, \mathcal{A}, P)$. Then, in accordance with Puri and Ralescu (1986), a *fuzzy random variable* is defined as follows:

Definition 5. A mapping $\mathcal{X} : \Omega \to \mathcal{F}_c(\mathbb{R})$ is said to be a *fuzzy random variable* (also called random fuzzy set) associated with $(\Omega, \mathcal{A})$ if, and only if, the α-level mapping $\mathcal{X}_\alpha : \Omega \to \mathcal{K}_c(\mathbb{R})$ (where $\mathcal{X}_\alpha(\omega) = (\mathcal{X}(\omega))_\alpha$ for all $\omega \in \Omega$) is a convex compact random set (i.e., $(\mathcal{A}, \mathcal{B}_{d_H})$-measurable, $\mathcal{B}_{d_H}$ being the σ-field generated by the topology induced from d_H on $\mathcal{K}_c(\mathbb{R})$) whatever $\alpha \in [0, 1]$ may be.

The *expected value* of a fuzzy random variable (in Kudo-Aumman's sense) has been introduced by Puri and Ralescu (1986) as follows:

Definition 6. If $\mathcal{X} : \Omega \to \mathcal{F}_c(\mathbb{R})$ is a fuzzy random variable, the *expected value* of $\mathcal{X}$ is the unique fuzzy subset of $\mathbb{R}$ (if it exists), $\widetilde{E}(\mathcal{X})$, such that for all $\alpha \in (0, 1]$ we have that $\left(\widetilde{E}(\mathcal{X})\right)_\alpha = E(\mathcal{X}_\alpha)$, that is, $\left(\widetilde{E}(\mathcal{X})\right)_\alpha$ equals the Aumann integral of the convex compact random set $\mathcal{X}_\alpha$.

When a fuzzy random variable $\mathcal{X} : \Omega \to \mathcal{F}_c(\mathbb{R})$ is *integrably bounded* (which means that $\sup_{x \in \mathcal{X}_0(\cdot)} |x| \in L^1(\Omega, \mathcal{A}, P)$), the expected value of $\mathcal{X}$ is unique and it is given by the compact interval $\left[E(\inf \mathcal{X}_\alpha), E(\sup \mathcal{X}_\alpha)\right]$ for all $\alpha \in [0, 1]$.

It should be emphasized that Yager's ranking criterion is especially operational when it is combined with the (fuzzy) expected value of a fuzzy random variable. More precisely, and on the basis of the results from López-Díaz and Gil (1998) (which have been really established for the more general criterion of Campos and González, 1989) one can conclude that if $\mathcal{X}$ is an integrably bounded fuzzy random variable with expected value $\widetilde{E}(\mathcal{X})$, then

$$F\left(\widetilde{E}(\mathcal{X})\right) = E(F \circ \mathcal{X}),$$

so that the value of the ranking function F for the (fuzzy) expected value of the fuzzy random variable $\mathcal{X}$ reduces to the expected value of the real-valued random variable $F \circ \mathcal{X}$.

Some special cases of fuzzy random variables, we will refer sometimes to in the present work, are the following:

Definition 7. A fuzzy random variable $\mathcal{X} : \Omega \to \mathcal{F}_c(\mathbb{R})$ is said to be *degenerate*, if there exists $\widetilde{V} \in \mathcal{F}_c(\mathbb{R})$ such that $\mathcal{X} = \widetilde{V}$ almost surely $[P]$. In particular, if $\widetilde{V}$ reduces to the indicator function of an interval in $\mathcal{K}_c(\mathbb{R})$, $\mathcal{X}$ is said to be a *fuzzy random variable degenerate at an interval value*, and if $\widetilde{V}$ reduces to the indicator function of a singleton in $\mathbb{R}$, $\mathcal{X}$ is said to be a *fuzzy random variable degenerate at a real value.*

Definition 8. A fuzzy random variable $\mathcal{X} : \Omega \to \mathcal{F}_c((0, +\infty))$ is said to be a *positive fuzzy random variable.*

Finally, we would like to comment that in the literature on the *quantification of the inequality of a population* with respect to a certain real-valued random variable, many indices have been proposed. Several of them have been widely accepted by both the international mathematical community and the international scientific communities in the fields of application. During recent years, some of the best known and most used inequality measures are those coinciding with (or being increasing functions of) the additively decomposable indices of order α (see, for instance, Bourguignon, 1979, Cowell, 1980, Shorrocks, 1980, Cowell and Kuga, 1981, Eichhorn and Gehrig, 1982, Zagier, 1983, Gil *et al.*, 1989b). Theil's index, the index of the Shannon type, and the hyperbolic index are additively decomposable indices and have been examined in detail in previous works (see, for instance, Gil and Gil, 1989, Gil *et al.*, 1989ab, Martínez, 1991).

The studies on inequality usually assume that the values of the considered attribute are positive, since the most common attributes are monetary (like income, wealth, etc.) or correspond to the size of a subpopulation. In addition, variables for which inequality is measured are usually ratio scale ones.

Recently (see Alonso *et al.*, 2001), a generalized family of inequality indices, including all the additively decomposable indices has been introduced. These indices are based on the generalized family of the directed-divergence measures stated by Csiszár (1967). The interest in using families of measures rich enough to allow us to find a proper measure in the family to handle each problem, supports the idea of constructing *generalized inequality measures.*

The indices we present in the following section are those introduced and analyzed for fuzzy random variables by Lubiano (1999) and also examined with respect to some special aspects by Alonso *et al.* (2001).

3 The f-inequality indices for fuzzy random variables

Assume that we consider a general population Ω and let $(\Omega, \mathcal{A}, P)$ be a probability space defined on it. Let $\mathcal{X} : \Omega \to \mathcal{F}_c((0, +\infty))$ be an integrably bounded fuzzy random variable associated with $(\Omega, \mathcal{A}, P)$ (that is, $\mathcal{X}$ being a positive fuzzy random variable such that $\sup \mathcal{X}_0 \in L^1(\Omega, \mathcal{A}, P)$). Let $f : (0, +\infty) \to \mathbb{R}$ be a strictly convex (intended as convex downward) and monotonic function satisfying that $f(u) + f(1/u) \geq 0$ for all $u \in (0, +\infty)$ and $f(1) = 0$, and let F be the Yager ranking function in $\mathcal{F}_c(\mathbb{R})$.

To quantify the inequality associated with $\mathcal{X}$ in Ω by means of a real-valued measurement, we suggest the following indices:

Definition 9. The *f-inequality index* associated with $\mathcal{X}$ in the population Ω given by the value (if it exists)

$$\mathrm{I}_f(\mathcal{X}) = F\left(\widetilde{\mathrm{I}}_f(\mathcal{X})\right),$$

where

$$\widetilde{\mathrm{I}}_f(\mathcal{X}) = \widetilde{E}\left[f\left(\mathcal{X} \oslash \widetilde{E}(\mathcal{X})\right)\right],$$

is the fuzzy f-inequality index associated with $\mathcal{X}$ in Ω (Colubi *et al.*, 1997, Colubi Cervero, 1997), where $f\left(\mathcal{X} \oslash \widetilde{E}(\mathcal{X})\right)$ denotes the image of $\mathcal{X} \oslash \widetilde{E}(\mathcal{X})$ induced from f on the basis of Zadeh's extension principle (that is,

$$\left(f\left(\mathcal{X} \oslash \widetilde{E}(\mathcal{X})\right)\right)_\alpha = \left[\min\left\{f\left(\frac{\inf \mathcal{X}_\alpha}{E(\sup \mathcal{X}_\alpha)}\right), f\left(\frac{\sup \mathcal{X}_\alpha}{E(\inf \mathcal{X}_\alpha)}\right)\right\},\right.$$
$$\left.\max\left\{f\left(\frac{\inf \mathcal{X}_\alpha}{E(\sup \mathcal{X}_\alpha)}\right), f\left(\frac{\sup \mathcal{X}_\alpha}{E(\inf \mathcal{X}_\alpha)}\right)\right\}\right]$$

for all $\alpha \in [0,1]$).

Occasionally, and when the probability measure P has to be specified, we will denote $\mathrm{I}_f(\mathcal{X})$ alternatively by $\mathrm{I}_f(\mathcal{X} \mid P)$.

The conditions assumed for f are satisfied by the functions associated with the additively decomposable indices for $\alpha \neq 0, 1$, and the function $f(x) = -\log x$ associated with the *index of the Shannon type* (see Alonso *et al.* 2001). However, the function associated with Theil's index ($f(x) = x\log(x)$) is nonmonotonic and it will be removed from the present study.

The f-inequality index is not necessarily defined for a fuzzy random variable in any population, and conditions guaranteeing the existence of $\widetilde{\mathrm{I}}_f$ depend on the function f.

The following conditions, which ensure the existence of $\widetilde{\mathrm{I}}_f(\mathcal{X})$ (see Colubi Cervero, 1997), would also guarantee that $\mathrm{I}_f(\mathcal{X}) \in \mathbb{R}$.

Consider a probabilistic space $(\Omega, \mathcal{A}, P)$, and assume that $\mathcal{X} : \Omega \to \mathcal{F}_c((0,+\infty))$ is a positive integrably bounded fuzzy random variable associated with $(\Omega, \mathcal{A}, P)$. Let $f : (0,+\infty) \to \mathbb{R}$ be a strictly convex and monotonic function belonging to C^1 and satisfying that $f(u) + f(1/u) \geq 0$ for all $u \in (0,+\infty)$ and $f(1) = 0$.

Then, the f-inequality index $\mathrm{I}_f(\mathcal{X})$ is well-defined and belongs to $\mathcal{F}_c(\mathbb{R})$ if, and only if,

(1) $f(\inf \mathcal{X}_0 / E(\sup \mathcal{X}_0)) \in L^1(\Omega, \mathcal{A}, P)$ if f is nonincreasing;
(2) $f(\sup \mathcal{X}_0 / E(\inf \mathcal{X}_0)) \in L^1(\Omega, \mathcal{A}, P)$ if f is nondecreasing.

On the other hand, the fuzzy f-inequality index associated with $\mathcal{X}$ in the population (when it exists under the conditions above assumed for f), is the fuzzy number $\widetilde{\mathrm{I}}_f \in \mathcal{F}_c(\mathbb{R})$ such that for all $\alpha \in [0,1]$

$$\left(\widetilde{\mathrm{I}}_f(\mathcal{X})\right)_\alpha = \left[\inf\left(\widetilde{\mathrm{I}}_f(\mathcal{X})\right)_\alpha, \sup\left(\widetilde{\mathrm{I}}_f(\mathcal{X})\right)_\alpha\right],$$

where

$$\inf\left(\widetilde{\mathrm{I}}_f(\mathcal{X})\right)_\alpha = E\left(\min\left\{f\left(\frac{\inf \mathcal{X}_\alpha}{E(\sup \mathcal{X}_\alpha)}\right), f\left(\frac{\sup \mathcal{X}_\alpha}{E(\inf \mathcal{X}_\alpha)}\right)\right\}\right),$$

$$\sup\left(\widetilde{\mathrm{I}}_f(\mathcal{X})\right)_\alpha = E\left(\max\left\{f\left(\frac{\inf \mathcal{X}_\alpha}{E(\sup \mathcal{X}_\alpha)}\right), f\left(\frac{\sup \mathcal{X}_\alpha}{E(\inf \mathcal{X}_\alpha)}\right)\right\}\right).$$

On the basis of the latter assertion, and because of the properties of the ranking function F, we have that

Theorem 1. *Let $(\Omega, \mathcal{A}, P)$ be a probabilistic space and let $\mathcal{X} : \Omega \to \mathcal{F}_c((0, +\infty))$ be an integrably bounded fuzzy random variable. Consider a mapping $f : (0, +\infty) \to \mathbb{R}$ strictly convex and monotonic, belonging to C^1, and satisfying $f(u) + f(1/u) \geq 0$ for all $u \in (0, +\infty)$ and $f(1) = 0$.*

If either (1) *or* (2) *are also satisfied, then,*

$$\mathrm{I}_f(\mathcal{X}) = \frac{1}{2}\int_{[0,1]} E\left[f\left(\frac{\inf \mathcal{X}_\alpha}{E(\sup \mathcal{X}_\alpha)}\right) + f\left(\frac{\sup \mathcal{X}_\alpha}{E(\inf \mathcal{X}_\alpha)}\right)\right] d\alpha.$$

Obviously, if we forget about either condition for the values of $\mathcal{X}$ or the monotonicity of the function f, we could not characterize the f-indices as presented in Theorem 1, which would mean an important (both, practical and theoretical) inconvenience for the later studies in this paper.

In the following section we will examine several properties being convenient and desirable in the measurement of the inequality of a population wih respect to a fuzzy random variable.

4 Properties of the f-inequality indices

From now on, and without mentioning it for each property, we will consider a probability space $(\Omega, \mathcal{A}, P)$, and a function $f : (0, +\infty) \to \mathbb{R}$ which is strictly convex and monotonic, belongs to C^1, and satisfies that $f(u) + f(1/u) \geq 0$ for all $u \in (0, +\infty)$ and $f(1) = 0$. We also will suppose that if f is nonincreasing Condition (1) in Section 3 is satisfied, and if f is nondecreasing Condition (2) in Section 3 is satisfied. The strict convexity of f could be weakened by assuming f is convex, but in such a case we could not establish the conditions under which the equality would hold in several of the properties below.

Proofs of some these properties can be found in Alonso *et al.* (2001), and all of them are gathered in Lubiano (1999).

The indices introduced in Section 3 are not changed by an equiproportional real-valued variation in the values of the fuzzy random variables. In other words, and in accordance with Kölm's terminology (1976ab) these indices are "rightist measures", and following Blackorby and Donaldson (1978) they are measures of "relative inequality". In this way, the following result extends to fuzzy random variables, the *mean independence property* (also referred to as *scale invariance* or *homogeneity of degree 0*) of most of the classical inequality indices.

Proposition 1 (Mean independence). *If $\mathcal{X} : \Omega \to \mathcal{F}_c\big((0,+\infty)\big)$ is an integrably bounded fuzzy random variable, then for all $k \in (0,+\infty)$ we have that $\mathrm{I}_f(k \odot \mathcal{X}) = \mathrm{I}_f(\mathcal{X})$.*

As commented in Alonso *et al.* (2001), as a consequence from Proposition 1 and whenever $\sup \mathcal{X}_0$ is assumed to be "uper bounded" in Ω, the extension of Theil's index could be carried out and properties in this section would be also applicable.

The *sign-preserving* (or *nonnegativeness*) holds for all the indices in Definition 9. Thus,

Proposition 2 (Nonnegativeness). *Let $\mathcal{X} : \Omega \to \mathcal{F}_c\big((0,+\infty)\big)$ be an integrably bounded fuzzy random variable. Then, we have that $\mathrm{I}_f(\mathcal{X}) \geq 0$.*

The *positiveness* (or *sensitivity out of equality*) is formalized in the following result:

Proposition 3 (Sensitivity out of equality). *Let $\mathcal{X} : \Omega \to \mathcal{F}_c\big((0,+\infty)\big)$ be an integrably bounded fuzzy random variable. If $\mathrm{I}_f(\mathcal{X}) = 0$, then $\mathcal{X}$ has to be a degenerate fuzzy random variable (that is, if $\mathcal{X}$ is nondegenerate we can ensure that $\mathrm{I}_f(\mathcal{X})$ is positive).*

The *insensitivity* or *nullity* of the f-inequality indices cannot be guaranteed for a degenerate fuzzy random variable whatever f may be. The following result states that for fuzzy random variables degenerate at a positive real number this insensitivity always holds.

Proposition 4 (Insensitivity). *Let $\mathcal{X} : \Omega \to \mathcal{F}_c\big((0,+\infty)\big)$ be an integrably bounded fuzzy random variable. If $\mathcal{X}$ is degenerate at a positive real value, then $\mathrm{I}_f(\mathcal{X}) = 0$.*

The next results present two different *minimality* properties for the f-indices, depending on certain conditions the function f satisfies.

Proposition 5 (Minimality I). *Let $\mathcal{X} : \Omega \to \mathcal{F}_c\big((0,+\infty)\big)$ be an integrably bounded fuzzy random variable. If f satisfies that $f(u) + f(1/u) = 0$ if, and only if, $u = 1$, then $\mathrm{I}_f(\mathcal{X}) = 0$ if, and only if, $\mathcal{X}$ is a fuzzy random variable degenerate at a positive real number.*

The additional condition $f(u) + f(1/u) = 0$ if, and only if, $u = 1$, is satisfied by many functions f (in particular, for those serving to extend the additively decomposable indices of order $\alpha \neq 0, 1$), although other valuable functions like $f(x) = -\log x$ (which is the basis of the Shannon type index) satisfy that $f(u) + f(1/u) = 0$ for all $u \in (0,+\infty)$ (see Alonso *et al.* 2001). In this latter case, the necessary and sufficient condition for $\mathrm{I}_f(\mathcal{X})$ being null is gathered in the following result:

Proposition 6 (Minimality II). *If $\mathcal{X} : \Omega \to \mathcal{F}_c\big((0,+\infty)\big)$ is an integrably bounded fuzzy random variable, and $f(u) + f(1/u) = 0$ for all $u \in (0,+\infty)$, then $\mathrm{I}_f(\mathcal{X}) = 0$ if, and only if, $\mathcal{X}$ is degenerate at an element in $\mathcal{F}_c\big((0,+\infty)\big)$.*

In accordance with Propositions 5 and 6, if $\mathcal{X}$ is a fuzzy random variable degenerate at a fuzzy number in $\mathcal{F}_c\big((0,+\infty)\big)$, the f-inequality index does not necessarily equal 0. Thus, for instance, if $\mathcal{X}$ equals almost surely the value $\tilde{x} = \text{Tri}(1,2,3)$ on Ω and we consider $f(x) = x^{-1} - 1$ for all $x \in (0,1)$, then we obtain that $\text{I}_f(\mathcal{X}) = .099$.

Reasons justifying that some f-indices do not vanish for degenerate fuzzy random variables, lie in the fact that several of these indices (in particular, those associated with functions f such that $f(u)+f(1/u) = 0$ if, and only if, $u = 1$) in addition to quantifying the *inter*values inequality also measures the *intra*values inequality. In this sense, and as a special case, one can prove the following *additive descomposition* property for the *hyperbolic index* I_H (that is, I_f with $f(x) = x^{-1} - 1$ for all $x \in (0,+\infty)$):

Proposition 7 (Additive decomposition of the hyperbolic index). *If $\Omega = \{\omega_1,\ldots,\omega_N\}$, $\mathcal{X}(\Omega) = \{\tilde{x}_1^*,\ldots,\tilde{x}_r^*\}$ and $p_l = P(\{\omega \in \Omega \,|\, \mathcal{X}(\omega) = \tilde{x}_l^*\})$, $l = 1,\ldots,r$, we have that*

$$\text{I}_H(\mathcal{X}) = \sum_{l=1}^{r} p_l^2\, \text{I}_H(\{\tilde{x}_l^*\}) + \text{I}_H^{bv}(\mathcal{X})$$

(where $\text{I}_H^{bv}(\mathcal{X})$ represents a kind of intervalues inequality index and $\text{I}_H(\{\tilde{x}_l^*\})$ denotes the intravalue inequality index for value $\tilde{x}_l^*$) with $\text{I}_H^{bv}(\mathcal{X}) = 0$ if, and only if, $\mathcal{X}$ is a degenerate fuzzy random variable.

When we deal with finite populations, there are more properties having a clear meaning and application. For this reason, for the remaining properties in this section we will consider a probability space $(\Omega, \mathcal{P}(\Omega), P)$ where Ω is the finite population $\{\omega_1,\ldots,\omega_N\}$ and P is the uniform distribution on Ω.

Proposition 8 (Expression for finite populations). *Let $\mathcal{X}$ be a positive fuzzy random variable defined on the population $\Omega = \{\omega_1,\omega_2,\ldots,\omega_N\}$. Then, we have that*

$$\text{I}_f(\mathcal{X}) = \frac{1}{2}\int_{[0,1]} \left[\frac{1}{N}\sum_{j=1}^{N} f\left(\frac{\inf \mathcal{X}_\alpha(\omega_j)}{\sup E(\mathcal{X}_\alpha)}\right) + \frac{1}{N}\sum_{j=1}^{N} f\left(\frac{\sup \mathcal{X}_\alpha(\omega_j)}{\inf E(\mathcal{X}_\alpha)}\right)\right] d\alpha.$$

The *symmetry* of the f-inequality indices formalizes the fact that they do not depend on the identity or numbering of individuals in the population (that is, the f-inequality indices are objective measures). In this way,

Proposition 9 (Symmetry). *Let $\mathcal{X} : \Omega \to \mathcal{F}_c\big((0,+\infty)\big)$ be a fuzzy random variable. Then, $\text{I}_f(\mathcal{X} \circ \sigma) = \text{I}_f(\mathcal{X})$ for any permutation σ on Ω.*

The *principle of population* (or *population homogeneity*) underlines the fact that the value of an f-inequality index of a given population Ω with respect to a positive fuzzy random variable coincides with that of the population $\Omega^{(r)} = \{\omega_{11}, \ldots, \omega_{1N}, \ldots, \omega_{r1}, \ldots, \omega_{rN}\}$ obtained from Ω by replicating it an arbitrary finite number r of times (i.e., $\omega_{ij} = \omega_j$ for all i, j), with respect to the immediate extension of this variable. In a more concise way, the f-inequality index only depends on the frequency of each possible value $\mathcal{X}$ in the population (that is, on the population structure) irrespectively of its size.

Proposition 10 (Population homogeneity). *If $\mathcal{X} : \Omega \to \mathcal{F}_c\big((0,+\infty)\big)$ is a fuzzy random variable and $\mathcal{X}^{(r)} : \Omega^{(r)} \to \mathcal{F}_c\big((0,+\infty)\big)$ is the fuzzy random variable extending $\mathcal{X}$ to $\Omega^{(r)}$, then $\mathrm{I}_f(\mathcal{X}^{(r)}) = \mathrm{I}_f(\mathcal{X})$.*

Another relevant property of the family of f-inequality indices is the *continuity,* in accordance with which "small" (fuzzy) changes in the values of the fuzzy random variables entail "small" (real-valued) variations in the f-inequality indices. To state this property we will make use of the d_∞ metric to formalize te idea that the "fuzzy change tends to" 0. Thus,

Proposition 11 (Continuity). *Let $\mathcal{X} : \Omega \to \mathcal{F}_c\big((0,+\infty)\big)$ be a fuzzy random variable and let $\mathcal{X}_{l,\widetilde{\mathcal{E}}}$ be a fuzzy random variable defined on Ω such that $\mathcal{X}_{l,\widetilde{\mathcal{E}}}(\omega_j) = \mathcal{X}(\omega_j)$ for all $j \in \{1, \ldots, N\} \setminus \{l\}$ and $\mathcal{X}_{l,\widetilde{\mathcal{E}}}(\omega_l) = \mathcal{X}(\omega_l) \oplus \widetilde{\mathcal{E}}$ with $\widetilde{\mathcal{E}} \in \mathcal{F}_c\big(\mathbb{R}\big)$ such that $\mathcal{X}(\omega_l) \oplus \widetilde{\mathcal{E}} \in \mathcal{F}_c\big((0,+\infty)\big)$. Then,*

$$\lim_{d_\infty(\widetilde{\mathcal{E}}, \mathbf{1}_{\{0\}}) \to 0} \mathrm{I}_f(\mathcal{X}_{l,\widetilde{\mathcal{E}}}) = \mathrm{I}_f(\mathcal{X}).$$

Some of the main and desirable properties of the inequality indices are those concerning the reaction of these indices to a "redistribution" of the values of the considered attribute.

The *strict Schur-convexity* of I_f, formalizes the fact that when the redistribution is carried out by considering convex linear combinations, the f-inequality index of the fuzzy random variable cannot increase in the population. Thus,

Proposition 12 (Strict Schur-convexity). *If $\mathcal{X} : \Omega \to \mathcal{F}_c\big((0,+\infty)\big)$ is a fuzzy random variable, (μ_{jl}) is an $N \times N$ doubly stochastic matrix and $\mathcal{X}' : \Omega \to \mathcal{F}_c\big((0,+\infty)\big)$ is another fuzzy random variable defined from $\mathcal{X}$ as follows:*

$$\begin{pmatrix} \mathcal{X}'(\omega_1) \\ \mathcal{X}'(\omega_2) \\ \vdots \\ \mathcal{X}'(\omega_N) \end{pmatrix} = \begin{pmatrix} \mu_{11} & \mu_{12} & \cdots & \mu_{1N} \\ \mu_{21} & \mu_{22} & \cdots & \mu_{2N} \\ \vdots & \vdots & \ddots & \vdots \\ \mu_{N1} & \mu_{N2} & \cdots & \mu_{NN} \end{pmatrix} \odot \begin{pmatrix} \mathcal{X}(\omega_1) \\ \mathcal{X}(\omega_2) \\ \vdots \\ \mathcal{X}(\omega_N) \end{pmatrix},$$

then, we have that $\mathrm{I}_f(\mathcal{X}) \geq \mathrm{I}_f(\mathcal{X}')$ with equality if, and only if, $\mathcal{X}' = \mathcal{X} \circ \sigma$ for certain permutation σ on Ω.

The well-known Lorenz criterion of the real-valued case could be extended to ordering the vectors of the values of a fuzzy random variable on a finite population by considering the Ramík and Římánek (1985) ranking. Thus, if we denote $V \succ_S W$ if, and only if, $V \succeq_S W$ but not $V \preceq_S W$, the *compatibility with Lorenz's criterion* of the f-inequality indices can be stated as follows:

Proposition 13 (Compatibility with Lorenz's criterion). *Consider two fuzzy random variables* $\mathcal{X} : \Omega \to \mathcal{F}_c((0,+\infty))$ *and* $\mathcal{X}' : \Omega \to \mathcal{F}_c((0,+\infty))$ *such that* $\mathcal{X}(\omega_N) \succeq_S \ldots \succeq_S \mathcal{X}(\omega_1), \mathcal{X}'(\omega_N) \succeq_S \ldots \succeq_S \mathcal{X}'(\omega_1)$, $\widetilde{E}(\mathcal{X}) = \widetilde{E}(\mathcal{X}')$ *and* $\mathcal{X}_L\, \mathcal{X}'$ *(which will mean that* $\mathcal{X}(\omega_1) \oplus \cdots \oplus \mathcal{X}(\omega_k) \succeq_S \mathcal{X}'(\omega_1) \oplus \cdots \oplus \mathcal{X}'(\omega_k)$ *for all* $k \in \{1, \ldots, N\}$ *with* $\succ_S$ *for at least one* k*). Then,* $\mathrm{I}_f(\mathcal{X}) < \mathrm{I}_f(\mathcal{X}')$.

The *progressive (and regressive) principles of transfers* formalize the reaction to redistributions which are carried out by means of transfers from a higher value to a lower one (or conversely) when preserving the expected value of the fuzzy random variable. More precisely, the progressive principle of transfers indicates that

Proposition 14 (Progressive principle of transfers). *If* $\mathcal{X} : \Omega \to \mathcal{F}_c((0, +\infty))$ *is a fuzzy random variable, and* $\mathcal{X}' : \Omega \to \mathcal{F}_c((0,+\infty))$ *is another fuzzy random variable such that* $\mathcal{X}(\omega_j) = \mathcal{X}'(\omega_j)$ *for all* $j \in \{1, \ldots, N\} \setminus \{l, l'\}$, $\mathcal{X}(\omega_l) \succeq_S \mathcal{X}'(\omega_l) \succeq_S \mathcal{X}(\omega_{l'})$, $\mathcal{X}(\omega_l) \succeq_S \mathcal{X}'(\omega_{l'}) \succeq_S \mathcal{X}(\omega_{l'})$ *and* $\widetilde{E}(\mathcal{X}) = \widetilde{E}(\mathcal{X}')$*. Then, we have that* $\mathrm{I}_f(\mathcal{X}) \geq \mathrm{I}_f(\mathcal{X}')$.

Furthermore, $\mathrm{I}_f(\mathcal{X}) = \mathrm{I}_f(\mathcal{X}')$ *if, and only if, either* $\mathcal{X} = \mathcal{X}'$ *on* Ω, *or* $\mathcal{X}' = \mathcal{X} \circ \sigma_{ll'}$ *for the permutation* $\sigma_{ll'}$ *on* Ω *which exchanges* ω_l *and* $\omega_{l'}$, *or for all* $\alpha \in [0, 1]$ *it happens that* $\inf \mathcal{X}_\alpha(\omega_l) = \inf \mathcal{X}'_\alpha(\omega_{l'})$, $\inf \mathcal{X}_\alpha(\omega_{l'}) = \inf \mathcal{X}'_\alpha(\omega_l)$, $\sup \mathcal{X}_\alpha(\omega_l) = \sup \mathcal{X}'_\alpha(\omega_l)$ *and* $\sup \mathcal{X}_\alpha(\omega_{l'}) = \sup \mathcal{X}'_\alpha(\omega_{l'})$, *or* $\inf \mathcal{X}_\alpha(\omega_l) = \inf \mathcal{X}'_\alpha(\omega_l)$, $\inf \mathcal{X}_\alpha(\omega_{l'}) = \inf \mathcal{X}'_\alpha(\omega_{l'})$, $\sup \mathcal{X}_\alpha(\omega_l) = \sup \mathcal{X}'_\alpha(\omega_{l'})$ *and* $\sup \mathcal{X}_\alpha(\omega_{l'}) = \sup \mathcal{X}'_\alpha(\omega_l)$.

On the other hand, the regressive principle of transfers can be derived from the last one by exchanging the roles of $\mathcal{X}$ and $\mathcal{X}'$ in it. Thus,

Proposition 15 (Regressive principle of transfers). *If* $\mathcal{X} : \Omega \to \mathcal{F}_c((0, +\infty))$ *is a fuzzy random variable, and* $\mathcal{X}' : \Omega \to \mathcal{F}_c((0,+\infty))$ *is another fuzzy random variable such that* $\mathcal{X}(\omega_j) = \mathcal{X}'(\omega_j)$ *for all* $j \in \{1, \ldots, N\} \setminus \{l, l'\}$, $\mathcal{X}'(\omega_l) \succeq_S \mathcal{X}(\omega_l) \succeq_S \mathcal{X}'(\omega_{l'})$, $\mathcal{X}'(\omega_l) \succeq_S \mathcal{X}(\omega_{l'}) \succeq_S \mathcal{X}'(\omega_{l'})$ *and* $\widetilde{E}(\mathcal{X}) = \widetilde{E}(\mathcal{X}')$*. Then, we have that* $\mathrm{I}_f(\mathcal{X}') \geq \mathrm{I}_f(\mathcal{X})$.

Furthermore, $\mathrm{I}_f(\mathcal{X}) = \mathrm{I}_f(\mathcal{X}')$ *if, and only if, either* $\mathcal{X} = \mathcal{X}'$ *on* Ω, *or* $\mathcal{X}' = \mathcal{X} \circ \sigma_{ll'}$ *(with the notation in Proposition 14), or for all* $\alpha \in [0, 1]$ *we have that either* $\inf \mathcal{X}'_\alpha(\omega_l) = \inf \mathcal{X}_\alpha(\omega_{l'})$, $\inf \mathcal{X}'_\alpha(\omega_{l'}) = \inf \mathcal{X}_\alpha(\omega_l)$, $\sup \mathcal{X}'_\alpha(\omega_l) = \sup \mathcal{X}_\alpha(\omega_l)$ *and* $\sup \mathcal{X}'_\alpha(\omega_{l'}) = \sup \mathcal{X}_\alpha(\omega_{l'})$, *or* $\inf \mathcal{X}'_\alpha(\omega_l) = \inf \mathcal{X}_\alpha(\omega_l)$, $\inf \mathcal{X}'_\alpha(\omega_{l'}) = \inf \mathcal{X}_\alpha(\omega_{l'})$, $\sup \mathcal{X}'_\alpha(\omega_l) = \sup \mathcal{X}_\alpha(\omega_{l'})$ *and* $\sup \mathcal{X}'_\alpha(\omega_{l'}) = \sup \mathcal{X}_\alpha(\omega_l)$.

The following result corresponds to a property formalizing the *effects* of the *"grouping" of fuzzy data* in quantifying the f-inequality index. More precisely, this property expresses the "ordering relation" between the inequality of the population and the inequality between the groups of a given (classical) partition of the population, when each of the groups is represented by the expected value of the fuzzy random variable in it. From this result we can conclude that grouping entails an increase in inequality.

Proposition 16 (Grouping effects). *Consider a finite population* $\Omega = \{\omega_{11}, \ldots, \omega_{1N_1}, \ldots\ldots, \omega_{M1}, \ldots, \omega_{MN_M}\}$ *(with* $N = N_1 + \cdots + N_M$*) which is divided into* M *subpopulations* $\Omega_m = \{\omega_{m1}, \ldots, \omega_{mN_m}\}$, $m = 1, \ldots, M$, *and assume that* $(\Omega, \mathcal{P}(\Omega))$ *is endowed with the uniform distribution* P *and that* $\mathcal{P} = \{\Omega_m\}_{m=1}^M$ *denotes the above partition. If* $\mathcal{X} : \Omega \to \mathcal{F}_c((0, +\infty))$ *is a fuzzy random variable associated with* $(\Omega, \mathcal{P}(\Omega), P)$, *and* $\mathcal{X}_{\mathcal{P}} : \mathcal{P} \to \mathcal{F}_c((0, +\infty))$ *is the fuzzy random variable such that* $\mathcal{X}_{\mathcal{P}}(\Omega_m)$ = *expected value of* $\mathcal{X}$ *on* Ω_m $(m = 1, \ldots, M)$, *and* $\mathcal{X}_{\Omega_m}$ *denotes the restriction of* $\mathcal{X}$ *from* Ω *to* Ω_m $(m = 1, \ldots, M)$, *then we have that*

$$\mathrm{I}_f(\mathcal{X}) \geq \mathrm{I}_f(\mathcal{X}_{\mathcal{P}}).$$

On the other hand, $\mathrm{I}_f(\mathcal{X}) = \mathrm{I}_f(\mathcal{X}_{\mathcal{P}})$ *if, and only if, for each* $m \in \{1, \ldots, M\}$ *the fuzzy random variable* $\mathcal{X}_{\Omega_m}$ *is degenerate in* Ω_m.

The *additive decomposability* of the indices for real-valued random variables is lost in the extension to the fuzzy case, except for Shannon's index. In this way,

Proposition 17 (Additive decomposability of the index of the Shannon type). *If in Proposition* 16 *we consider the function* $f(x) = -\log x$ *for all* $x \in (0, +\infty)$, *then we have that*

$$\mathrm{I}_{Sh}(\mathcal{X}) = \mathrm{I}_{Sh}(\mathcal{X}_{\mathcal{P}}) + \sum_{m=1}^{M} \frac{N_m \mathrm{I}_{Sh}(\mathcal{X}_{\Omega_m})}{N},$$

that is, the inequality in the population coincides with the sum of the inequality between groups (more concretely, between the expected values of $\mathcal{X}$ *in different groups) and the average of the inequality within groups.*

5 Estimating the hyperbolic index in random samplings from finite populations

In this section we consider the problem of estimating the population inequality index associated with a fuzzy random variable in a finite population in random samplings from finite populations.

To this purpose, we are going to check that it is possible to construct an unbiased estimator of the population hyperbolic index from samples of any size in samplings with and without replacement. However, for the other inequality indices introduced in the previous section, the construction of unbiased estimators is either very complex or even unfeasible.

Consider a finite population Ω of N units, $\omega_1, \ldots, \omega_N$, and let $\mathcal{X} : \Omega \to \mathcal{F}_c\big((0, +\infty)\big)$ be a fuzzy random variable associated with a probabilistic space defined on Ω which is endowed with the uniform distribution.

Assume that a sample of size n is chosen at random and without replacement from Ω, υ denotes a generic simple random sample of size n, and $\omega_{\upsilon 1}, \ldots, \omega_{\upsilon n}$ are the units in it. Then, the *sample hyperbolic index* of $\mathcal{X}$ in υ is given by

$$\mathrm{I}_H\left(\mathcal{X}[\upsilon]\right) = F\left(\frac{1}{n^2} \odot \bigoplus_{i=1}^{n} \bigoplus_{i'=1}^{n} \left(\mathcal{X}(\omega_{\upsilon i}) \oslash \mathcal{X}(\omega_{\upsilon i'})\right) \ominus 1\right)$$

$$= \frac{1}{n^2} \sum_{i=1}^{n} \sum_{i'=1}^{n} F\left(\mathcal{X}(\omega_{\upsilon i}) \oslash \mathcal{X}(\omega_{\upsilon i'})\right) - 1.$$

$\mathrm{I}_H\left(\mathcal{X}[\cdot]\right)$ is a real-valued random variable associated with the probability space $\big(\Upsilon_n, \mathcal{P}(\Upsilon_n), p\big)$ (Υ_n being the space of the $C_{N,n} = \binom{N}{n}$ distinct possible random samples without replacement of size n from the given population, $\mathcal{P}(\Upsilon_n)$ being the associated power set, and $p[\upsilon] = 1/C_{N,n}$ for all $\upsilon \in \Upsilon_n$), and hence, defines a real-valued estimator of the *population hyperbolic index*, which is given by

$$\mathrm{I}_H(\mathcal{X} \,|\, P) = F\left(\frac{1}{N^2} \odot \bigoplus_{j=1}^{N} \bigoplus_{j'=1}^{N} \left(\mathcal{X}(\omega_j) \oslash \mathcal{X}(\omega_{j'})\right) \ominus 1\right)$$

$$= \frac{1}{N^2} \sum_{j=1}^{N} \sum_{j'=1}^{N} F\left(\mathcal{X}(\omega_j) \oslash \mathcal{X}(\omega_{j'})\right) - 1.$$

To obtain from the sample index an unbiased estimator of the population index, the first one has to be revised. The correction which we will apply is based on the following result about the expected value of the sample index:

Proposition 18. *In random sampling without replacement of size n from the population Ω, if $f = n/N$ we have that*

$$E\left(\mathrm{I}_H\left(\mathcal{X}[\cdot]\right)\right) = \frac{(n-1)N}{n(N-1)} \mathrm{I}_H(\mathcal{X} \,|\, P) + \frac{1-f}{n(N-1)} \sum_{j=1}^{N} \mathrm{I}_H(\{\mathcal{X}(\omega_j)\}),$$

where $\mathrm{I}_H(\{\mathcal{X}(\omega_j)\})$ represents the (intra)hyperbolic inequality index of the fuzzy random variable degenerate at the value $\mathcal{X}(\omega_j) \in \mathcal{F}_c\big((0, +\infty)\big)$.

On the basis of the above result, we can conclude that

Theorem 2. *In random sampling without replacement of size n from the population Ω, if $f = n/N$ we have that the estimator $\widehat{\mathrm{I}_H}(\mathcal{X}[\cdot])$ such that*

$$\widehat{\mathrm{I}_H}(\mathcal{X}[v]) = \frac{n(N-1)}{N(n-1)}\mathrm{I}_H\left(\mathcal{X}[v]\right) - \frac{1-f}{n}\left\{\frac{1}{n-1}\sum_{i=1}^{n}\mathrm{I}_H\left(\{\mathcal{X}(\omega_{vi})\}\right)\right\},$$

is an unbiased estimator of $\mathrm{I}_H(\mathcal{X}\,|\,P)$.

To establish the accuracy of the preceding estimator of $\mathrm{I}_H(\mathcal{X}\,|\,P)$, we determine the associated "mean squared error", which in this case coincides with the variance $\mathrm{Var}\left(\widehat{\mathrm{I}_H}(\mathcal{X}[\cdot])\right)$ in the following result:

Proposition 19. *In random sampling without replacement of size n from Ω, if $f = n/N$ we have that*

$$\mathrm{Var}\left(\widehat{\mathrm{I}_H}(\mathcal{X}[\cdot])\right) = \frac{(1-f)}{n(n-1)N^2(N-1)(N-2)(N-3)}$$

$$\cdot\Bigg\{[N(6-4n)+6(n-1)]N^3(N-1)\left(\mathrm{I}_H(\mathcal{X}\,|\,P)\right)^2$$

$$+[4N^2(3-2n)+13N(n-1)+3(n-3)]N^2(N-1)\,\mathrm{I}_H(\mathcal{X}\,|\,P)$$

$$+[N^2(7-3n)+N(5n-7)-4(n-2)]\left(\sum_{j=1}^{N}\mathrm{I}_H(\{\mathcal{X}(\omega_j)\})\right)^2$$

$$+[N^2(n-5)+N(5n+1)-10(n-2)]\sum_{j=1}^{N}\left(\mathrm{I}_H(\{\mathcal{X}(\omega_j)\})\right)^2$$

$$+2[N^3(3-n)+N^2(3n-8)+N(n+9)-10(n-2)]\sum_{j=1}^{N}\mathrm{I}_H(\{\mathcal{X}(\omega_j)\})$$

$$+4(n-2)N^2(N-1)\mathrm{I}_H(\mathcal{X}\,|\,P)\sum_{j=1}^{N}\mathrm{I}_H(\{\mathcal{X}(\omega_j)\})$$

$$+(n-2)(N-1)(N-2)\sum_{j=1}^{N}\left(\sum_{l=1}^{N}[F(\mathcal{X}(\omega_j)\oslash\mathcal{X}(\omega_l))+F(\mathcal{X}(\omega_l)\oslash\mathcal{X}(\omega_j))]\right)^2$$

$$+(N-n+1)(N-1)(N-3)\sum_{j=1}^{N}\sum_{l=1}^{N}F(\mathcal{X}(\omega_j)\oslash\mathcal{X}(\omega_l))$$

$$\cdot[F(\mathcal{X}(\omega_j)\oslash\mathcal{X}(\omega_l))+F(\mathcal{X}(\omega_l)\oslash\mathcal{X}(\omega_j))]$$

$$-[N(3n-7)-3(n-3)](N-1)\sum_{j=1}^{N}\sum_{l=1}^{N}\mathrm{I}_H(\{\mathcal{X}(\omega_l)\})F\left(\mathcal{X}(\omega_j)\oslash\mathcal{X}(\omega_l)\right)$$

$$+[N^5(6-4n)+8N^4n+N^3(n-9)-6N^2(n-2)+N(5n+1)-10(n-2)]\Bigg\}.$$

If, instead of adopting a choice at random and without replacement, we consider a choice at random and with replacement of n units from $\Omega = \{\omega_1, \ldots, \omega_N\}$, and v represents a generic random sample with replacement of size n, and $\omega_{v1}, \ldots, \omega_{vn}$ are the units in v, then, the *sample hyperbolic index* of $\mathcal{X}$ in v is given now by

$$\mathrm{I}_H\left(\mathcal{X}[v]\right)=\frac{1}{N^2}\sum_{j=1}^{N}\sum_{j'=1}^{N}F\left(\mathcal{X}(\omega_j)\oslash\mathcal{X}(\omega_{j'})\right)\cdot t_j[v]t_{j'}[v]-1,$$

where t_j is the real-valued random variable defined on the probabilistic space $(\Upsilon_n^w, \mathcal{P}(\Upsilon_n^w), p^w)$ (Υ_n^w being the space of the $CR_{N,n} = \binom{N+n-1}{n}$ distinct possible random samples with replacement of size n from the considered population, and $p^w[v]$ is the probability of choosing the sample $v \in \Upsilon_n^w$, which does not determine a uniform distribution on Υ_n^w for the considered sample) so that $t_j[v]$ is the "number of times that ω_j appears in v".

$\mathrm{I}_H\left(\mathcal{X}[\cdot]\right)$ is a real-valued random variable associated with the probability space$(\Upsilon_n^w, \mathcal{P}(\Upsilon_n^w), p^w)$, and, hence, it defines an estimator of the population hyperbolic index.

As for the simple random sampling, to obtain an unbiased estimator of $\mathrm{I}_H(\mathcal{X})$ from the sampling one, we first examine the expected value of the sample hyperbolic index in the sampling with replacement.

Proposition 20. *In random sampling with replacement of size n from the population Ω, we have that*

$$E\left(\mathrm{I}_H\left(\mathcal{X}[\cdot]\right)\right)=\frac{(n-1)}{n}\mathrm{I}_H(\mathcal{X}\,|\,P)+\frac{1}{nN}\sum_{j=1}^{N}\mathrm{I}_H\left(\{\mathcal{X}(\omega_j)\}\right).$$

On the basis of the above result, we can conclude that

Theorem 3. *In random sampling with replacement of size n from the population Ω, we have that the estimator $\widehat{\mathrm{I}_H}^w\left(\mathcal{X}[\cdot]\right)$ which for each sample v takes on the value*

$$\widehat{\mathrm{I}_H}^w\left(\mathcal{X}[v]\right)=\frac{n}{(n-1)}\mathrm{I}_H\left(\mathcal{X}[v]\right)-\frac{1}{n}\left\{\frac{1}{n-1}\sum_{i=1}^{n}\mathrm{I}_H\left(\{\mathcal{X}(\omega_{vi})\}\right)\right\},$$

is an unbiased estimator of $\mathrm{I}_H(\mathcal{X}\,|\,P)$.

The mean squared error associated with the estimator in Theorem 3, coincides with the variance $\mathrm{Var}\left(\widehat{\mathrm{I}_H}^w(\mathcal{X}[\cdot])\right)$, and will be given by

Proposition 21. *In random sampling with replacement of size n from the population Ω, we have that*

$$\mathrm{Var}\left(\widehat{\mathrm{I}_H}^w(\mathcal{X}[\cdot])\right) = \frac{1}{n(n-1)N^3}$$

$$\cdot\Bigg\{(6-4n)N^3\left(\mathrm{I}_H(\mathcal{X}\mid P)\right)^2 + 2(6-4n)N^3\mathrm{I}_H(\mathcal{X}\mid P)$$

$$-[N-4(n-1)]\left(\sum_{j=1}^{N}\mathrm{I}_H(\{\mathcal{X}(\omega_j)\})\right)^2 + [N-4(n-1)]\sum_{j=1}^{N}\left(\mathrm{I}_H(\{\mathcal{X}(\omega_j)\})\right)^2$$

$$-2[N-4(n-1)](N-1)\sum_{j=1}^{N}\mathrm{I}_H(\{\mathcal{X}(\omega_j)\})$$

$$+(n-2)\sum_{j=1}^{N}\left(\sum_{l=1}^{N}[F(\mathcal{X}(\omega_j)\oslash\mathcal{X}(\omega_l)) + F(\mathcal{X}(\omega_l)\oslash\mathcal{X}(\omega_j))]\right)^2$$

$$+N\sum_{j=1}^{N}\sum_{l=1}^{N}F(\mathcal{X}(\omega_j)\oslash\mathcal{X}(\omega_l))\,[F(\mathcal{X}(\omega_j)\oslash\mathcal{X}(\omega_l)) + F(\mathcal{X}(\omega_l)\oslash\mathcal{X}(\omega_j))]$$

$$+[N^2(5-4n) + N(4n-3) - 4(n-1)]N\Bigg\}.$$

The results we have just stated can be used to compare the accuracy in the estimation of the population hyperbolic index associated with different variables, to derive confidence intervals and testing hypotheses, and determining adequate sample sizes, although as usual this would lead to conservative procedures and we would need to estimate the population parameters in the variance of the estimators.

6 Examples

In the following examples we illustrate the computation and use of certain f-inequality indices in comparing populations and in estimating population inequality.

Example 1. Consider the variable ANNUAL INCOME, $\mathcal{X}$, in accordance with the classification which is adopted in some credit assessment systems. Following Cox (1994), this variable can be viewed as a variable whose (fuzzy) values are $\tilde{x}_1$ = SOMEWHAT HIGH, $\tilde{x}_2$ = MODERATELY HIGH, $\tilde{x}_3$ = HIGH and $\tilde{x}_4$ = VERY HIGH, where $\tilde{x}_1$, $\tilde{x}_2$, $\tilde{x}_3$ and $\tilde{x}_4$ are described (in US thousand dollars) by means of the following S- and Π-curves:

$$\tilde{x}_1 = 1 - S(100, 125),$$
$$\tilde{x}_2 = \Pi(100, 125, 150),$$
$$\tilde{x}_3 = \Pi(125, 147.5, 170),$$
$$\tilde{x}_4 = S(147.5, 170),$$

and $\operatorname{supp} \tilde{x}_i \subset [90, 180]$, $i = 1, 2, 3, 4$, for all candidates for a credit in the considered system (see Figure 1).

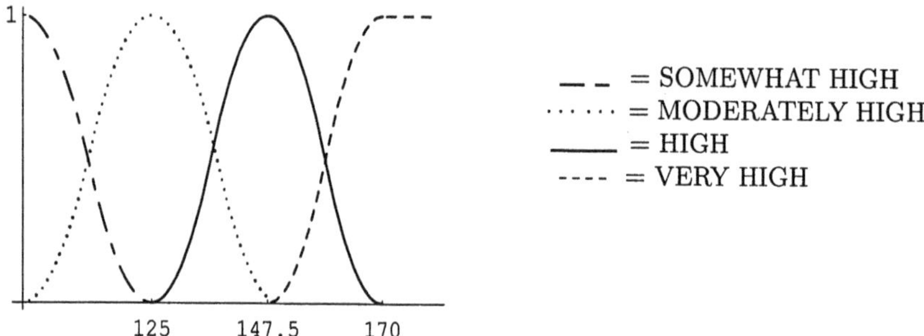

Fig. 1. Fuzzy values of the variable ANNUAL INCOME

Assume that a bank adopting the above system wishes to compare two different towns by means of the income inequality, and to this purpose we observe the values of $\mathcal{X}$ in the central offices of these two towns.

If there are 125 candidates for a credit in one of the offices (Ω_1) during a certain period, 28 of them having a SOMEWHAT HIGH annual income, 43 MODERATELY HIGH, 31 HIGH and 23 VERY HIGH, whereas there are 178 candidates for a credit in the other office (Ω_2) during the same period, 63 of them having SOMEWHAT HIGH, 79 MODERATELY HIGH, 27 HIGH and 9 VERY HIGH, and we employ the f-inequality index with $f(x) = -\log x$, we obtain that

$$\mathrm{I}_{Sh}(\mathcal{X}_{\Omega_1}) = 4.892,$$
$$\mathrm{I}_{Sh}(\mathcal{X}_{\Omega_2}) = 4.815,$$

whence we can conclude that the two towns have a close inequality of annual income.

Example 2. In Klir and Yuan (1995) it has been pointed out that there is a large number of situations in Civil Engineering to which Fuzzy Set Theory has already proven to be especially valuable, like those consisting of problems of assessing or evaluating existing constructions. Typical examples of these problems are the assessment of fatigue in metal structure, the assessment of quality of highway pavements, the assessment of damage to a building after an earthquake, etc.

Klir and Yuan have mentioned as an example in the study of the physical conditions of highway bridges, the variable CURRENT CONDITION OF THE PIERS ($\mathcal{X}$) of a bridge, whose values are POOR ($\tilde{x}_1$), FAIR ($\tilde{x}_2$) and GOOD ($\tilde{x}_3$), and which have been assumed to be characterized (Klir and Yuan, 1995) by means of the triangular fuzzy numbers, $\tilde{x}_1 = \mathrm{Tri}(1,2,3)$, $\tilde{x}_2 = \mathrm{Tri}(2,3,5)$ and $\tilde{x}_3 = \mathrm{Tri}(3,5,5)$ (see Figure 2).

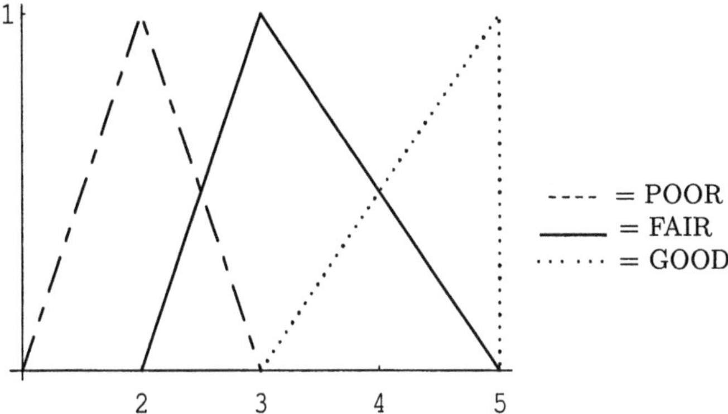

Fig. 2. Fuzzy values of the CURRENT CONDITION OF THE PIERS of the highway bridges

In the United States there are approximately 600,000 highway bridges, about one half of which were built before 1940. To estimate by point and testing the value of the f-inequality index associated with $\mathcal{X}$ in the population Ω of the 600,000 US bridges, with $f = x^{-1} - 1$, we can consider a simple random sample v of $n = 400$ bridges in the US. Assume that the above sample provides us with the following data

$\tilde{x}_l$	POOR	FAIR	GOOD
absol. freq.	15	56	329

In virtue of Theorem 2, $\mathrm{I}_H(\mathcal{X})$ can be estimated by means of the value

$$\widehat{\mathrm{I}_H}\left(\mathcal{X}[v]\right) = \frac{400(599,999)}{600,000(399)}\mathrm{I}_H\left(\mathcal{X}[v]\right)$$

$$-\frac{1-\dfrac{400}{600,000}}{400}\left\{\frac{1}{399}\sum_{i=1}^{400}\mathrm{I}_H\left(\{\mathcal{X}(\omega_{vi})\}\right)\right\} = .099.$$

Acknowledgements

The research in this paper has been partially supported by DGESIC Grant No. DGE-99-PB98-1534, and this financial support is gratefully acknowledged to the Spanish MECD. The authors are deeply grateful to their colleagues Professors Ana Colubi, Norberto Corral, María Teresa López and Miguel López-Díaz for their valuable comments and suggestions in connection with this paper.

References

1. Adamo, J.M. (1980) Fuzzy decision trees. *Fuzzy Sets and Systems* **4**, 207-219.
2. Alonso, M.C., Bertoluzza, C., Brezmes, T. and Lubiano, M.A. (2001) A generalized real-valued measure of the inequality associated with a fuzzy random variable. *Int. J. Approx. Reas.* (accepted, in press).
3. Blackorby, C. and Donaldson, D. (1978). Measures of relative equality and their meaning in terms of social welfare. *J. Econom. Theory* **18**, 59-80.
4. Bortolan, G. and Degani, P. (1985). A review of some methods for ranking fuzzy subsets. *Fuzzy Sets and Systems* **15**, 1-19.
5. Bourguignon, F. (1979). Decomposable income inequality measures. *Econometrica* **47**, 901-920.
6. Campos, L.M. de and González, A. (1989). A subjective approach for ranking fuzzy numbers. *Fuzzy Sets and Systems* **29**, 145-153.
7. Colubi, A. (1997). *Los f-índices de desigualdad difusos asociados a una variable aleatoria difusa.* Trabajo de Investigación de Tercer ciclo. Universidad de Oviedo.
8. Colubi, A., Gil, M.A. and López-García, H. (1997). Measuring the inequality associated with a fuzzy random variable in terms of statistical f-divergences. *Proc. EUFIT'97* **3**, 1764-1768.
9. Cowell, F.A. (1980). On the structure of additive inequality measures. *Rev. Econ. Stud.* **47**, 521-531.
10. Cowell, F.A. and Kuga, K. (1981). Additivity and the entropy concept. An axiomatic approach. *J. Econom. Theory* **25**, 131-143.
11. Cox, E. (1994). *The Fuzzy Systems Handbook.* Academic Press, Cambridge.
12. Csiszár, I. (1967). Information-type measures of difference of probability distributions and indirect observations. *Studia Scient. Math. Hung.* **2**, 299-318.
13. Delgado, M., Verdegay, J.L. and Vila, M.A. (1988). A procedure for ranking fuzzy numbers using fuzzy relations. *Fuzzy Sets and Systems* **26**, 49-62.
14. Diamond, P. and Kloeden, P. (1994). *Metric Spaces of Fuzzy Sets: Theory and Applications.* World Scientific, Singapore.
15. Dubois, D. and Prade, H. (1980). *Fuzzy Sets and Systems: Theory and applications.* Academic Press, New York.
16. Eichhorn, W. and Gehrig, W. (1982). Measurement of inequality in Economics. In *Applied Mathematics-Optimization and Operations Research* (B. Korte, Eds.), 657-693. North-Holland, Amsterdam.
17. Gil, M.A. and Gil, P. (1989). On some information measures of degree $\beta = 2$. Estimation in simple-stage cluster sampling. *Statist. Probab. Lett.* **8**, 157-162.
18. Gil, M.A., Caso, C. and Gil, P. (1989a). Estudio asintótico de una clase de índices de desigualdad muestrales. *Trab. de Est.* **4**, 95-109.

19. Gil, M.A., Pérez, R. and Gil, P. (1989b). A family of measures of uncertainty involving utilies: definitions, properties, applications and statistical inferences. *Metrika* **36**, 129-147.
20. González, A. and Vila, M.A. (1992). Dominance relations on fuzzy numbers. *Inform. Sci.* **64**, 1-16.
21. Klir, G.J. and Yuan, B. (1995). *Fuzzy sets and Fuzzy Logic. Theory and Applications.* Prentice Hall, New Jersey.
22. Kölm, S. Ch. (1976). Unequal inequalities I. *J. Econom. Theory* **12**, 416-442.
23. Kölm, S. Ch. (1976). Unequal inequalities II. *J. Econom. Theory* **13**, 82-111.
24. Kołodziejczyk, W. (1986). Orlovsky's concept of Decision-Making with fuzzy preference relation-further results. *Fuzzy Sets and Systems* **19**, 11-20.
25. López Díaz, M. (1996). *Medibilidad e integración de variables aleatorias difusas. Aplicación a problemas de decisión.* PhD Thesis. Universidad de Oviedo.
26. López-Díaz, M. and Gil, M.A. (1998). The λ-average value and the fuzzy expectation of a fuzzy random variable. *Fuzzy Sets and Systems* **99**, 347-352.
27. Lubiano, M.A. (1999). *Medidas de variación para elementos aleatorios imprecisos.* PhD Thesis. Universidad de Oviedo.
28. Martínez López, I. (1991). *Aproximación de algunas familias de índices de variación en poblaciones finitas.* PhD. Thesis. Universidad de Oviedo.
29. Nakamura, K. (1986). Preference relations on a set of fuzzy utilities as a basis for decision making. *Fuzzy Sets and Systems* **20**, 147-162.
30. Norwich, A.M. and Turksen, I.B. (1984). A model for the measurement of membership and the consequences of its empirical implementation. *Fuzzy Sets and Systems* **12**, 1-25.
31. Nguyen, H.T. (1978). A note on the extension principle for fuzzy sets. *J. Math. Anal. Appl.* **64**, 369-380.
32. Puri, M.L. and Ralescu, D. (1981). Différentielle d'une fonction floue. *C.R. Acad. Sci. Paris, Sér. I* **293**, 237-239.
33. Puri, M.L. and Ralescu, D. (1983). Differentials of fuzzy functions. *J. Math. Annal. Appl.* **91**, 552-558.
34. Puri, M.L. and Ralescu, D. (1986). Fuzzy random variables. *J. Math. Anal. Appl.* **114**, 409-422.
35. Ramík, J. and Římánek, J. (1985). Inequality relation between fuzzy numbers and its use in fuzzy optimization. *Fuzzy Sets and Systems* **16**, 123-138.
36. Sánchez de Posada Martínez, C. (1998). *Estudio matemático de los números difusos.* Trabajo Integrado y Académicamente Dirigido del Plan de Estudios de la Licenciatura de Matemáticas. Universidad de Oviedo.
37. Shorrocks, A.F. (1980). The class of additively decomposable inequality measures. *Econometrica* **48**, 613-625.
38. Tseng, T.V. and Klein, C.M. (1989). New algorithm for the ranking procedure in fuzzy decision-making. *IEEE Trans. Syst. Man and Cybernet.* **19**, 1289-1296.
39. Yager, R.R. (1981) A procedure for ordering fuzzy subsets of the unit interval. *Inf. Sci.* **24**, 143-161.
40. Zadeh, L.A. (1975). The concept of a linguistic variable and its application to approximate reasoning. *Inform. Sci.*, Part 1 **8**, 199-249; Part 2 **8**, 301-353; Part 3 **9**, 43-80.
41. Zagier, D. (1983). On the decomposability of the Gini coefficient and other indices of inequality. Discussion paper No. 108, University of Bonn.

Traditional techniques to prove some limit theorems for fuzzy random variables

Ana Colubi[1]

Departamento de Estadística e I.O. y D.M., Universidad de Oviedo, 33071 Oviedo, Spain

Abstract. In the last years, some limit theorems for fuzzy random variables have been proven by means of different techniques developed for this purpose. In this work we deal with the cadlag representation of a kind of fuzzy sets to show that these limit results can be also proved by applying well-known techniques in Probability Theory (specifically, the ones which make valid the analogous theorems for $D[0,1]$-valued random elements). In this context, we will study a strong law of large numbers (whose proof will suggest a characterization of the uniform convergence) and a strong law of the iterated logarithm. Furthermore, we will check the relationships between these techniques and the ones used by Molchanov to prove a SLLN for the same random elements.

1 Introduction

One of the most useful metrics defined on the wide class $\mathcal{F}(B)$ of the normal fuzzy sets with closed and bounded α-level sets is the supremum one (see Puri and Ralescu, 1981,1986). This metric is not separable which forces us to develop complex techniques to prove limit theorems for fuzzy set-valued random elements (see, for instance, Proske, 1997, and Colubi *et al.* 1997, 1999). Proske's proof of the strong law of large numbers makes use of the studies concerning nonseparable spaces by Hoffman-Jørgensen (1985). The proof of the same result developed independently by Colubi *et al.* (1997, 1999) is based on a result by ópez-Díaz and Gil (1998) stating an approximation of fuzzy random variables and on the relationship found between the Glivenko-Cantelli Theorem and a certain particular case of the classical strong law. The aim of this work is to point out that some well-known results in the space $D[0,1]$ of the right-continuous functions having left-limits at every point can be useful in the space $\mathcal{F}(B)$. In this way, the study developed in this paper is based, on one hand, on the connection between the fuzzy sets of $\mathcal{F}(B)$ and the cadlag functions (see Colubi *et al.*, 2000) and, on the other hand, on the proof of the strong law of large numbers in terms of the supremum metric for D[0,1]-valued random elements stated by Daffer and Taylor (1979). The cadlag representation will allow us to apply well-known techniques of the spaces $D_E[0,1]$ to the space $\mathcal{F}(B)$. More precisely, the scheme of Daffer and Taylor's proof is used to demonstrate the analogous strong law for fuzzy random variables. We will compare this technique with the one applied by

Molchanov (1999). As in Molchanov's work, a characterization of uniform convergence in $\mathcal{F}(B)$ is then deduced from the employed technique. Moreover, it is possible to use this method to find a law of the iterated logarithm for fuzzy random variables.

2 Preliminaries

Let $\mathcal{K}(B)$ be the class of non-empty closed and bounded subsets of a separable Banach space $(B, |\cdot|)$. On $\mathcal{K}(B)$ the *Hausdorff metric* is defined in such a way that for all $R, S \in \mathcal{K}(B)$

$$d_H(R,S) = \max\left\{\sup_{r\in R}\inf_{s\in S}|r-s|, \sup_{s\in S}\inf_{r\in R}|r-s|\right\}.$$

The metric space $(\mathcal{K}(B), d_H)$ is complete and separable (see Debreu, 1967). On the other hand, if $R_1 \subset R \subset R_2$ and $S_1 \subset S \subset S_2$ are sets belonging to $\mathcal{K}(B)$, the Hausdorff metric satisfies that

$$d_H(R,S) \leq \max\{d_H(R_1,S_2), d_H(R_2,S_1)\}.$$

The space $\mathcal{K}(B)$ can be endowed with a semilinear structure induced by the Minkowski sum and the multiplication by a scalar, that is,

$$R+S = \{r+s \,|\, r \in R, s \in S\}, \quad \lambda R = \{\lambda r \,|\, r \in R\},$$

with $R, S \in \mathcal{K}(B)$, $\lambda \in \mathbb{R}$. The space $(\mathcal{K}(B), +, \cdot)$ is not vectorial. Let $(\Omega, \mathcal{A}, P)$ be a probability space. A mapping $X : \Omega \to \mathcal{K}(B)$ is a *random set* if it is Borel-measurable. The *expected value of a random set* X will be defined by means of the Aumann integral (if it exists), that is, $E(X) = \{Ef | f \in L^1(\Omega, \mathcal{A}, P), f(\omega) \in X(\omega)\, a.s.[P]\}$, where Ef is the Bochner integral of the random vector f (see Aumann, 1965). We will denote by $\mathcal{F}(B)$ the class of normal fuzzy sets $V : B \to [0,1]$ with bounded support, that is, $\mathcal{F}(B) = \{V : B \to [0,1] \,|\, V_\alpha \in \mathcal{K}(B)$, for all $\alpha \in [0,1]\}$ where $V_\alpha = \{b \in B \,|\, V(b) \geq \alpha\}$ if $\alpha \in (0,1]$ and $V_0 = \text{cl}\{b \in B | V(b) > 0\}$, the sets V_α being called the α-level sets of V. The set $\{b \in B \,|\, V(b) > \alpha\}$ will be called the strict α-level of V. The *supremum metric* d_∞ is defined on $\mathcal{F}(B)$, in such a way that for all $U, V \in \mathcal{F}(B)$,

$$d_\infty(U,V) = \sup_{\alpha\in[0,1]} d_H(U_\alpha, V_\alpha).$$

The above operations on $\mathcal{K}(B)$ are inherited by $\mathcal{F}(B)$ through the α-levels and in this way a sum and a multiplication by scalars can be defined as the fuzzy sets belonging to $\mathcal{F}(B)$ satisfying that $(U+V)_\alpha = U_\alpha + V_\alpha$ and $(\lambda U)_\alpha = \lambda U_\alpha$ for all $\alpha \in [0,1]$, these operations being equivalent to those defined through Zadeh's extension principle (1975). The metric space $(\mathcal{F}(B), d_\infty)$ is non-separable (see Klement *et al.*, 1986). If $(\Omega, \mathcal{A}, P)$ is a probability space, a *fuzzy random variable* is a mapping $X : \Omega \to \mathcal{F}(B)$ such that the *α-level mappings* $X_\alpha : \Omega \to \mathcal{K}(B)$ are random sets for all $\alpha \in [0,1]$ (see Puri and Ralescu, 1986). A fuzzy random variable X is said to be *integrably bounded* if $\|X_0\| \in L^1(\Omega, \mathcal{A}, P)$ where $\|X_0(\omega)\| = d_H(\{0\}, X_0(\omega))$ for all $\omega \in \Omega$. If X is an integrably bounded fuzzy random variable, its *expected value* is the unique element of $\mathcal{F}(B)$, $E(X)$ satisfying that $(E(X))_\alpha = E(X_\alpha)$ for all $\alpha \in [0,1]$ (see Puri and Ralescu, 1986). The convex hull of a fuzzy random variable is the random element $\mathrm{Co}X : \Omega \to \mathcal{F}(B)$ such that $(\mathrm{Co}X(\omega))_\alpha = \mathrm{Co}((X(\omega)_\alpha)$ for all $\alpha \in [0,1]$. If (E, d) is a complete and separable metric space, we denote by $D_E[0,1]$ the class of the *cadlag functions*, that is, the functions $f : [0,1] \to E$ being right-continuous, having left-limits at every point and being continuous at $t = 1$. We denote by $D_E^{\uparrow}[0,1]$ the class of non-decreasing cadlag functions (whenever it makes sense). On $D_E[0,1]$ we define the metric m_∞, so that for all $x, y \in D_E[0,1]$

$$m_\infty(x,y) = \sup_{t \in [0,1]} d(x(t), y(t)).$$

In Billingsley (1968) the following result is proved for elements in $D[0,1]$ (and the same arguments preserve it for elements in $D_E[0,1]$). For each $x \in D_E[0,1]$ and each $\epsilon > 0$, there exists a discretization of $[0,1]$, $0 = t_0 < t_1 < \ldots < t_r = 1$, so that

$$w_x([t_i, t_{i+1})) = \sup_{t,s \in [t_i, t_{i+1})} d(x(t), x(s)) < \epsilon$$

for all $i \in \{0, 1, \ldots, r-1\}$. The metric space $(D_E[0,1], m_\infty)$ is non-separable (see, for example, Billingsley, 1968). A random element with values in $D_E[0,1]$ is a mapping $X : \Omega \to D_E[0,1]$ satisfying that $X(t) : \Omega \to E$ is Borel measurable for all $t \in [0,1]$ (see, for example, Ethier and Kurtz, 1986). The cadlag representation of a fuzzy set will allow us to connect the space $\mathcal{F}(B)$ with $D_E[0,1]$. If V is a fuzzy set, the function $x_V : [0,1] \to \mathcal{K}(B)$ defined in such a way that $x_V(\alpha) = V_{1-\alpha}$ for all $\alpha \in [0,1]$ will be referred to as the *cadlag representation* of V. If a fuzzy set $V \in \mathcal{F}(B)$, then, $x_V \in D^{\uparrow}_{\mathcal{K}(B)}[0,1]$ and, conversely, if $x \in D^{\uparrow}_{\mathcal{K}(B)}[0,1]$ there exists a fuzzy set $V \in \mathcal{F}(B)$ so that $x_V = x$ (see Colubi *et al.*, 2000). Note that from this representation it is possible to identify completely the concepts of fuzzy random variable and $D^{\uparrow}_{\mathcal{K}(B)}[0,1]$-valued random element.

3 Results

The cadlag representation of elements of $\mathcal{F}(B)$ allows us to apply on them the techniques for the spaces $D_E[0,1]$. In this paper, we will consider the proof given by Daffer and Taylor (1979) for the strong law in $D[0,1]$, which is based on the proof of the Glivenko-Cantelli Theorem. On $D^{\uparrow}_{\mathcal{K}(B)}[0,1]$ we can employ step by step the same reasoning to prove the *strong law in* $\mathcal{F}(B)$.

Theorem 1. *Let* $X_1 : \Omega \to \mathcal{F}(B)$ *an integrably bounded fuzzy random variable. If we assume that* $X_1, X_2, \ldots$ *is a sequence of independent and identically distributed random elements and* $S_n = X_1 + \ldots + X_n$*, then*

$$\lim_{n\to\infty} d_\infty\left(\frac{S_n}{n}, E(\mathrm{Co}X_1)\right) = 0 \quad a.s.[P].$$

Proof. Given that $E(\mathrm{Co}X_1) \in D_{\mathcal{K}(B)}[0,1]$, as we have commented in the preliminaries we can assure that for every $m \in N$ there exists a partition of $[0,1]$, $0 = t_0 < t_1 < \ldots < t_{k(m)} = 1$, so that

$$\sup_{t,s\in[t_i,t_i+1)} d_H(E(\mathrm{Co}X_1)(t), E(\mathrm{Co}X_1)(s)) < 1/m.$$

If $t \in [0,1]$, then it is verified that either $t \in [t_{i-1}, t_i)$ for some $i \in \{1,2,\ldots,$ $k(m)\}$ or $t = 1$. Since the functions are non-decreasing we have that

$$E(\mathrm{Co}X_1)(t_{i-1}) \subset E(\mathrm{Co}X_1)(t) \subset E(\mathrm{Co}X_1)(t_i^-)$$

and, in the same way, if we denote $S_n^\omega = S_n(\omega)$ for each $\omega \in \Omega$, it is satisfied that

$$\frac{1}{n}S_n^\omega(t_{i-1}) \subset \frac{1}{n}S_n^\omega(t) \subset \frac{1}{n}S_n^\omega(t_i^-).$$

In virtue of the properties of the Hausdorff metric we can guarantee that

$$d_H\left(\frac{1}{n}S_n^\omega(t), E(\mathrm{Co}X_1)(t)\right)$$

$$\leq \max\left\{d_H\left(\frac{1}{n}S_n^\omega(t_{i-1}), E(\mathrm{Co}X_1)(t_i^-)\right), d_H\left(\frac{1}{n}S_n^\omega(t_i^-), E(\mathrm{Co}X_1)(t_{i-1})\right)\right\}.$$

The triangle inequality allows us to conclude that

$$d_H\left(\frac{1}{n}S_n^\omega(t_{i-1}), E(\mathrm{Co}X_1)(t_i^-)\right)$$

$$\leq d_H\left(\frac{1}{n}S_n^\omega(t_{i-1}), E(\mathrm{Co}X_1)(t_{i-1})\right) + d_H\left(E(\mathrm{Co}X_1)(t_{i-1}), E(\mathrm{Co}X_1)(t_i^-)\right)$$

$$\leq d_H\left(\frac{1}{n}S_n^\omega(t_{i-1}), E(\mathrm{Co}X_1)(t_{i-1})\right) + \frac{1}{m}$$

and

$$d_H\left(\frac{1}{n}S_n^\omega(t_i^-), E(\mathrm{Co}X_1)(t_{i-1})\right)$$

$$\leq d_H\left(\frac{1}{n}S_n^\omega(t_i^-), E(\mathrm{Co}X_1)(t_i^-)\right) + d_H\left(E(\mathrm{Co}X_1)(t_i^-), E(\mathrm{Co}X_1)(t_{i-1})\right)$$

$$\leq d_H\left(\frac{1}{n}S_n^\omega(t_i^-), E(\text{Co}X_1)(t_i^-)\right) + \frac{1}{m},$$

whence, for each $t \in [0,1]$ we can write

$$d_H\left(\frac{1}{n}S_n^\omega(t), E(\text{Co}X_1)(t)\right)$$

$$\leq \max_{1\leq i\leq k(m)} \max\left\{d_H\left(\frac{1}{n}S_n^\omega(t_i^-), E(\text{Co}X_1)(t_i^-)\right),\right.$$

$$\left.d_H\left(\frac{1}{n}S_n^\omega(t_{i-1}), E(\text{Co}X_1)(t_{i-1})\right)\right\} + \frac{1}{m}.$$

By taking the supremum on $[0,1]$, we have that

$$d_\infty\left(\frac{1}{n}S_n(\omega), E(\text{Co}X_1)\right)$$

$$\leq \max_{1\leq i\leq k(m)} \max\left\{d_H\left(\frac{1}{n}S_n^\omega(t_i^-), E(\text{Co}X_1)(t_i^-)\right),\right.$$

$$\left.d_H\left(\frac{1}{n}S_n^\omega(t_{i-1}), E(\text{Co}X_1)(t_{i-1})\right)\right\} + \frac{1}{m}.$$

Given that m is arbitrary and the strong law for random sets (see Artstein and Hansen, 1985, Hiai, 1985) guarantees that:

$$\lim_{n\to\infty} d_H\left(\frac{1}{n}S_n(t^-), E(\text{Co}X_1)(t^-)\right) = 0 \quad a.s.[P]$$

and

$$\lim_{n\to\infty} d_H\left(\frac{1}{n}S_n(t), E(\text{Co}X_1)(t)\right) = 0 \quad a.s.[P]$$

for all $t \in [0,1]$, we obtain that

$$\lim_{n\to\infty} d_\infty\left(\frac{1}{n}S_n, E(\text{Co}X_1)\right) = 0 \quad a.s.[P]. \qquad \blacksquare$$

In the same way, it is possible to prove a *law of the iterated logarithm in* $\mathcal{F}(\mathbb{R}^p)$) from the LIL for random sets (see, for instance, Giné *et al.*, 1983).

Theorem 2. *Let* $X_1 : \Omega \to \mathcal{F}(\mathbb{R}^p)$ *a random element so that* $E\|X_1\|^2 < \infty$. *If we suppose that* $X_1, X_2, \ldots$ *is a sequence of independent and identically distributed fuzzy random variables and* $S_n = X_1 + \ldots + X_n$*, then:*

$$\limsup_{n\to\infty} \frac{\sqrt{n}}{\sqrt{2\log\log n}}\, d_\infty\left(\frac{1}{n}S_n, E(\text{Co}X_1)\right) \leq \sqrt{E\|\text{Co}X\|^2} \quad a.s.[P].$$

It is possible to check that the method used in the proof of Theorem 1 is equivalent to the one used in Molchanov (1999). Whereas Molchanov takes into account the distances between the α-levels and the closure of the strict α-levels, in the technique by Daffer and Taylor (1979) the considered distances are those between the α-levels and the left-limit at each point α of the β-levels. The next proposition shows that both viewpoints are equivalent.

Proposition 1. *If $V \in \mathcal{F}(B)$, and x_V is its cadlag representation, it is satisfied that $x_V((1-\alpha)^-) = \mathrm{cl}\{a \in B \,|\, V(a) > \alpha\}$ for all $\alpha \in [0,1]$.*

Proof. If $\alpha = 1$, the equality is verified as a consequence of the continuity of x_V at this point and by the definition of the 0-level. If $\alpha \in [0,1)$, we have to prove that for every non-decreasing sequence $\{\beta_n\}_n$ so that $\beta_n \uparrow 1-\alpha$ as n tends to ∞, it is verified that $\lim_{\beta_n \uparrow 1-\alpha} x_V(\beta_n) = \mathrm{cl}\{a \in B \,|\, V(a) > \alpha\}$. Since $\{x_V(\beta_n)\}_n$ is a non-decreasing sequence, we have that

$$\lim_{\beta_n \uparrow 1-\alpha} x_V(\beta_n) = \mathrm{cl}\left(\bigcup_{n=1}^{\infty} x_V(\beta_n)\right).$$

Besides, $x_V(\beta_n) = \{a \in B \,|\, V(a) \geq 1-\beta_n\}$ for all $n \in N$, and hence the sequence $1-\beta_n \downarrow \alpha$ as n tends to ∞, whence

$$\bigcup_{n=1}^{\infty} \{a \in B \,|\, V(a) \geq 1-\beta_n\} = \{a \in B \,|\, V(a) > \alpha\}. \qquad \blacksquare$$

As we can observe in the proof of Theorem 1, and Molchanov pointed out except for the formal differences above remarked, there exists a characterization of convergence of fuzzy sets which, due to the bijection between $\mathcal{F}(B)$ and $D^{\uparrow}_{\mathcal{K}(B)}[0,1]$, can be expressed as follows:

Corollary 1. *Let $x, x_n \in D^{\uparrow}_{\mathcal{K}(B)}[0,1]$, with $n \in N$ be a sequence of fuzzy sets. Then, $\lim_{n\to\infty} m_\infty(x_n, x) = 0$ if, and only if, $\lim_{n\to\infty} d_H(x_n(\alpha), x(\alpha)) = 0$ and $\lim_{n\to\infty} d_H(x_n(\alpha^-), x(\alpha^-)) = 0$ for all $\alpha \in [0,1] \cap \mathbb{Q}$.*

4 Concluding remarks

In the methodology followed to prove Theorem 1, the equality of expected values to find the adequate discretization of the interval $[0,1]$ becomes crucial. Thus, the removal of the hypothesis of equality of distributions following this scheme requires to introduce some condition to ensure certain uniformity at the discontinuity points of the expected values (see, for other approach, Colubi *et al.*, 2001). There are some strong laws for sequences of independent random sets in which the equality of distributions is not assumed and that could be applied in this case (see, for example, Lyashenko, 1982, Hiai, 1985).

On the other hand, there are strong laws of large numbers in the space $D^{\uparrow}(\mathbb{R}^+, \mathbb{R})$ (see, for example, Daffer and Schiopu-Kratina, 1988). A similar reasoning could be useful to prove the strong law for toll sets, an interesting extension of fuzzy sets (see, for example, Aubin, 1999).

Acknowledgements

The research in this paper has been partially supported by DGESIC Grant No. DGE-98-PB97-1282. This financial support is gratefully acknowledged to the Spanish MECD.

References

1. Artstein, Z. and Hansen, J.C. (1985). Convexification in limit laws of random sets in Banach spaces. *Ann. Probab.* **13**, 307-309.
2. Aubin, J.P. (1999). *Mutational and Morphological Analysis.* Birkhäuser, Boston.
3. Aumann, R.J. (1965). Integrals of set-valued functions. *J. Math. Anal. Appl.* **12**, 1-12.
4. Billingsley, P. (1968). *Convergence of Probability Measures.* John Wiley & Sons, New York.
5. Colubi, A., López-Díaz, M., Domínguez-Menchero, J.S. and Gil, M.A. (1997). *A generalized strong law of large numbers.* Technical Report. Universidad de Oviedo.
6. Colubi, A., López-Díaz, M., Domínguez-Menchero, J.S. and Gil, M.A. (1999). A generalized strong law of large numbers. *Probab. Theory Relat. Fields* **114**, 401-417.
7. Colubi, A., Domínguez-Menchero, J.S., López-Díaz, M. and Ralescu D. A. (2001). A $D_E[0,1]$ representation of random upper semicontinuous functions, (submitted for publication)
8. Colubi, A., Domínguez-Menchero, J.S., López-Díaz, M. and Körner R. (2001). A method to derive strong laws of large numbers for random upper semicontinuous functions, *Statist. Probab. Let.* (accepted for publication).
9. Daffer, P.Z. and Taylor, R.L. (1979). Laws of large numbers for $D[0,1]$. *Ann. Prob.* **7**, 85-95.
10. Daffer, P. and Schiopu-Kratina, I. (1988). L^1-tightness and the law of large numbers in $D(\mathbb{R})$. *Can. J. Stat.* **16**, 393-397.
11. Debreu, G. (1967). Integration of correspondences. *Proc. Fifth Berkeley Symp. Math. Statist. Prob.* 1965/66 **2**, Part 1. Univ. of California Press, Berkeley, 351-372.
12. Ethier, S.N. and Kurtz, T.G. (1986). *Markov Processes. Characterizations and Convergence.* John Wiley & Sons.
13. Giné, E., Hahn, M. and Zinn, J. (1983). Limit theorems for random sets: an application of probability in Banach space results. *Probability in Banach spaces IV. Berlin, Springer-Verlag* **990**, 112-135.
14. Hiai, F. (1985). Convergence of conditional expectations and strong laws of large numbers for multivalued random variables. *Trans. Amer. Math. Soc.* **291**, 613-627.

15. Hoffman-Jørgensen, J. (1985a). The law of large numbers for non-measurable and non-separable random elements. *Asterisque.* **131** 299-356.
16. Klement, E.P., Puri M.L. and Ralescu, D.A. (1986). Limit theorems for fuzzy random variables. *Proc. R. Soc, Lond. A* **407**, 171-182.
17. López-Díaz, M. and Gil, M.A. (1998a). Approximating integrably bounded fuzzy random variables in terms of the "generalized" Hausdorff metric. *Inform. Sci.* **104**, 279-291.
18. Lyashenko, N.N. (1982). Limit theorems for sums of independent compact random subsets of euclidean space. *J. Soviet. Math.* **20**, 2187-2196.
19. Molchanov, I. (1999). On strong laws of large numbers for random upper semicontinuous functions. *J. Math. Anal. Appl.* **235**, 349-355.
20. Proske, F. (1997). *Grenzwertsätze für Fuzzy-Zufallsvariablen unter dem Gesichspunkt der Wahrscheinlichkeitstheorie ouf inseparablen semigruppen.* PhD Thesis. Univ. Ulm.
21. Puri, M.L. and Ralescu, D.A. (1981). Différentielle d'une fonction floue. *C.R. Acad. Sci. Paris Sér. A* **293**, 237-239.
22. Puri, M.L. and Ralescu, D. (1986). Fuzzy random variables. *J. Math. Anal. Appl.* **114**, 409-422.

Convergence in graph for fuzzy valued martingales and smartingales

Shoumei Li[1] and Yukio Ogura[2]

[1] Department of Applied Mathematics, Beijing Polytechnic University, 100 Ping Le Yuan, Chao Yang District, Beijing, 100022, P. R. China
[2] Department of Mathematics, Saga University, 1 Honjo-Machi, Saga, 840-8502, Japan

Abstract. In this paper, we introduce the concept of convergence in graph for fuzzy-valued random variables, give an equivalent definition and then obtain convergence theorems for fuzzy-valued martingales, submartingales and supermartingales based on the results of our previous papers (Li and Ogura, 1996, 1998, 1999).

1 Introduction

Puri and Ralescu (1986) introduced the concept of fuzzy-valued random variable by exploiting the theory of set-valued random variables (or called random sets). There are very rich mathematical properties in the theory of set-valued random variables (cf. Aumann, 1965, Kendall, 1974, Hiai and Umegaki, 1977, Hiai, 1985, Hess, 1991, 1998). Following their work, many authors have obtained many important results such as approximative theorem, ergodic theorems, strong law of large numbers and central limit theorems for fuzzy-valued random variables (Klement *et al.*, 1986, Ralescu, 1986, Ban, 1991, López-Díaz and Gil, 1998). In Puri and Ralescu (1991) built a concept of fuzzy-valued martingale and got a convergence theorem of fuzzy-valued martingale. There the basic space was the n-dimensional Euclidean space $\mathbb{R}^n$ and convergence was in the Hausdorff distance. The typical method was by using Rådström's embedding theorem (1952) and their results focused on the fuzzy valued random variables whose cut sets are compact set-valued random variables.

In our previous papers (Li and Ogura, 1996, 1997, 1998, 1999), we focused on closed set-valued random variables and fuzzy-valued random variables whose cut sets are closed subsets of general Banach space, in contrast with most of above mentioned works, whose cut sets are compact subsets of $\mathbb{R}^n$. Our basic method was by using selection method. Among the results in 1996, we used the method of martingale selections to obtain a regularity theorem (cf. [Li and Ogura, 1996; Lemma 5.7]) for closed convex set valued martingales based on the results of Hiai and Umegaki (1977), and then applied it to fuzzy-valued martingales (cf. [Li and Ogura, 1996; Theorem 5.1]). In Li and Ogura (1999), we made use of Kuratowski-Mosco convergence (cf. Mosco, 1969, Salinetti and Roger, 1981, e.g.) in place of Hausdorff convergence, which was a main tool in Li and Ogura (1996) (and in most of former works of other authors) and got convergence theorems both for closed convex set-valued martingales (cf. [Li and Ogura, 1999; Theorem 3.1]) and fuzzy-valued martingales (cf. [Li and Ogura, 1999; Theorem 3.3]). In Li and Ogura (1998), we proved convergence theorems for closed convex sub-martingales (cf. [Li and Ogura, 1998; Theorem 4]) and super-martingales (cf. [Li and Ogura, 1998; Theorem 5]) in the Kuratowski-Mosco sense, after discussing the closedness of Aumman integrals and conditional expectations of closed set-valued random variable.

Hess discussed closed convex set-valued random variable whose values may be unbounded, proved the existence of integrable martingale selections and got convergence theorems for martingales and supermartingales in Hess (1991, 1998) in the senses of Kuratowski-Mosco convergence and Wijsman convergence.

In this paper, we first compare three convergences briefly, that is, the Wijsman convergence, the Kuratowski-Mosco convergence and the Hausdorff convergence, and then state some results for set-valued martingales. We then introduce a new convergence, i.e. convergence in graph for fuzzy-valued random variables. We can consider this convergence for the fuzzy-valued case as the Kuratowski-Mosco convergence for the set-valued case in the product space. Thus this convergence has very clear topology. We firstly give an equivalent definition, and then prove convergence theorems for fuzzy-valued martingales, supermartingales and submartingales based on the results of our previous papers (Li and Ogura, 1996, 1998, 1999). We notice that above results are for integrably bounded fuzzy-valued random variables. Finally we extend some results to unbounded case.

2 Topologies and convergences on $\mathcal{K}_*(\mathbb{X})$

Throughout this paper, assume that $(\mathbb{X}, \|\cdot\|_{\mathbb{X}})$ is a real separable Banach space, $\mathcal{K}_*(\mathbb{X})$ is the family of all nonempty, closed subsets of $\mathbb{X}$, $\mathcal{K}_{c^*}(\mathbb{X})$ is the family of all closed, convex subsets of $\mathbb{X}$ and $\mathcal{K}_c(\mathbb{X})$ is the family of all compact, convex subsets of $\mathbb{X}$. For $B \in \mathcal{K}_*(\mathbb{X})$, $\|B\|_{\mathcal{K}_*(\mathbb{X})}$ denotes the *norm* of B defined as $\|B\|_{\mathcal{K}_*(\mathbb{X})} = \sup_{a \in B} \|a\|_{\mathbb{X}}$.

Two operations are defined in $\mathcal{K}_*(\mathbb{X})$:

$$A + B = \text{cl}\{a + b : a \in A, b \in B\},$$

$$\lambda A = \{\lambda a : a \in A\},$$

where λ is a real number.

For any $B \in \mathcal{K}_*(\mathbb{X})$, the distance function $d(\cdot, B)$ is defined by

$$d(x, B) = \inf_{y \in B} \|x - y\|_{\mathbb{X}}, \qquad x \in \mathbb{X}.$$

In our papers (Li and Ogura, 1998, 1999), we mainly discussed the Kuratowski-Mosco convergence for the sequences of set-valued random variables. Here we briefly compare two topologies and relative convergences on $\mathcal{K}_*(\mathbb{X})$ or $\mathcal{K}_{c^*}(\mathbb{X})$, namely, the Wijsman topology and the Mosco topology.

The *Wijsman topology* on $\mathcal{K}_*(\mathbb{X})$ (cf. Wijsman, 1966, and Beer, 1993), denoted by τ_W, is the topology determined by the family of distance functionals

$$\{B \to d(x, B) : x \in \mathbb{X}\}.$$

It is known that (e.g. cf. Beer, 1993) that the Wijsman topology is metrizable and separable. We say that a sequence $\{B_n\}$ in $\mathcal{K}_*(\mathbb{X})$ *converges* to B *in the Wijsman topology* if, for any $x \in \mathbb{X}$, $d(x, B_n) \to d(x, B)$, denoted by $B_n \xrightarrow{\tau_W} B$.

Now we recall some known facts concerning Kuratowski-Mosco convergence and Mosco topology on $\mathcal{K}_*(\mathbb{X})$. Given a topology τ on $\mathbb{X}$ and a sequence $\{B_n\}$ in $\mathcal{K}_*(\mathbb{X})$, we define two subsets

$$\tau\text{-}LiB_n := \{x = \tau\text{-}\lim x_n : x_n \in B_n, n \in \mathbb{N}\},$$

$$\tau\text{-}LsB_n := \{x = \tau\text{-}\lim x_{n_k} : x_{n_k} \in B_{n_k}, k \in \mathbb{N}\},$$

where $\{B_{n_k}\}$ is a subsequence of $\{B_n\}$. τ-LiB_n and τ-LsB_n are respectively called the lower limit and the upper limit of sequence $\{B_n\}$ relatively to the topology τ. The sequence $\{B_n\}$ is said to converge to B with respect to the topology τ if

$$\tau\text{-}LsB_n = B = \tau\text{-}LiB_n.$$

This is denoted by $B_n \xrightarrow{\tau} B$. When we consider two different topologies τ_1, τ_2 at the same time on $\mathbb{X}$, we can get the lower limit τ_1-LiB_n and τ_2-LsB_n. According to this idea, Mosco introduced the following convergence in Mosco (1969).

Let s be the strong topology and w be the weak topology of $\mathbb{X}$. We call B_n *convergence* to B *in the Kuratowski-Mosco sense*, denoted by $B_n \xrightarrow{K-M} B$, if

$$w\text{-}LsB_n = B = s\text{-}LiB_n.$$

Beer (1993) showed that the Mosco convergence arises from a hit-and-miss topology on $\mathcal{K}_{c^*}(\mathbb{X})$ in the following way. For any open subset O of $\mathbb{X}$, let

$$\mathcal{K}_*(O) := \{B \in \mathcal{K}_{c^*}(\mathbb{X}) : B \cap O \neq \emptyset\},$$

and for any weakly compact subset K of $\mathbb{X}$, write

$$\mathcal{K}_*(K) := \{B \in \mathcal{K}_{c^*}(\mathbb{X}) : B \cap K = \emptyset\}.$$

The *Mosco topology* τ_M on $\mathcal{K}_{c^*}(\mathbb{X})$ is the topology generated by all sets of the form $\mathcal{K}_*(O)$ and $\mathcal{K}_*(K)$.

When $\mathbb{X}$ is finite dimensional, the Kuratowski-Mosco convergence and Wijsman convergence coincide. When $\mathbb{X}$ is infinite dimensional, the Kuratowski-Mosco convergence implies Wijsman convergence in general. However if $\mathbb{X}$ is separable reflexive Banach space having a Fréchet differentiable norm, the Kuratowski-Mosco convergence and the Wijsman convergence are equivalent. On the other hand, every reflexive Banach space can be renormed so that the new norm is Fréchet differentiable. This allow us to assume that the Mosco and Wijsman topologies coincide if $\mathbb{X}$ is reflexive.

There is another topology and relative convergence that often to be used, that is, the Hausdorff topology and the Hausdorff convergence, (e.g. cf. Hiai and Umegaki, 1977, Hiai, 1985, Puri and Ralescu, 1986, 1991, for its extension), derived from the Hausdorff distance d_H,

$$d_H(A, B) = \max\{\sup_{a \in A} d(a, B), \sup_{b \in B} d(b, A)\}$$

where $A, B \in \mathcal{K}_*(\mathbb{X})$. But if A or B are unbounded, $d_H(A, B)$ may be infinite. It is well-known (cf. [Kuratowski, 1965, p.214, p.407]) that the family of all bounded elements in $\mathcal{K}_*(\mathbb{X})$ is a complete space with respect to the Hausdorff metric d_H, and the family of all bounded elements in $\mathcal{K}_{c^*}(\mathbb{X})$ is a closed subset of this complete space. Generally speaking, the Hausdorff convergence is stronger than the Kuratowski-Mosco and the Wijsman convergences. However, if $\mathbb{X}$ is finite dimensional and we limit in $\mathcal{K}_c(\mathbb{X})$, then all the convergences are equivelent (cf. Klein and Thompson, 1984). From the discussion above, we can see that the Kuratowski-Mosco and the Wijsman convergences are also available for sequences of unbounded closed subsets but the Hausdorff convergence is not even in the finite dimensional case.

3 Set valued martingales and convergence

Throughout this paper, assume that $(\Omega, \mathcal{A}, \mu)$ is a complete probability space.

A *set-valued random variable* $F : \Omega \to \mathcal{K}_*(\mathbb{X})$ is a measurable mapping, that is, for every $B \in \mathcal{K}_*(\mathbb{X})$, $F^{(-1)}(B) := \{\omega \in \Omega; F(\omega) \cap B \neq \emptyset\} \in \mathcal{A}$ (cf. Hiai and Umegaki, 1977). It is equivalent to that, for each $x \in \mathbb{X}$, distance function $d(x, F(\omega))$ is measurable.

A measurable mapping $f : \Omega \to \mathbb{X}$ is called *a measurable selection* of F if $f(\omega) \in F(\omega)$ for all $\omega \in \Omega$. Denote by $\mathcal{L}^1[\Omega, \mathbb{X}]$ the Banach space of all measurable mappings $g : \Omega \to \mathbb{X}$ such that the norm $\|g\|_L = \int_\Omega \|g(\omega)\|_{\mathbb{X}} d\mu$ is finite. For a measurable set-valued random variable F, define the set

$$S_F = \{f \in \mathcal{L}^1[\Omega, \mathbb{X}] : f(\omega) \in F(\omega) \quad a.e.(\mu)\}.$$

For a sub-σ-field $\mathcal{A}_0$, denote by $S_F(\mathcal{A}_0)$ the set of all $\mathcal{A}_0$-measurable mappings in S_F. S_F is closed in $\mathcal{L}^1[\Omega, \mathbb{X}]$.

S_F is non empty if, and only if, the distance function $d(0, F(\omega)) \in \mathcal{L}^1[\Omega, \mathbb{X}]$. In this case, F is called *integrable* (cf. Hess, 1998). On the other hand, a set random variable $F : \Omega \to \mathcal{K}_*(\mathbb{X})$ is called *integrably bounded* iff the real-valued random variable $\|F(\omega)\|_{\mathcal{K}_*(\mathbb{X})}$ is integrable. Hiai and Umegaki (1977) has shown that F is integrably bounded if and only if S_F is bounded in $\mathcal{L}^1[\Omega, \mathbb{X}]$.

From the definitions, we can see that an integrable random variable may take unbounded-valued. But an intergable bounded random variable only takes bounded set values *a.e.*(μ).

Let $U^1[\Omega, \mathcal{A}, \mu; \mathcal{K}_*(\mathbb{X})]$ denote the space of all integrable set-valued random variables and $\mathcal{L}^1[\Omega, \mathcal{A}, \mu; \mathcal{K}_*(\mathbb{X})]$ denote the space of all integrably bounded set random variables where two set random variables $F_1, F_2 \in \mathcal{L}^1[\Omega, \mathcal{A}, \mu; \mathcal{K}_*(\mathbb{X})]$ are considered to be identical if $F_1(\omega) = F_2(\omega)$, a.e.$(\mu)$. $\mathcal{L}^1[\Omega, \mathcal{A}, \mu; \mathcal{K}_{c^*}(\mathbb{X})]$ (resp. $\mathcal{L}^1[\Omega, \mathcal{A}, \mu; \mathcal{K}_c(\mathbb{X})]$) denote the space of all closed (resp. compact) convex integrably bounded set random variables.

For each $F \in U^1[\Omega, \mathcal{A}, \mu; \mathcal{K}_*(\mathbb{X})]$, Aumann integral of F is given by

$$\int_\Omega F d\mu = \mathrm{cl}\{\int_\Omega f d\mu : f \in S_F\}, \tag{1}$$

where $\int_\Omega f d\mu$ is the usual Bochner integral and the closure is taken in $\mathbb{X}$. Define $\int_A F d\mu = \mathrm{cl}\{\int_A f d\mu : f \in S_F\}$, for $A \in \mathcal{A}$.

Remark 1. Here the definition is a little difference with the original one of Aumann (cf. Aumann, 1965). He defined $\int_\Omega F d\mu = \{\int_\Omega f d\mu : f \in S_F\}$. The set $\{\int_\Omega f d\mu : f \in S_F\}$, however, is not closed in general (cf. Example in Li and Ogura, 1998). But if $\mathbb{X}$ is reflexive, it is closed (cf. Theorems 1, 2 by Li and Ogura, 1998).

Let $\mathcal{A}_0$ be a sub-σ-field of $\mathcal{A}$. The *conditional expectation* $E[F|\mathcal{A}_0]$ of an $F \in U^1[\Omega, \mathcal{A}, \mu; \mathcal{K}_*(\mathbb{X})]$ is determined as a $\mathcal{A}_0$-measurable element of $U^1[\Omega, \mathcal{A}, \mu; \mathcal{K}_*(\mathbb{X})]$ by

$$S_{E[F|\mathcal{A}_0]}(\mathcal{A}_0) = \mathrm{cl}\{E(f|\mathcal{A}_0) : f \in S_F\}, \tag{2}$$

where the closure is taken in the $\mathcal{L}^1[\Omega, \mathbb{X}]$ (cf. Hiai and Umegaki, 1977). We have to notice that Hiai and Umegaki gave the definition on $\mathcal{L}^1[\Omega, \mathcal{A}, \mu; \mathcal{K}_*(\mathbb{X})]$. But we can extend it to $U^1[\Omega, \mathcal{A}, \mu; \mathcal{K}_*(\mathbb{X})]$ and prove the existence theorem without any difficulty. If $\mathbb{X}^*$ is separable, this is equivalent to the formula

$$\mathrm{cl} \int_A F d\mu = \mathrm{cl} \int_A E[F|\mathcal{A}_0] d\mu \quad \text{for} \quad A \in \mathcal{A}_0. \tag{3}$$

Let $\{\mathcal{A}_n : n \in \mathbb{Z}\}$ be a family of complete sub-σ-fields of $\mathcal{A}$ such that $\mathcal{A}_n \subset \mathcal{A}_{n+1}$ for all $n \in \mathbb{Z}$, $\mathcal{A}_\infty$ the σ-field generated by $\bigcup_{n=1}^{\infty} \mathcal{A}_n$, and $\mathcal{A}_{-\infty} = \bigcap_{n=1}^{\infty} \mathcal{A}_{-n}$.

A system $\{F_n, \mathcal{A}_n : n \in \mathbb{N}\}$ is called a *set-valued martingale* (resp. *set-valued submartingale, supermartingale*) iff

1) $F_n \in \mathcal{L}^1[\Omega, \mathcal{A}_n, \mu : \mathcal{K}_*(\mathbb{X})], \quad n \in \mathbb{N}$,
2) $F_n = E[F_{n+1}|\mathcal{A}_n], \quad n \in \mathbb{N}$, *a.e.*$(\mu)$. (resp. $\subset$, $\supset$).

Remark 2. In our paper (1998), the definition of set-valued submartingale is a little different with above one, since the subset $\{E(f|\mathcal{A}_{n+1}) : f \in S_{F_n}(\mathcal{A}_n)\}$ is not closed in general. However, if $\mathbb{X}$ is reflexive, the set $E[F_{n+1}|\mathcal{A}_n]$ is closed (cf. Theorems 1, 2 in Li and Ogura, 1998). Thus both definitions are equivalent.

By using embedding theorem, Hiai and Umegaki (1977) obtained the following results:

(R.1) *Let $\{F_n, \mathcal{A}_n; n \geq 1\}$ be a compact convex-valued martingale such that $F_n = E[F|\mathcal{A}_n], n \geq 1$, where $F \in \mathcal{L}^1[\Omega, \mathcal{A}, \mu; \mathcal{K}_c(\mathbb{X})]$. Then*

$$d_H(F_n, F_\infty) \to 0,$$

where $F_\infty = E[F|\mathcal{A}_\infty]$.

(R.2) *Let $\{F_n, \mathcal{A}_n; n \leq -1\}$ be a set-valued martingale such that $F_n = E[F|\mathcal{A}_n], n \leq -1$, where $F_{-1} \in \mathcal{L}^1[\Omega, \mathcal{A}, \mu; \mathcal{K}_c(\mathbb{X})]$. Then*

$$d_H(F_n, F_{-\infty}) \to 0$$

as $n \to -\infty$, where $F_{-\infty} = E[F|\mathcal{A}_{-\infty}]$, (cf. Hiai and Umegaki, 1977).

Remark 3. The $\mathbb{X}$-valued martingales with regular property, as we know, imply convergence in almost everywhere with respect to μ. In the case of compact convex-valued martingales, it is also right according to above results. But in the case of bounded closed convex set-valued martingales, regular property does not imply convergence in the Hausdorff metric (cf. [Li and Ogura, 1998, Example 4.2]). We can see from this fact that the Hausdorff topology is too strong for the study of set-valued random variables. Then we used the Kuratowski-Mosco convergence and got the following theorems (cf. Li and Ogura, 1998, 1999).

A Banach space $\mathbb{X}$ is said to have the *Radon-Nikodym property* (RNP) with respect to a finite measure space $(\Omega, \mathcal{A}, \mu)$ if for each μ-continuous $\mathbb{X}$-valued measure $m : \mathcal{A} \to \mathbb{X}$ of bounded variation, there exists an integrable mapping $f : \Omega \to \mathbb{X}$ such that $m(A) = \int_A f d\mu$ for all $A \in \mathcal{A}$. It is known that every separable dual space and every reflexive space has the RNP (cf. Chatterji, 1968).

Theorem 1. (Li and Ogura, 1999) *Assume that* $\mathbb{X}$ *is a Banach space satisfying the RNP with the separable dual* $\mathbb{X}^*$. *Then, for every uniformly integrably set-valued martingale* $\{F_n, \mathcal{A}_n : n \in \mathbb{N}\}$, *there exists a unique* $F_\infty \in \mathcal{L}^1[\Omega, \mathcal{A}, \mu; \mathcal{K}_{c^*}(\mathbb{X})]$ *such that* $F_n = E[F_\infty|\mathcal{A}_n]$ *for each* $n \in \mathbb{N}$ *and* $F_n \overset{K-M}{\longrightarrow} F_\infty$, *a.e.*$(\mu)$.

Theorem 2. (Li and Ogura, 1998) *Assume that* $\mathbb{X}$ *is a Banach space satisfying the RNP with the separable dual* $\mathbb{X}^*$, $\{F_n, \mathcal{A}_n : n \in \mathbb{N}\}$ *is uniformly integrably set-valued supermartingale and*

$$M = \bigcap_{n=1}^{\infty} \{f \in \mathcal{L}^1[\Omega, \mathcal{A}_\infty, \mu; \mathbb{X}] : E(f|\mathcal{A}_n) \in S_{F_n}(\mathcal{A}_n)\} \tag{4}$$

is a nonempty set. Then there exists a unique $F_\infty \in \mathcal{L}^1[\Omega, \mathcal{A}_\infty, \mu; \mathcal{K}_{c^*}(\mathbb{X})]$ *such that* $F_n \overset{K-M}{\longrightarrow} F_\infty$, *a.e.*$(\mu)$.

Theorem 3. (Li and Ogura, 1998) *Assume that* $\mathbb{X}$ *is a reflexive Banach space. Then, for every uniformly integrably set-valued submartingale* $\{F_n, \mathcal{A}_n : n \in \mathbb{N}\}$, *there exists a unique* $F_\infty \in \mathcal{L}^1[\Omega, \mathcal{A}_\infty, \mu; \mathcal{K}_{c^*}(\mathbb{X})]$ *such that* $F_n \overset{K-M}{\longrightarrow} F_\infty$, *a.e.*$(\mu)$.

4 Convergence in graph for sequences of fuzzy sets and fuzzy valued random variables

Let $\mathcal{F}_*(\mathbb{X})$ denote the family of all fuzzy sets $\nu : \mathbb{X} \to [0,1]$, which satisfy the following two conditions:

1) each ν is upper semicontinous function, i.e. for each $\alpha \in (0,1]$, the cut set $\nu_\alpha = \{x \in \mathbb{X} : \nu(x) \geq \alpha\}$ is a closed subset of $\mathbb{X}$,
2) the cut set $\nu_1 = \{x \in \mathbb{X} : \nu(x) = 1\} \neq \emptyset$.

For two fuzzy sets $\nu^1, \nu^2 \in \mathcal{F}_*(\mathbb{X})$, we denote $\nu^1 \leq \nu^2$ iff $\nu^1_\alpha \subset \nu^2_\alpha$ for every $\alpha \in [0,1]$. Obviously, $(\mathcal{F}_*(\mathbb{X}), \leq)$ is a partial ordered set. Similarly, we define $\mathcal{F}_{c^*}(\mathbb{X})$ and $\mathcal{F}_c(\mathbb{X})$.

For any two fuzzy sets $\nu^1, \nu^2 \in \mathcal{F}_*(\mathbb{X})$, define two operators $\oplus, \odot$ by using Zadeh's extension principle,

$$(\nu^1 \oplus \nu^2)(z) = \sup_{z=x+y} \min\{\nu^1(x), \nu^2(y)\} \qquad \text{for all } z \in \mathbb{X}$$

and for any $x \in \mathbb{X}$

$$(\lambda \odot \nu^1)(x) = \begin{cases} \nu^1(x/\lambda), & \text{if } \lambda \neq 0, \\ I_0(x), & \text{if } \lambda = 0, \end{cases}$$

where I_0 is the indicator function of 0.

For example, let $\nu^1(x) = \exp\{-(x-m)^2\}$, $\nu^2(x) = \exp\{-(x-n)^2\}$, then $(\nu^1 \oplus \nu^2)(z) = \exp\{-(z-m-n)^2/4\}$.

It is known that for any $\alpha \in [0,1], (\nu^1 \oplus \nu^2)_\alpha = \nu^1_\alpha + \nu^2_\alpha$, and $(\lambda \odot \nu^1)_\alpha = \lambda \nu^1_\alpha$. We can see that the operators $\oplus, \odot$ on $\mathcal{F}_*(\mathbb{X})$ are the extension of the operators of plus and multiplication on $\mathcal{K}_*(\mathbb{X})$.

A fuzzy set ν is called convex if,

$$\nu(\lambda x + (1-\lambda)y) \geq \min\{\nu(x), \nu(y)\}, \qquad \text{for any} \quad x, y \in \mathbb{X}, \quad \lambda \in [0,1].$$

It is known that ν is convex in above sense if and only if, for any $\alpha \in [0,1]$, the cut set ν_α is a convex subset of $\mathbb{X}$.

Now we introduce convergence in graph in $\mathcal{F}_*(\mathbb{X})$. Let $\nu \in \mathcal{F}_*(\mathbb{X})$, write

$$Gr\ \nu = \{(x,y) \in \mathbb{X} \times [0,1], \nu(x) \geq y\}.$$

It is clear that $Gr\ \nu$ denotes the area between the curve of ν and $\mathbb{X}$-axis if $\mathbb{X}$ is $\mathbb{R}$. We call it *the graph of* ν. Since all the cut set of ν are nonempty closed set, the graph of ν is a closed set of space $\mathbb{X} \times [0,1]$.

For $\nu_n,\ \nu \in \mathcal{F}_*(\mathbb{X})$, ν_n is called to *converge to* ν *in graph* (denoted by $\nu_n \xrightarrow{Gr} \nu$) iff $Gr\ \nu_n$ converges to $Gr\ \nu$ in $\mathbb{X} \times [0,1]$ in the Kuratowski-Mosco sense.

Lemma 1. *If, for any* $\alpha \in [0,1]$, $(\nu_n)_\alpha \xrightarrow{K-M} (\nu)_\alpha$, *then we have that* $\nu_n \xrightarrow{Gr} \nu$.

Proof. Firstly we will prove that

$$Gr\ \nu \subset s\text{-}\liminf_{n\to\infty} Gr\ \nu_n, \qquad \text{in } \mathbb{X} \times [0,1].$$

Indeed, for any $(x_0, y_0) \in Gr\ \nu$, we have $x_0 \in \mathbb{X}$, and $\nu(x_0) \geq y_0$ according to the definition. Thus $x_0 \in \nu_{y_0} = \{x \in \mathbb{X} : \nu(x) \geq y_0\}$. Since $(\nu_n)_{y_0} = \{x \in \mathbb{X} : \nu_0(x) \geq y_0\}$ converges to ν_{y_0} in the Kuratowski-Mosco sense, there exist $x_n \in (\nu_n)_{y_0}$, such that $x_n \to x$ in strong topology in $\mathbb{X}$. Notice that $\nu_n(x_n) \geq y_0$ and let $y_n = y_0$, then we have $(x_n, y_n) \in Gr\nu_n$, $x_n \to x_0$ in $\|\cdot\|$ and $y_n \to y_0$. Thus $(x_0, y_0) \in s\text{-}\liminf_{n\to\infty} Gr\ \nu_n$.

Now we will prove that

$$w\text{-}\limsup_{n\to\infty} Gr\ \nu_n \subset Gr\ \nu, \qquad \text{in } \mathbb{X} \times [0,1].$$

For any $(x_0, y_0) \in w\text{-}\limsup_{n\to\infty} Gr\ \nu_n$, there exist a subsequence $(x_{n_k}, y_{n_k}) \in Gr\ \nu_{n_k}, k \in \mathbb{N}$ such that $w\text{-}\lim_{k\to\infty}(x_{n_k}, y_{n_k}) = (x_0, y_0)$. Since $(x_{n_k}, y_{n_k}) \in Gr\nu_{n_k}$, we have $\nu_{n_k}(x_{n_k}) \geq y_{n_k}$.

From $\lim_{k\to\infty} y_{n_k} = y_0$, we get $\limsup_{k\to\infty} \nu_{n_k}(x_{n_k}) \geq y_0$. Thus there exists a subsequence of $\{n_k : k \in \mathbb{N}\}$, denoted by $\{n'_k : k \in \mathbb{N}\}$ such that

$\lim_{k\to\infty} \nu_{n'_k}(x_{n'_k}) \geq y_0$. Since $(\nu_{n'_k})_{y_0} = \{x \in \mathbb{X} : \nu_{n'_k}(x) \geq y_0\} \xrightarrow{K-M} (\nu)_{y_0} = \{x \in \mathbb{X} : \nu(x) \geq y_0\}$, $x_{n'_k} \in (\nu_{n'_k})_{y_0}$ and w- $\lim_{k\to\infty} x_{n'_k} = x_0$ imply $x_0 \in (\nu)_{y_0}$ i.e. $\nu(x_0) \geq y_0$. Thus $(x_0, y_0) \in Gr\ \nu$. ∎

Remark 4. But the opposite of Lemma 1 is not correct. We can see it from the following example.

Example 1. Let $\mathbb{X} = \mathbb{R}$, $a < b < c$ and

$$\nu(x) = \begin{cases} 0, & x < a, x > c, \\ 1/2, & a \leq x < b, \\ 1, & b \leq x \leq c, \end{cases}$$

and

$$\nu_n(x) = \begin{cases} 0, & x < a, x > c, \\ 1/2 - 1/2n, & a \leq x < b, \\ 1, & b \leq x \leq c. \end{cases}$$

Then $\nu_n \xrightarrow{Gr} \nu$, but $(\nu_n)_{1/2}$ does not converge to $\nu_{1/2}$ in the Kuratowski-Mosco sense.

Now we give an equivalent definitions for the convergence in graph in the following theorem.

Theorem 4. *Let ν_n, $\nu \in \mathcal{F}_*(\mathbb{X})$, then $\nu_n \xrightarrow{Gr} \nu$ iff the following two conditions are satisfied,*

(*i*) *for any $x \in \mathbb{X}$, there exists a sequence $\{x_n, n \in \mathbb{N}\}$ of $\mathbb{X}$ converging to x in strong topology of X such that*

$$\liminf_{n\to\infty} \nu_n(x_n) \geq \nu(x),$$

(*ii*) *For any given subsequence $\{\nu_{n_k}\}$ of ν_n and any sequence $\{x_{n_k}\}$ which converges to x in the weak topology of $\mathbb{X}$, we have*

$$\limsup_{k\to\infty} \nu_{n_k}(x_{n_k}) \leq \nu(x).$$

Proof. *Step 1* We will prove that (*i*) is equivalent to

$$Gr\ \nu \subset s\text{-}\liminf_{n\to\infty} Gr\ \nu_n, \qquad \text{in } \mathbb{X} \times [0,1]. \tag{5}$$

Suppose that we have (*i*). Since $Gr\ \nu \neq \emptyset$, take $(x, y) \in Gr\ \nu$, i.e. $x \in \mathbb{X}, y \in [0, 1]$ with $\nu(x) \geq y$. By (*i*), there exists a sequence $\{x_n\}$ of $\mathbb{X}$, such that $x_n \to x$, as $n \to \infty$, in strong topology of $\mathbb{X}$, and

$$\liminf_{n\to\infty} \nu_n(x_n) \geq \nu(x).$$

If let $y_n = \min\{\nu_n(x_n), y\}$, then we have $y = \lim_{n\to\infty} y_n$. Therefore $(x_n, y_n) \in Gr\ \nu_n$ for all $n \in \mathbb{N}$ and $(x_n, y_n) \to (x, y)$, as $n \to \infty$, in strong topology of $\mathbb{X} \times [0,1]$. Thus $(x,y) \in s\text{-}\liminf_{n\to\infty} Gr\ \nu_n$.

Conversely, suppose that (5) is satisfied. Let $x \in \mathbb{X}$, then $(x, \nu(x)) \in Gr\ \nu \subset s\text{-}\liminf_{n\to\infty}$
$Gr\ \nu_n$. Therefore by the definition, there exists $(x_n, y_n) \in Gr\ \nu_n$, for all $n \in \mathbb{N}$ such that $x_n \to x$ in strong topology of $\mathbb{X}$ and $y_n \to \nu(x)$ in $[0,1]$, as $n \to \infty$.

Since $(x_n, y_n) \in Gr\ \nu_n$, we have $\nu_n(x_n) \geq y_n$. Thus

$$\liminf_{n\to\infty} \nu_n(x_n) \geq \liminf_{n\to\infty} y_n = \lim_{n\to\infty} y_n = \nu(x),$$

i.e. (5) is satisfied.

Step 2 We will prove that (ii) is equivalent to

$$w\text{-}\limsup_{n\to\infty} Gr\ \nu_n \subset Gr\ \nu, \qquad \text{in } \mathbb{X} \times [0,1]. \tag{6}$$

Assume now that (ii) holds. Let $(x,y) \in w\text{-}\limsup_{n\to\infty} Gr\ \nu_n$, then $(x,y) \in \mathbb{X} \times [0,1]$ is the weak limit of a subsequence (x_{n_k}, y_{n_k}) in $\mathbb{X} \times [0,1]$ with $(x_{n_k}, y_{n_k}) \in Gr\ \nu_{n_k}$ for each $k \in \mathbb{N}$. Since $(x_{n_k}, y_{n_k}) \in Gr\ \nu_{n_k}$ implies $\nu_{n_k}(x_{n_k}) \geq y_{n_k}$. With

$$x_{n_k} \xrightarrow{w} x, \quad \text{and} \quad \limsup_{k\to\infty} \nu_{n_k}(x_{n_k}) \leq \nu(x),$$

we have

$$y = \lim_{k\to\infty} y_{n_k} \leq \limsup_{k\to\infty} \nu_{n_k}(x_{n_k}) \leq \nu(x).$$

This implies $(x,y) \in Gr\ \nu$.

On the other hand, let (6) be satisfied. Suppose $\{\nu_{n_k}\}$ be any subsequence of $\{\nu_n\}$ and $x_{n_k} \xrightarrow{w} x$. Denote $y = \limsup_{k\to\infty} \nu_{n_k}(x_{n_k})$. There exists a subsequence $\{\nu_{n'_k}(x_{n'_k})\}$ of $\{\nu_{n_k}(x_{n_k})\}$ such that $y_{n'_k} := \nu_{n'_k}(x_{n'_k}) \to y$. Since $(x_{n'_k}, y_{n'_k}) \in Gr\nu_{n'_k}$, and $(x_{n'_k}, y_{n'_k}) \xrightarrow{w} (x,y)$, we have $(x,y) \in Gr\ \nu$ by (6). This implies $\nu(x) \geq y$ with $y = \limsup_{n\to\infty} \nu_n(x_n)$. Thus we have the condition (ii). ■

Remark 5. The condition (ii) in Theorem 4 is equivalent to the condition $(ii)'$ as follows.

$(ii)'$ For any given any sequence $\{x_n\}$ which converges to x in the weak topology of $\mathbb{X}$, we have

$$\limsup_{n\to\infty} \nu_n(x_n) \leq \nu(x). \tag{7}$$

It is clear that (ii) implies $(ii)'$. Now we prove that $(ii)'$ also implies (ii). As a matter of fact, take $\{n_k\}$, $\{x_{n_k}\}$, such that $x_{n_k} \to x$ weakly in $\mathbb{X}$. Let

$$x_n = \begin{cases} x_{n_k}, & n = n_k, \\ x, & n \notin \{n_k\}. \end{cases}$$

Then $x_n \to x$ weakly. Hence, by $(ii)'$, we have (7). But $\{\nu_n(x_n) : n \in \mathbb{N}\} \supset \{\nu_{n_k}(x_{n_k}) : k \in \mathbb{N}\}$, therefore

$$\limsup_{k\to\infty} \nu_{n_k}(x_{n_k}) \le \limsup_{n\to\infty} \nu_n(x_n).$$

This with (7) implies

$$\limsup_{k\to\infty} \nu_{n_k}(x_{n_k}) \le \nu(x).$$

Now we prove the following theorem that will be used in the next section.

Theorem 5. *Let $\{\nu^n : n \in \mathbb{N}\} \subset \mathcal{F}_*(\mathbb{X})$, and for any $\alpha \in \mathbb{Q} \cap [0,1]$, $\nu^n_\alpha \xrightarrow{K-M} \nu_\alpha$. Define $\tilde{\nu}_\alpha = \bigcap_{\beta<\alpha,\beta\in\mathbb{Q}\cap[0,1)} \nu_\beta$, then $\nu^n \xrightarrow{Gr} \tilde{\nu}$, where $\tilde{\nu} \in \mathcal{F}_*(\mathbb{X})$ determined by $\tilde{\nu}(x) = \sup\{\alpha \in [0,1] : x \in \tilde{\nu}_\alpha\}$ for any $x \in \mathbb{X}$.*

Proof. *Step 1* Firstly we prove

$$Gr\ \tilde{\nu} \subset s\text{-}\liminf_{n\to\infty} Gr\ \nu_n.$$

For any $(x_0, y_0) \in Gr\ \tilde{\nu}$, it holds that $x_0 \in \tilde{\nu}_{y_0} = \bigcap_{y<y_0,y\in\mathbb{Q}\cap[0,1)} \nu_y$. Taking $\{y_k\} \subset \mathbb{Q} \cap [0,1]$, such that, $y_k \uparrow y_0$, we have $x_0 \in \nu_{y_k}$, $k \in \mathbb{N}$. Since the fact $\nu^n_{y_k} \xrightarrow{K-M} \nu_{y_k}$ implies $\nu_{y_k} \subset s\text{-}\liminf_{n\to\infty} \nu^n_{y_k}$ for each $k \in \mathbb{N}$, we have $x_0 \in s\text{-}\liminf_{n\to\infty} \nu^n_{y_1}$. Thus

$$\lim_{n\to\infty} d(x_0, \nu^n_{y_1}) = 0. \tag{8}$$

By using the property of fuzzy cut sets $\nu^n_{y_k} \supset \nu^n_{y_1}$, $k, n \in \mathbb{N}$, we have $d(x_0, \nu^n_{y_k}) \le d(x_0, \nu^n_{y_1})$ for $k, n \in \mathbb{N}$. Especially, $d(x_0, \nu^n_{y_n}) \le d(x_0, \nu^n_{y_1})$ for $n \in \mathbb{N}$. Thus we have from (8) that

$$\lim_{n\to\infty} d(x_0, \nu^n_{y_n}) = 0,$$

i.e. there exists $x_n \in \nu^n_{y_n}$ such that $x_0 = s\text{-}\lim_{n\to\infty} x_n$. This implies that $(x_n, y_n) \in Gr\ \nu^n$ and $(x_0, y_0) = s\text{-}\lim_{n\to\infty}(x_n, y_n)$.

Step 2 Take $(x_0, y_0) \in w\text{-}\limsup_{n\to\infty} Gr\ \nu^n$. By the definition, there exist $\{n_k\}$, and $(x_{n_k}, y_{n_k}) \in Gr\ \nu^{n_k}$ such that

$$x_0 = w\text{-}\lim_{k\to\infty} x_{n_k}, \qquad y_0 = \lim_{k\to\infty} y_{n_k}. \tag{9}$$

Let $\beta < y_0$, $\beta \in \mathbb{Q} \cap [0,1]$. Then, $\nu^{n_k}(x_{n_k}) \ge y_{n_k}$ with (9) implies that there exists a $k_0 \in \mathbb{N}$ such that

$$\nu^{n_k}(x_{n_k}) \ge \beta, \quad \text{for all} \quad k \ge k_0,$$

i.e. $x_{n_k} \in \nu^{n_k}_\beta$, for all $k \ge k_0$. Since $\nu^{n_k}_\beta \xrightarrow{K-M} \nu_\beta$, we have $x_0 \in \nu_\beta$ from (9). Due to the arbitrariness of β, we have $x_0 \in \tilde{\nu}_{y_0} = \bigcap_{\beta<y_0,\beta\in\mathbb{Q}\cap[0,1)} \nu_\beta$. Therefore $(x_0, y_0) \in Gr\ \tilde{\nu}$. ■

A *fuzzy-valued random variable* (cf. Puri and Ralescu, 1986) or *fuzzy random variable* is a function $X : \Omega \to \mathcal{F}_*(\mathbb{X})$, such that $X_\alpha(\omega) = \{x \in \mathbb{X} : X(\omega)(x) \geq \alpha\}$ is a set random variable for every $\alpha \in (0,1]$.

A fuzzy random variable X is called *integrable* if the real-valued random variable $d(0, X_1(\omega))$ is integrable. Since for any $\alpha \in (0,1]$, any $\omega \in \Omega$, $X_1(\omega) \subset X_\alpha(\omega)$, we have $d(0, X_\alpha(\omega)) \leq d(0, X_1(\omega))$ from the definition of d. Thus, if X is integrable, then for any $\alpha \in (0,1]$, X_α is an integrable set-valued random variable. On the otehr hand, a fuzzy random variable X is called *integrably bounded* if for every $\alpha \in (0,1]$, the real-valued random variable $\|X_\alpha(\omega)\|_{\mathcal{K}}$ is integrable.

Let $U^1[\Omega, \mathcal{A}, \mu; \mathcal{F}_*(\mathbb{X})]$ be the set of all integrable fuzzy random variables. Two fuzzy random variables $X, Y \in U^1[\Omega, \mathcal{A}, \mu; \mathcal{F}_*(\mathbb{X})]$ are considered to be identical if for any $\alpha \in [0,1]$, $X_\alpha(\omega) = Y_\alpha(\omega)$, *a.e.*$(\mu)$. Let $\mathcal{L}^1[\Omega, \mathcal{A}, \mu; \mathcal{F}_*(\mathbb{X})]$ be the set of all integrably bounded fuzzy random variables and $\mathcal{L}^1[\Omega, \mathcal{A}, \mu; \mathcal{F}_{c^*}(\mathbb{X})]$ denote the set of all fuzzy random variables whose cut sets are closed convex-valued random variables. Similarly we have the notation $\mathcal{L}^1[\Omega, \mathcal{A}, \mu; \mathcal{F}_c(\mathbb{X})]$ denotes the set of all fuzzy random variables whose cut sets are compact convex-valued random variables.

Let $X^n, X \in U^1[\Omega, \mathcal{A}, \mu; \mathcal{F}_*(\mathbb{X})]$, $n = 1, 2, ...$, it is called X^n converges to X in Graph if $X^n(\omega) \xrightarrow{Gr} X(\omega)$, *a.e.*$(\mu)$.

The *expected value* of a fuzzy random variable X, denoted by $E[X]$, is a fuzzy set such that, for every $\alpha \in (0,1]$,

$$(E[X])_\alpha = \int_\Omega X_\alpha d\mu = \mathrm{cl}\{E(f); f \in S_{X_\alpha}\}, \tag{10}$$

where the closure is taken in $\mathbb{X}$. From the existence theorem, we can get an equivalent definition as follows,

$$E(X)(x) = \sup\{\alpha \in [0,1] : x \in E(X_\alpha)\}.$$

From now on, we limit ourselves to deal with fuzzy random variable in $U^1[\Omega, \mathcal{A}, \mu; \mathcal{F}_{c^*}(\mathbb{X})]$ only for simplicity.

The *conditional expectation* of fuzzy-valued random variable X, denoted by $E[X|\mathcal{A}_0]$, is a fuzzy random variable satisfying the following two conditions.

$$E[X|\mathcal{A}_0] \in U^1[\Omega, \mathcal{A}_0, \mu; \mathcal{F}_{c^*}(\mathbb{X})], \tag{11}$$

$$\int_A E[X|\mathcal{A}_0] d\mu = \int_A X d\mu \quad \text{for every} \quad A \in \mathcal{A}_0. \tag{12}$$

Similar to the proof of in Li and Ogura (1996), the conditional expectation $E[X|\mathcal{A}_0]$ uniquely exists for $X \in U^1[\Omega, \mathcal{A}, \mu; \mathcal{F}_{c^*}(\mathbb{X})]$.

$\{X^n, \mathcal{A}_n : n \in \mathbb{N}\}$ is called a *fuzzy-valued martingale* (resp. *fuzzy-valued submartingale, supermartingale*) iff

1) $X^n \in \mathcal{L}^1[\Omega, \mathcal{A}_n, \mu; \mathcal{F}_{c^*}(\mathbb{X})]$, for all $n \in \mathbb{N}$,
2) $X^n = E[X^{n+1}|\mathcal{A}_n]$, for $n \in \mathbb{N}$ (resp. "$\leq$", "$\geq$" instead of "=" in the formula).

5 Convergence in graph for fuzzy valued martingales, submartingales and supermatingales

In this section, we give the convergence in graph for fuzzy-valued martingales, submartingales and supermatingales.

Theorem 6. *Assume that* $\mathbb{X}$ *is a Banach space satisfying the RNP with the separable dual* $\mathbb{X}^*$. *Then, for every uniformly integrably fuzzy valued martingales* $\{X^n, \mathcal{A}_n : n \in \mathbb{N}\}$, *there exists a unique* $X^\infty \in \mathcal{L}^1[\Omega, \mathcal{A}_\infty, \mu; \mathcal{F}_{c^*}(\mathbb{X})]$ *such that* $X^n = E[X^\infty|\mathcal{A}_n]$ *for each* $n \in \mathbb{N}$ *and* $X^n \xrightarrow{Gr} X^\infty$ *a.e.*(μ).

Proof. From the proof of Theorem 3.3 in Li and Ogura (1999), we have that there exists a unique $X^\infty \in \mathcal{L}^1[\Omega, \mathcal{A}_\infty, \mu; \mathcal{F}_{c^*}(\mathbb{X})]$ such that $X^n = E[X^\infty|\mathcal{A}_n]$ for each $n \in \mathbb{N}$ and $X^n_\alpha \xrightarrow{K-M} X^\infty_\alpha$, *a.e.*$(\mu)$ and for every $\alpha \in (0,1]$. By using Lemma 1, we get $X^n \xrightarrow{Gr} X^\infty$ *a.e.*(μ). ∎

To prove the following theorem, we need Lemma 2.

Lemma 2. (Li and Ogura (1996)) *Let* $\{S_\alpha : \alpha \in [0,1]\}$ *be a family of subsets of* $\mathcal{L}^1[\Omega, \mathbb{X}]$, S_α *be nonempty, closed, bounded and decomposable for every* $\alpha > 0$ *and satisfy the following conditions:*

(*i*) $\alpha \leq \beta \Longrightarrow S_\beta \subset S_\alpha$,

(*ii*) $\alpha_1 \leq \alpha_2 \leq ... \leq \alpha_n \leq ...,\ \lim_{n\to\infty} \alpha_n = \alpha \Longrightarrow S_\alpha = \bigcap_{n=1}^{\infty} S_{\alpha_n}$.

Then, there exists a unique $Y \in \mathcal{L}^1[\Omega, \mathcal{A}, \mu; \mathcal{F}_*(\mathbb{X})]$ *such that for every* α,

$$S_\alpha = \{f \in \mathcal{L}^1[\Omega, \mathbb{X}]; f(\omega) \in Y_\alpha(\omega),\ a.e.(\mu)\}. \tag{13}$$

If $\{S_\alpha : \alpha \in [0,1]\}$ *have the above conditions* (*i*) - (*ii*) *and*
(*iii for any* $\alpha \in (0,1], S_\alpha$ *is convex.*
Then, there exists a unique $Y \in \mathcal{L}^1[\Omega, \mathcal{A}, \mu; \mathcal{F}_{c^*}(\mathbb{X})]$ *which satisfies (13).*

Theorem 7. *Assume that* $\mathbb{X}$ *is a Banach space satisfying the RNP with the separable dual* $\mathbb{X}^*$, $\{X^n, \mathcal{A}_n : n \in \mathbb{N}\}$ *is uniformly integrable fuzzy-valued supermartingale and for,*

$$M_1 = \bigcap_{n=1}^{\infty} \{f \in \mathcal{L}^1[\Omega, \mathcal{A}_\infty, \mu; \mathbb{X}] : E(f|\mathcal{A}_n) \in S_{X^n_1}(\mathcal{A}_n)\}$$

is a nonempty set. Then there exists a $X^\infty \in \mathcal{L}^1[\Omega, \mathcal{A}_\infty, \mu; \mathcal{F}_{c^*}(\mathbb{X})]$ *such that* $X^n \xrightarrow{Gr} X^\infty$ *a.e..*

Proof. Since fuzzy-valued submartingale $\{X^n, \mathcal{A}_n : n \in \mathbb{N}\}$ is uniformly integrable, there exists a unique set-valued random variable $X_\alpha^\infty \in \mathcal{L}^1[\Omega, \mathcal{A}_\infty, \mu; \mathcal{K}_{c^*}(\mathbb{X})]$ such that $X_\alpha^n \stackrel{K-M}{\longrightarrow} X_\alpha^\infty$, *a.e.*$(\mu)$ by using Theorem 2 (also cf. Theorem 5 in Li and Ogura, 1998).

Now we prove that $\{X_\alpha^\infty : \alpha \in (0,1]\}$ can form a fuzzy-valued random variable $X^\infty \in \mathcal{L}^1[\Omega, \mathcal{A}_\infty, \mu; \mathcal{F}_{c^*}(\mathbb{X})]$. From the proof of Theorem 5 in Li and Ogura, 1998, we have

$$S_{X_\alpha^\infty} = M_\alpha = \bigcap_{n=1}^{\infty} \{f \in \mathcal{L}^1[\Omega, \mathcal{A}_\infty, \mu; \mathbb{X}] : E(f|\mathcal{A}_n) \in S_{X_\alpha^n}(\mathcal{A}_n)\}. \tag{14}$$

From assumption that M_1 is nonempty, we have that M_α is nonempty for any $\alpha \in [0,1]$. It is easy to see that $S_{X_\alpha^\infty}$ is closed, convex, bounded and decomposable subset in $\mathcal{L}^1[\Omega; \mathbb{X}]$ for each $\alpha \in (0,1]$. Thus we only need to check that $\{S_{X_\alpha^\infty} : \alpha \in (0,1]\}$ satisfies conditions (i) and (ii) of Lemma 2.

(i) Since for any $n \in \mathbb{N}$, X^n is a set-valued random variable, we have $X_\beta^n \subset X_\alpha^n$, for any $\alpha, \beta \in (0,1]$ and $\alpha \leq \beta$. Thus $S_{X_\beta^\infty} \subset S_{X_\alpha^\infty}$ from (14).

To show condition (ii), we take a sequence $\alpha_1 \leq \alpha_2 \leq ... \leq \alpha_k \leq ...$ such that $\lim_{k \to \infty} \alpha_k = \alpha$. It is then clear that the sequence $\{X_{\alpha_k}^n\}_{k \in \mathbb{N}}$ is decreasing a.e. (μ), and $\bigcap_{k=1}^{\infty} X_{\alpha_k}^n(\omega) = X_\alpha^n(\omega)$ a.e. (μ). Hence, we see that $\{S_{X_{\alpha_k}^n}\}_{k \in \mathbb{N}}$ is decreasing and $\bigcap_{k=1}^{\infty} S_{X_{\alpha_k}^n} = S_{X_\alpha^n}$. With (14), we have $\{S_{X_{\alpha_k}^\infty}\}_{k \in \mathbb{N}}$ is decreasing and $\bigcap_{k=1}^{\infty} S_{X_{\alpha_k}^\infty} = S_{X_\alpha^\infty}$. ■

Theorem 8. *Assume that $\mathbb{X}$ is reflexive. Then, for every uniformly integrable fuzzy-valued submartingale $\{X^n, \mathcal{A}_n : n \in \mathbb{N}\}$, there exists a $X^\infty \in \mathcal{L}^1[\Omega, \mathcal{A}_\infty, \mu; \mathcal{F}_{c^*}(\mathbb{X})]$ such that $X^n \stackrel{Gr}{\longrightarrow} X^\infty$ a.e.(μ).*

Proof. Since fuzzy-valued submartingale $\{X^n, \mathcal{A}_n : n \in \mathbb{N}\}$ is uniformly integrable, $\{X_\alpha^n, \mathcal{A}_n : n \in \mathbb{N}\}$ is uniformly integrable set-valued submartingale for every $\alpha \in (0,1]$. There exists a unique set-valued random variable $Y_\alpha^\infty \in \mathcal{L}^1[\Omega, \mathcal{A}_\infty, \mu; \mathcal{K}_{c^*}(\mathbb{X})]$ such that $X_\alpha^n \stackrel{K-M}{\longrightarrow} Y_\alpha^\infty$, *a.e.*$(\mu)$ for every $\alpha \in [0,1]$ by using Theorem 3. Since $\mathbb{Q}$ is countable, there exists a null set A, such that $X_\alpha^n(\omega) \stackrel{K-M}{\longrightarrow} Y_\alpha^\infty(\omega)$, for each $\omega \in \Omega \setminus A$ and for all $\alpha \in \mathbb{Q} \cap [0,1]$. Define

$$X_\alpha^\infty(\omega) = \begin{cases} \bigcap_{\beta<\alpha, \beta \in \mathbb{Q} \cap [0,1)} Y_\beta^\infty(\omega), & \omega \in \Omega \setminus A, \\ 0, & \omega \in A, \end{cases}$$

for any $\alpha \in [0,1]$. It is obvious that $X_\alpha^\infty \in \mathcal{L}^1[\Omega, \mathcal{A}_\infty, \mu; \mathcal{K}_{c^*}(\mathbb{X})]$. By Theorem 5, we have $X^\infty(\omega) \in \mathcal{F}_{c^*}(\mathbb{X})$ determined by $X^\infty(\omega)(x) = \sup\{\alpha \in [0,1] : x \in X_\alpha^\infty(\omega)\}$ for any $x \in \mathbb{X}$ and $X^n(\omega) \stackrel{Gr}{\longrightarrow} X^\infty(\omega)$ for any $\omega \in \Omega \setminus A$, i.e. $X^n \stackrel{Gr}{\longrightarrow} X^\infty$, *a.e.*$(\mu)$. ■

Remark 6. 1) In the proof of Theorem 8, we cannot prove directly that $\{X_\alpha^\infty : \alpha \in (0,1]\}$ can form a fuzzy-valued random variable $X^\infty \in \mathcal{L}^1[\Omega, \mathcal{A}_\infty, \mu; \mathcal{F}_{c^*}(\mathbb{X})]$ as we did in the proof of Theorem 7. It is because, from Theorem 3 (cf. the proof of Theorem 4 in Li and Ogura, 1998), we have

$$S_{X_\alpha^\infty} = \mathrm{cl}[\bigcup_{m=1}^{\infty} \bigcap_{n=m}^{\infty} \{f \in \mathcal{L}^1[\Omega, \mathcal{A}_\infty, \mu; \mathbb{X}] : E(f|\mathcal{A}_n) \in S_{X_\alpha^n}(\mathcal{A}_n)\}], \tag{15}$$

and $S_{X_\alpha^\infty}$ is a nonempty, closed, convex, bounded and decomposable subset in $\mathcal{L}^1[\Omega; \mathbb{X}]$ for each $\alpha \in (0,1]$. It is easy to check that $\{S_{X_\alpha^\infty} : \alpha \in (0,1]\}$ satisfies condition (i) of Lemma 2. But it is difficult to prove that it satisfies condition (ii) of Lemma 2, since it takes closure in (15).

2) We also can prove Theorem 7 by using Theorem 5. But since $\{S_{X_\alpha^\infty} : \alpha \in (0,1]\}$ satisfies conditions (i) and (ii) of Lemma 2 there, we have already had $X_\alpha^\infty(\omega) = \bigcap_{\beta<\alpha, \beta \in \mathbb{Q}\cap[0,1)} X_\beta^\infty(\omega)$.

6 The extension on unbounded set valued and relative fuzzy valued martingales

In our papers (Li and Ogura, 1996, 1997, 1998, 1999) and Section 5 above, we mainly focused on integral bounded case, i.e. in $\mathcal{L}^1[\Omega, \mathcal{A}_\infty, \mu; \mathcal{K}_{c^*}(\mathbb{X})]$ or $\mathcal{L}^1[\Omega, \mathcal{A}_\infty, \mu; \mathcal{F}_{c^*}(\mathbb{X})]$. Even in the set-valued case, this demanded that almost every $\omega \in \Omega$ with respect to μ, $F(\omega)$ is a bounded closed subset of Banach space $\mathbb{X}$, where F is a set-valued random variable. But from the following simple example, we can see that it is necessary to consider the unbounded case for the set-valued and fuzzy-valued random variables, i.e. in $U^1[\Omega, \mathcal{A}_\infty, \mu; \mathcal{K}_{c^*}(\mathbb{X})]$ or $U^1[\Omega, \mathcal{A}_\infty, \mu; \mathcal{F}_{c^*}(\mathbb{X})]$.

Example 2. Let $\xi : \Omega \to \mathbb{R}^1$ be a real-valued integrable random variable, that is, the expectation of ξ, $E(|\xi|) < \infty$. We define a set-valued random variable $F(\omega) = [\xi(\omega), +\infty)$ (or $(-\infty, \xi(\omega)]$) for each $\omega \in \Omega$. We can see that F is integrable but not integrably bounded-valued random variable. It is easy to prove that $E[F] = [E(\xi), +\infty)$ (or $(-\infty, E(\xi)]$). If $\mathcal{A}_0$ is a sub-σ-field, the $E[F|\mathcal{A}_0] = [E(\xi|\mathcal{A}_0), +\infty)$ (or $(-\infty, E(\xi|\mathcal{A}_0)]$).

If ξ_n, ξ are real-valued random variables such that $\xi_n \to \xi$ $a.s.(\mu)$ in $\mathbb{R}^1$. It is easy to prove that $F_n := [\xi_n, +\infty)$ or $(-\infty, \xi_n]$ converges to $F := [\xi, +\infty)$ (or $(-\infty, \xi]$) $a.e.(\mu)$ in the Kuratowski-Mosco sense. Especially if $\{\xi_n, \mathcal{A}_n : n \in \mathbb{N}\}$ is a real-valued martingale with uniformly integrable property, then there exists unique ξ_∞ such that $\xi_n \to \xi_\infty$, $a.e.(\mu)$. we can prove that $F_n := [\xi_n, +\infty)$ (or $(-\infty, \xi_n]$) is a set-valued martingale and converges to $F := [\xi, +\infty)$ (or $(-\infty, \xi]$) $a.e.(\mu)$ in the Kuratowski-Mosco sense. If $\{\xi_n, \mathcal{A}_n : n \in \mathbb{N}\}$ is a real-valued submartingale (or supermartingale) with uniformly integrable property, then $F_n := [\xi_n, +\infty)$ is a supermartingale (or submartingale) with $\sup_{n\in\mathbb{N}} Ed(0, F_n) < +\infty$, and $F_n \xrightarrow{K-M} F_\infty$, $a.e.(\mu)$.

Hess mainly discussed unbounded set-valued random variables and obtained convergence theorem for a set-valued supermartingales whose values may be unbounded in the Wijsman sense (cf. Hess, 1998). It is easy to see from the proof that it is also true for an unbounded set-valued martingales. For simplicity, we suppose in this section that $\mathbb{X}$ is finite dimensional. In this case, we can consider that the Wijsman convergecne and the Kuratowski-Mosco convergence coincide as we pointed in Section 2.

Theorem 9. *Assume that* $\{F_n, \mathcal{A}_n : n \in \mathbb{N}\}$ *is a set-valued martingale (or supermartingale) in* $U^1[\Omega, \mathcal{A}, \mu; \mathcal{K}_{c^*}(\mathbb{X})]$*, and satisfying the condition*

$$\sup_{n \in \mathbb{N}} Ed(0, F_n) < +\infty,$$

then, there exists an $F_\infty \in U^1[\Omega, \mathcal{A}, \mu; \mathcal{K}_{c^*}(\mathbb{X})]$ *such that* $F_n \stackrel{K-M}{\longrightarrow} F_\infty$*, a.e.*$(\mu)$*.*

And we can also extend it to the fuzzy-valued case. To do it, we have the following Lemma similar to Lemma 2 for integrable fuzzy random variable.

Lemma 3. *Let* $\{S_\alpha : \alpha \in [0,1]\}$ *be a family of subsets of* $\mathcal{L}^1[\Omega, \mathbb{X}]$*,* S_α *be nonempty, closed and decomposable for every* $\alpha > 0$ *and satisfy two conditions (i) and (ii)) of Lemma 2, then, there exists a unique* $Y \in U^1[\Omega, \mathcal{A}, \mu; \mathcal{F}_*(\mathbb{X})]$ *such that for every* α*,*

$$S_\alpha = \{f \in \mathcal{L}^1[\Omega, \mathbb{X}]; f(\omega) \in Y_\alpha(\omega), \ a.e.(\mu)\}. \tag{16}$$

If $\{S_\alpha : \alpha \in [0,1]\}$ *have the above conditions (i) - (ii) and (iii), then, there exists a unique* $Y \in U^1[\Omega, \mathcal{A}, \mu; \mathcal{F}_{c^*}(\mathbb{X})]$ *which satisfies (16).*

The proof is the similar to Lemma 2 (cf. Li and Ogura, 1996).

Theorem 10. *Assume that* $\{X^n, \mathcal{A}_n : n \in \mathbb{N}\}$ *is a set-valued martingale (or supermartingale) in* $U^1[\Omega, \mathcal{A}, \mu; \mathcal{F}_{c^*}(\mathbb{X})]$*, and satisfying the condition*

$$\sup_{n \in \mathbb{N}} Ed(0, X_1^n) < +\infty,$$

then, there exists an $X^\infty \in U^1[\Omega, \mathcal{A}_\infty, \mu; \mathcal{F}_{c^*}(\mathbb{X})]$ *such that* $X^n \stackrel{Gr}{\longrightarrow} X^\infty$*, a.e.*$(\mu)$*.*

Proof. Similar to the proof of Theorem 7, we can finish its proof by using Theorem 9 and Lemma 3 and Lemma 1. ∎

Acknowledgements

The research in this paper has been partially supported by BNSF and by Grant-in-Aid for Scientific Research No. 09440085.

References

1. Aumann, R. J. (1965). Integrals of set-valued functions, *J. Math. Anal. Appl.* **12**, 1-12.
2. Ban, J. (1991). Ergodic theorems for random compact sets and fuzzy variables in Banach spaces, *Fuzzy Sets and Systems* **44**, 71-82.
3. Beer, G. (1993). *Topologies on Closed and Closed Convex Sets, Mathematics and Its Applications.* Kluwer Academic Publishers, Dordrecht, Holland.
4. Chatterji, S. D.(1968). Martingale convergence and the RN-theorems, *Math. Scand.* **22**, 21-41.
5. Hess, C.(1991). On multivalued martingales whose values may be unbounded: martingale selectors and Mosco convergence, *J. Multivar. Anal.* **39**, 175-201.
6. Hess, C.(1998). On the almost sure convergence of sequences of random sets: martingales and extensions, (to appear).
7. Hiai, F. and Umegaki, H. (1977). Integrals, conditional expectations and martingales of multivalued functions, *J. Multivar. Anal.* **7**, 149-182.
8. Hiai, F. (1985). Convergence of conditional expectations and strong laws of large numbers for multivalued random variables, *Trans. Amer. Math. Soc.* . **291**, 613-627.
9. Kendall, D. G. (1974). Foundations of a Theory of Random Sets, in *Stochastic Geometry* (Harding, E. F. and Kendall D. G., Eds.), J. Wiley & Sons, New York.
10. Klement, E. P., Puri, L. M. and Ralescu, D. A. (1986). Limit theorems for fuzzy random variables, *Proc. R. Soc. Lond.* **407**, 171-182.
11. Klein, E. and Thompson, A. C. (1984). *Theory of Correspondences Including Applications to Mathematical Economics.* J. Wiley & Sons, New York.
12. Kim, B. K. and Kim, J. H. (1998) Stochastic integrals of set valued processes and fuzzy processes (preprint).
13. Kuratowski, K. (1965). *Topology.* **1**, New York-London-Warszawa (Tranl. from French).
14. Li, S. and Ogura, Y. (1996). Fuzzy random variables, conditional expectations and fuzzy martingales, *J. Fuzzy Math.* **4**, 905-927.
15. Li, S. and Ogura, Y. (1997). An optional sampling theorem for fuzzy valued martingales, *Proc. IFSA'97, Prague* **4**, 9-14.
16. Li, S. and Ogura, Y. (1999). Convergence of set valued and fuzzy valued martingales, *Fuzzy sets and Systems* **101**, 139-147.
17. Li, S. and Ogura, Y. (1998). Convergence of set valued sub- and super-martingales in the Kuratowski-Mosco sense, *Ann. Probab.* **26**, 1384-1402.
18. López-Diaz, M and Gil, M. A. (1998). Approximating integrably bounded fuzzy random variables in terms of the generalized Hausdorff metric, *Inform. Sci.* **104**, 279-291.
19. Luu, D. Q. (1981). Representations and regularity of multivalued martingales, *Acta Math. Vietn.* **6**, 29-40.
20. Matheron, G. (1975). Random Sets and Integral Geometry, J. Wiley & Sons.
21. Molchanov, I. S. (1993). *Limit Theorems for Unions of Random Closed Sets,* Lect. Notes in Math. **1561**, Springer-Verlag, Heidelberg.
22. Mosco, U. (1969). Convergence of convex set and of solutions of variational inequalities, *Advances Math.* **3**, 510-585.

23. Mosco, U. (1971). On the continuity of the Young-Fenchel transform, *J. Math. Anal. Appl.* **35**, 518-535.
24. Puri, M. L. and Ralescu, D. A. (1986). Fuzzy random variables, *J. Math. Anal. Appl.* **114**, 406-422.
25. Puri, M. L. and Ralescu, D. A. (1991). Convergence theorem for fuzzy martingales, *J. Math. Anal. Appl.* **160**, 107-121.
26. Rådström, H.(1952). An embedding theorem for spaces of convex sets, *Proc. Amer. Math. Soc.* **3**, 165-169.
27. Ralescu, D. A.(1986) Radon-Nikodym theorem for fuzzy set-valued measures. In *Fuzzy Sets Theory and Applications* (A. Jones *et al.*, Eds.), D. Reidel Pub., Dordrecht, 39-50.
28. Salinetti, G. and Roger J. B. Wets (1981). On the convergence of closed-valued measurable multifunctions, *Trans. Amer. Math. Soc.* **226**, 275-289.
29. Salinetti, G. and Roger J. B. Wets (1977). On the relations between two types of convergence for convex functions, *J. Math. Anal. Appl.* **60**, 211-226.
30. Thobie, C. G. (1974). Selections de multimesures, application a un theoreme de Radon-Nikodym multivoque. *C. R. Acad. Sci. Paris Ser. A* **279**, 603-606.
31. Wijsman, R. (1966) Convergence of sequences of convex sets, cones and fuctions, part 2, *Trans. Amer. Math. Soc.* **123**, 32-45
32. Zadeh, L. A. (1968). Probability measure of fuzzy events, *J. Math. Anal. Appl.* **23**, 421-427.

Remarks on Korovkin-type approximation of fuzzy random variables

Pedro Terán and Miguel López-Díaz

Departamento de Estadística e I.O. y D.M., Universidad de Oviedo
Facultad de Ciencias, C/ Calvo Sotelo s/n, 33007 Oviedo (Spain)

Abstract. In this paper we show how a technique used by R.A. Vitale to obtain a Korovkin-type approximation theorem for random sets can be exploited to develop a similar result for fuzzy random variables. A convergence theorem for positive linear operators is obtained, and consequences of this theorem in the Bernstein approximation of fuzzy random variables are analyzed.

1 Introduction

The aim of this paper is to introduce some aspects of the approximation of fuzzy random variables. Vitale (1979) established a first Korovkin-type theorem for random sets. Korovkin-type theorems deal with the following problem: given a sequence of operators in a function space, determine a family of test functions so that convergence of the images of that family determines convergence in all the function space.

For this purpose, Korovkin (1953, 1960) proved the first theorem of this kind for linear positive operators converging to the identity in the space of real-valued continuous functions on a compact interval, taking $\{1, x, x^2\}$ as test functions. Since then, this has become a fruitful subject in approximation theory (see Altomare and Campiti, 1994).

To obtain a Korovkin-type theorem for random sets, Vitale (1979) took advantage of a result for families of real-valued functions (which is proved by following a similar result by Shisha and Mond, 1968), by embedding the class of nonempty compact convex subsets of $\mathbb{R}^p$ into the class of real-valued continuous functions defined on S^{p-1}.

In this paper we show that the same result and technique can be used in order to obtain a Korovkin type-theorem for fuzzy random variables.

The main results of the paper are contained in Sections 3 and 4. In the former, a Korovkin type theorem for fuzzy random variables is obtained. In Section 4, the above result allows us to study the convergence of Bernstein approximants of those mappings. Concepts and results which are required in order to develop the paper are collected in Section 2.

2 Preliminaries

Let $\mathcal{K}(\mathbb{R}^p)$ denote the class of nonempty compact subsets of $\mathbb{R}^p$, let $\mathcal{K}_c(\mathbb{R}^p)$ be the subclass of nonempty compact convex subsets of $\mathbb{R}^p$, and let B denote the set $\{x \in \mathbb{R}^p \mid |x| \leq 1\}$, where $|\cdot|$ denotes the Euclidean norm in $\mathbb{R}^p$.

The space $\mathcal{K}(\mathbb{R}^p)$ can be endowed with a linear structure given by the Minkowski addition and the product by a scalar, that is,

$$A + C = \{a + c \,|\, a \in A,\, c \in C\}, \;\; \lambda A = \{\lambda a \,|\, a \in A\},$$

for all $A, C \in \mathcal{K}(\mathbb{R}^p)$, and $\lambda \in \mathbb{R}$.

The space $(\mathcal{K}(\mathbb{R}^p), +, \cdot)$ is not a vector space.

Recall that if $A \in \mathcal{K}_c(\mathbb{R}^p)$ and $\lambda_1, \lambda_2 \in [0, +\infty)$, then $\lambda_1 A + \lambda_2 A = (\lambda_1 + \lambda_2)A$. Moreover, $\mathcal{K}_c(\mathbb{R}^p)$ is closed under Minkowski addition and product by a scalar.

Given $A, C \in \mathcal{K}(\mathbb{R}^p)$, the *Hausdorff distance* between A and C is defined by

$$d_H(A, C) = \max\left\{\sup_{a \in A} \inf_{c \in C} |a - c|, \sup_{c \in C} \inf_{a \in A} |a - c|\right\},$$

then $(\mathcal{K}(\mathbb{R}^p), d_H)$ is a complete separable metric space and $(\mathcal{K}_c(\mathbb{R}^p), d_H)$ is a closed subspace (see Debreu, 1966).

If $A \in \mathcal{K}(\mathbb{R}^p)$, its *magnitude* is defined to be

$$\|A\| = d_H(\{0\}, A) = \sup_{x \in A} |x|.$$

Some well-known properties of the Hausdorff metric are the following:

i) $d_H(A + C, D + E) \leq d_H(A, D) + d_H(C, E)$,
ii) $d_H(aA, bA) \leq |a - b| \|A\|$,
iii) $d_H(\mathrm{co}\, A, \mathrm{co}\, C) \leq d_H(A, C)$,
iv) $d_H(A, C) = \inf\{\varepsilon > 0 \,|\, A \subset C + \varepsilon B,\, C \subset A + \varepsilon B)$,
where $A, C, D, E \in \mathcal{K}(\mathbb{R}^p)$, $a, b \in \mathbb{R}$, and co denotes the convex hull.

It is possible to embed the class $\mathcal{K}_c(\mathbb{R}^p)$ into the Banach space $C(S^{p-1})$ of continuous real functions on the unit sphere S^{p-1} of $\mathbb{R}^p$ by means of the *support function* $s : \mathcal{K}_c(\mathbb{R}^p) \to C(S^{p-1})$ with $K \mapsto s(\cdot, K) : S^{p-1} \to \mathbb{R}$ and $s(r, K) = \sup_{a \in K} < r, a >$ being $< \cdot, \cdot >$ the scalar product in $\mathbb{R}^p$. Obviously, $s(\cdot, B) = 1$.

For all $A, C \in \mathcal{K}_c(\mathbb{R}^p)$ we have that $d_H(A, C) = \|s(\cdot, A) - s(\cdot, C)\|_{S^{p-1}}$ and $\|A\| = \|s(\cdot, A)\|_{S^{p-1}}$ (from now on if $f : E \to \mathbb{R}$ is a bounded map, $\|f\|_E$ will denote the sup-norm). The mapping s is linear, that is, $s(\cdot, \lambda A) = \lambda s(\cdot, A)$ and $s(\cdot, A + C) = s(\cdot, A) + s(\cdot, C)$ for all $A, C \in \mathcal{K}(\mathbb{R}^p)$ and $\lambda \in [0, \infty)$.

Let $(\Omega, \mathcal{A})$ be a measurable space. A set-valued function $X : \Omega \to \mathcal{K}(\mathbb{R}^p)$ is called a *random set* if it is $\mathcal{A}$-$\mathcal{B}_{d_H}$ measurable, where $\mathcal{B}_{d_H}$ denotes the Borel σ-field in $\mathcal{K}(\mathbb{R}^p)$.

If $X : \Omega \longrightarrow \mathcal{K}(\mathbb{R}^p)$ is a random set, its *magnitude* is defined as the mapping $\|X\| : \Omega \longrightarrow \mathbb{R}$ with $\|X\|(\omega) = \|X(\omega)\|$ for all $\omega \in \Omega$. The measurability assumed for the random set X implies the Borel measurability of $\|X\|$ (see Hiai and Umegaki, 1977).

Some well-known properties of random sets are the following: If $X, Y : \Omega \rightarrow \mathcal{K}(R^p)$ are random sets, then $X+Y$, λX and $\operatorname{co} X$ are also random sets, where $(\operatorname{co} X)(\omega) = \operatorname{co}(X(\omega))$ for all $\omega \in \Omega$ and $\lambda \in \mathbb{R}$ (see Matheron, 1975).

We will denote by $\mathcal{F}(\mathbb{R}^p)$ (respectively, $\mathcal{F}_c(\mathbb{R}^p)$) the class of fuzzy sets V in $[0,1]^{\mathbb{R}^p}$ such that the *α-level sets* V_α belong to $\mathcal{K}(\mathbb{R}^p)$ (respectively, to $\mathcal{K}_c(\mathbb{R}^p)$) for all $\alpha \in [0,1]$, with $V_\alpha = \{x \in \mathbb{R}^p \mid V(x) \geq \alpha\}$ for $\alpha \in (0,1]$, and $V_0 = \operatorname{cl}\{x \in \mathbb{R}^p \mid V(x) > 0\}$ where cl denotes the topological closure. Elements of $\mathcal{F}_c(\mathbb{R}^p)$ are called convex, since $A \in \mathcal{F}_c(\mathbb{R}^p)$ if, and only if, $\lambda A + (1-\lambda)A = A$ for all $\lambda \in [0,1]$.

The class $\mathcal{F}(\mathbb{R}^p)$ can be endowed with a linear structure, for which addition and product by a scalar are defined as follows:

$$(U+V)(x) = \sup\{\alpha \in [0,1] \mid x \in U_\alpha + V_\alpha\},$$

$$(\lambda U)(x) = \begin{cases} U(\lambda^{-1}x) & \text{if } \lambda \neq 0 \\ 0 & \text{if } \lambda = 0 \text{ and } x \neq 0 \\ 1 & \text{if } \lambda = 0 \text{ and } x = 0 \end{cases}$$

for all $U, V \in \mathcal{F}(\mathbb{R}^p)$, $\lambda \in \mathbb{R}$.

It is possible to see (Puri and Ralescu, 1981) that these operations are inherited levelwise from those defined on $\mathcal{K}(\mathbb{R}^p)$, that is, for all $\alpha \in [0,1]$

$$(\lambda U)_\alpha = \lambda U_\alpha, \quad \text{and} \quad (U+V)_\alpha = U_\alpha + V_\alpha.$$

Obviously, $\mathcal{F}(\mathbb{R}^p)$ is not a vector space with these operations, and $\mathcal{F}_c(\mathbb{R}^p)$ is closed under them.

$\mathcal{F}(\mathbb{R}^p)$ can be endowed with the *d_q-metrics*, $q \in [1,+\infty]$, where

$$d_q(V,W) = \left(\int_0^1 d_H(V_\alpha, W_\alpha)^q \, d\alpha\right)^{1/q}, \quad \text{if } 1 \leq q < \infty, \text{ and}$$

$$d_\infty(V,W) = \sup_{\alpha \in [0,1]} d_H(V_\alpha, W_\alpha).$$

$(\mathcal{F}(\mathbb{R}^p), d_\infty)$ is a complete metric space (see Puri and Ralescu, 1981), but it is not separable (see Klement *et al.*, 1986), whereas $(\mathcal{F}(\mathbb{R}^p), d_q)$, $q \in [1,+\infty)$ are separable metric spaces but they are not complete (see Diamond and Kloeden, 1990, 1994).

$\mathcal{F}_c(\mathbb{R}^p)$ can be embedded into the Banach space $L^\infty(S^{p-1} \times [0,1])$ of bounded measurable real functions on $S^{p-1} \times [0,1]$ by means of the mapping

$s : \mathcal{F}_c(\mathbb{R}^p) \to L^\infty(S^{p-1} \times [0,1])$ with $V \mapsto s(\cdot,\cdot,V) : S^{p-1} \times [0,1] \to \mathbb{R}$ and $s(r,\alpha,V) = \sup_{a \in V_\alpha} < r, a >$, for all $V \in \mathcal{F}_c(\mathbb{R}^p)$ and $(r,\alpha) \in S^{p-1} \times [0,1]$.

Thus, $d_\infty(A,C) = \|s(\cdot,\cdot,A) - s(\cdot,\cdot,C)\|_{S^{p-1}\times[0,1]}$ for all $A, C \in \mathcal{F}_c(\mathbb{R}^p)$. Note that $\|s(\cdot,\cdot,A)\|_{S^{p-1}\times[0,1]} = \|A_0\| = d_\infty(A, I_{\{0\}})$ and $s(\cdot,\cdot,I_B) = 1$, where I_C is the indicator function of $C \in \mathcal{K}(\mathbb{R}^p)$ and $V \in \mathcal{F}(\mathbb{R}^p)$.

From the analogue properties of the Hausdorff metric it follows that

i) $d_\infty(A + C, D + E) \leq d_\infty(A,D) + d_\infty(C,E)$,
ii) $d_\infty(aA, bA) \leq |a - b|\,\|A_0\|$,
iii) $d_\infty(\operatorname{co} A, \operatorname{co} C) \leq d_\infty(A,C)$,
where $A, C, D, E \in \mathcal{F}(\mathbb{R}^p)$, $a, b \in \mathbb{R}$, and if $V \in \mathcal{F}(\mathbb{R}^p)$, then $\operatorname{co} V \in \mathcal{F}_c(\mathbb{R}^p)$ is the fuzzy set such that $(\operatorname{co} V)_\alpha = \operatorname{co}(V_\alpha)$ for all $\alpha \in [0,1]$.

To state a property similar to the fourth one for d_H we can consider a partial ordering on $\mathcal{F}(\mathbb{R}^p)$. Thus, for $V, W \in \mathcal{F}(\mathbb{R}^p)$ we will write $V \subset W$ if $V_\alpha \subset W_\alpha$ for all $\alpha \in [0,1]$ or, equivalently, $V(x) \leq W(x)$ for all $x \in \mathbb{R}^p$. Then,

iv) $d_\infty(A,C) = \inf\{\varepsilon > 0 \,|\, A \subset C + \varepsilon I_B,\ C \subset A + \varepsilon I_B\}$.

A mapping $X : \Omega \to \mathcal{F}(\mathbb{R}^p)$ will be said to be *a fuzzy random variable* if the *α-level function* $X_\alpha : \Omega \to \mathcal{K}(\mathbb{R}^p)$, with $X_\alpha(\omega) = (X(\omega))_\alpha$ for all $\omega \in \Omega$, is a random set for all $\alpha \in [0,1]$ (see Puri and Ralescu, 1986).

If $X, Y : \Omega \to \mathcal{F}(\mathbb{R}^p)$ are fuzzy random variables, then we have that $X + Y$, λX and $\operatorname{co} X$ are also fuzzy random variables, for all $\lambda \in \mathbb{R}$. These properties are inherited from random sets' analogue ones.

We will denote by $\mathcal{C}(\mathcal{F}(\mathbb{R}^p))$ (respectively $\mathcal{C}(\mathcal{F}_c(\mathbb{R}^p))$) the set of mappings $X : [0,1] \to \mathcal{F}(\mathbb{R}^p)$ (respectively in $\mathcal{F}_c(\mathbb{R}^p)$), being continuous with respect to the Euclidean and d_∞ metric.

Obviously, if $X \in \mathcal{C}(\mathcal{F}(\mathbb{R}^p))$, then for each $\alpha \in [0,1]$ the mapping $X_\alpha : [0,1] \to \mathcal{K}(\mathbb{R}^p)$ is continuous with respect to Euclidean and Hausdorff metric, and hence X is a fuzzy random variable.

On the class $\mathcal{C}(\mathcal{F}(\mathbb{R}^p))$ we can consider the D_∞ metric defined by

$$D_\infty(X,Y) = \sup_{t \in [0,1]} d_\infty(X(t), Y(t)).$$

Since $(\mathcal{F}(\mathbb{R}^p), d_\infty)$ is not a separable metric space, then $(\mathcal{C}(\mathcal{F}(\mathbb{R}^p)), D_\infty)$ is also non-separable.

Inherited from that on $\mathcal{F}(\mathbb{R}^p)$, a pointwise ordering can be considered on $\mathcal{C}(\mathcal{F}(\mathbb{R}^p))$, that is, if $X, Y \in \mathcal{C}(\mathcal{F}(\mathbb{R}^p))$ then $X \subset Y$ will denote $X(t) \subset Y(t)$ for all $t \in [0,1]$.

We have that

$$D_\infty(X,Y) = \inf\{\varepsilon > 0 \,|\, X \subset Y + \varepsilon I_B,\ Y \subset X + \varepsilon I_B\}$$

A mapping $T : \mathcal{C}(\mathcal{F}_c(\mathbb{R}^p)) \to \mathcal{C}(\mathcal{F}_c(\mathbb{R}^p))$ will be said to be $\mathcal{F}_c(\mathbb{R}^p)$-*linear* if for all $a, b \in [0,\infty)$ and $X, Y \in \mathcal{C}(\mathcal{F}_c(\mathbb{R}^p))$ we have that $T(aX +$

$bY) = aT(X) + bT(Y)$, and it will be called $\mathcal{F}_c(\mathbb{R}^p)$-*positive* if for all $X, Y \in \mathcal{C}(\mathcal{F}_c(\mathbb{R}^p))$ with $X \subset Y$ we have that $T(X) \subset T(Y)$.

The following result was established by Vitale (1979) by considering an argument similar to that of Shisha and Mond (1968). It will be essential to the development of Section 3.

Let P be an indexing set and let Σ be the collection of all $\sigma = \{\sigma_p\}_{p \in P}$ where $\{\sigma_p\}_{p \in P}$ is a bounded equicontinuous family of real functions on $[0,1]$, that is, given $\sigma = \{\sigma_p\}_{p \in P}$, then

i) there exists N_σ such that $\sup_{p \in P} \|\sigma_p\|_{[0,1]} \leq N_\sigma < \infty$.

ii) the modulus of continuity, $\omega_\sigma(\delta) = \sup_{|t-x| \leq \delta} \sup_{p \in P} |\sigma_p(t) - \sigma_p(x)|$, satisfies that $\omega_\sigma(0^+) = \lim_{\delta \to 0^+} \omega_\sigma(\delta) = 0$.

Given $\sigma^{(1)}, \sigma^{(2)} \in \Sigma$ and $a \in \mathbb{R}$ we can define $\sigma^{(1)} + \sigma^{(2)} = \{\sigma_p^{(1)} + \sigma_p^{(2)}\}_{p \in P}$, and $a\sigma^{(1)} = \{a\sigma_p^{(1)}\}_{p \in P}$, with these operations $(\Sigma, +, \cdot)$ is a vector space.

On the other hand, the mapping $\| \cdot \|^\Sigma : \Sigma \to \mathbb{R}$ with

$$\|\sigma\|^\Sigma = \sup_{t \in [0,1]} \sup_{p \in P} |\sigma_p(t)|$$

is a norm on Σ, and so $(\Sigma, +, \cdot, \| \ \|^\Sigma)$ is a normed linear space.

A partial ordering can be defined on Σ with $\sigma^{(1)} \prec \sigma^{(2)}$ if and only if $\sigma_p^{(1)}(t) \leq \sigma_p^{(2)}(t)$ for all $t \in [0,1]$ and $p \in P$.

Let Σ_0 be a subspace of Σ and consider a mapping $L : \Sigma_0 \to \Sigma$. L is said to be *linear* if $L(a\sigma^{(1)} + b\sigma^{(2)}) = aL(\sigma^{(1)}) + bL(\sigma^{(2)})$ for all $a, b \in \mathbb{R}$, $\sigma^{(1)}, \sigma^{(2)} \in \Sigma_0$, and it is called *positive* if $L(\sigma^{(1)}) \prec L(\sigma^{(2)})$ as $\sigma^{(1)} \prec \sigma^{(2)}$.

For the sake of convenience and according to Vitale (1979), Σ_0 will be said to be *full* if the following conditions hold:

i) for $i = 0, 1, 2$, ${}_i\sigma \in \Sigma_0$, where ${}_i\sigma_p(t) = t^i$ for all $p \in P$, $t \in [0,1]$

ii) if $\sigma = \{\sigma_p\}_{p \in P} \in \Sigma_0$, then for each fixed $x \in [0,1]$ ${}_{(x)}\sigma \in \Sigma_0$, being ${}_{(x)}\sigma_p(t) = \sigma_p(x)$ for all $p \in P$, $t \in [0,1]$.

Under the former definitions, the following holds:

Proposition 1. *Given $\sigma \in \Sigma$ and $L : \Sigma_0 \to \Sigma$ a positive linear mapping, let us define*

$$\gamma(\sigma, L) = \sup_{t \in [0,1]} \sup_{p \in P} |(L_{(t)}\sigma)_p(t) - \sigma_p(t)|.$$

If $\{L_n\}_n$ is a sequence of positive linear mappings from Σ_0 to Σ and $L_n \, {}_i\sigma \to {}_i\sigma$ for $i = 0, 1, 2$, then for each $\sigma \in \Sigma_0$, $\gamma(\sigma, L_n) \to 0$ implies that $L_n\sigma \to \sigma$.

3 A convergence theorem for positive linear operators

This section is devoted to show that the former result can be used to obtain an approximation theorem for positive linear operators on the class $\mathcal{C}(\mathcal{F}_c(\mathbb{R}^p))$. An application of that theorem will be developed in Section 4.

Theorem 1. *Let $\{T_n\}_n$ be a sequence of mappings $T_n : \mathcal{C}(\mathcal{F}_c(\mathbb{R}^p)) \to \mathcal{C}(\mathcal{F}_c(\mathbb{R}^p))$ which are $\mathcal{F}_c(\mathbb{R}^p)$-linear and $\mathcal{F}_c(\mathbb{R}^p)$-positive. Then, the following conditions are equivalent:*

i) $T_n X \overset{D_\infty}{\rightrightarrows} X$ *for all* $X \in \mathcal{C}(\mathcal{F}_c(\mathbb{R}^p))$,

ii) $T_n X^i \overset{D_\infty}{\rightrightarrows} X^i$, $i = 0, 1, 2$ *where* $X^i(t) = t^i I_B$, *and* $T_n A \overset{D_\infty}{\rightrightarrows} A$ *for all* $A \in \mathcal{F}_c(\mathbb{R}^p)$.

Proof. Obviously *i)* implies *ii)* since X^i $(i = 0, 1, 2)$ and A are in $\mathcal{C}(\mathcal{F}_c(\mathbb{R}^p))$ (we make the remark that throughout this section, constant functions and their only value will be denoted by the same symbol to simplify notations).

To obtain the converse result, Proposition 1 will be applied. Consider $P = S^{p-1} \times [0,1]$, and denote by $s(X)$ the bunch of functions $\{s(r, \alpha, X(\cdot))\}_{(r,\alpha) \in P}$, where

$$s(r, \alpha, X(t)) = s(r, X_\alpha(t)) = \sup_{a \in X_\alpha(t)} < r, a > \quad \text{for all } t \in [0,1],$$

and let $C = \{s(X)\}_{X \in \mathcal{C}(\mathcal{F}_c(\mathbb{R}^p))}$.

We can see now that C is a family of bounded equicontinuous bunches of real functions on $[0,1]$, that is, $C \subset \Sigma$. Thus, if $X \in \mathcal{C}(\mathcal{F}_c(\mathbb{R}^p))$, then

$$\omega_{s(X)}(\delta) = \sup_{|t-x| \leq \delta} \sup_{(r,\alpha) \in P} |s(r, \alpha, X(t)) - s(r, \alpha, X(x))|$$

$$= \sup_{|t-x| \leq \delta} \sup_{\alpha \in [0,1]} \sup_{r \in S^{p-1}} |s(r, X_\alpha(t)) - s(r, X_\alpha(x))|$$

$$= \sup_{|t-x| \leq \delta} \sup_{\alpha \in [0,1]} d_H(X_\alpha(t), X_\alpha(x)) = \sup_{|t-x| \leq \delta} d_\infty(X(t), X(x)) = \omega_X(\delta),$$

what proves the equicontinuity of $s(X)$ since X is continuous. On the other hand, $s(X)$ is bounded for each $X \in \mathcal{C}(\mathcal{F}_c(\mathbb{R}^p))$ since

$$\sup_{(r,\alpha) \in S^{p-1} \times [0,1]} \|s(r, \alpha, X(\cdot))\|_{[0,1]} = \sup_{(r,\alpha) \in S^{p-1} \times [0,1]} \sup_{t \in [0,1]} |s(r, \alpha, X(t))|$$

$$= \sup_{t \in [0,1]} \sup_{r \in S^{p-1}} |s(r, 0, X(t))| = \sup_{t \in [0,1]} \|s(\cdot, X_0(t))\|_{S^{p-1}} = \sup_{t \in [0,1]} \|X_0\|(t)$$

which is a bound since $\|X_0\|$ is a continuous mapping on $[0,1]$.

Given $T : \mathcal{C}(\mathcal{F}_c(\mathbb{R}^p)) \to \mathcal{C}(\mathcal{F}_c(\mathbb{R}^p))$, consider a mapping $L : C \to C$ with $L(s(X)) = s(T(X))$ for all $X \in \mathcal{C}(\mathcal{F}_c(\mathbb{R}^p))$.

Now define Σ_0 as the linear span of C,

$$\Sigma_0 = \left\{ \sum_{j=1}^{m} a_j s(X_j) \,|\, a_j \in \mathbb{R},\, X_j \in \mathcal{C}(\mathcal{F}_c(\mathbb{R}^p)),\, j \leq m,\, m \in \mathbb{N} \right\},$$

by putting $L(-s(X)) = -L(s(X))$, extend L to a positive linear mapping from Σ_0 to Σ.

The subspace Σ_0 is full since

- if $i \in \{0,1,2\}$, then $s(X^i) = s(t^i I_B) = \{s(r,\alpha,t^i I_B)\}_{(r,\alpha)\in P}$, but $s(r,\alpha, t^i I_B) = s(r,t^i B) = t^i s(r,B) = t^i$, and hence $s(X^i) = \{t^i\} = {}_i\sigma$, that is, ${}_i\sigma \in \Sigma_0$.
- let $\sigma = \sum_{j=1}^m a_j s(X_j) \in \Sigma_0$ and $x \in [0,1]$, then ${}_{(x)}\sigma_{(r,\alpha)}(t) = \sum_{j=1}^m a_j$ $s(r,\alpha,X_j(x))$ for all $(r,\alpha) \in P$, that is, ${}_{(x)}\sigma = \sum_{j=1}^m a_j s(X_j(x)) \in \Sigma_0$.

On the other hand, if $i \in \{0,1,2\}$ we have that

$$D_\infty(T_n X^i, X^i) = \sup_{t\in[0,1]} d_\infty((T_n X^i)(t), X^i(t))$$

$$= \sup_{t\in[0,1]} \|s(\cdot,\cdot,(T_n X^i)(t)) - s(\cdot,\cdot,X^i(t))\|_{S^{p-1}\times[0,1]}$$

$$= \sup_{t\in[0,1]} \sup_{(r,\alpha)\in P} |s(r,\alpha,(T_n X^i)(t)) - s(r,\alpha,X^i(t))|$$

$$= \|s(T_n X^i) - s(X^i)\|^\Sigma = \|L_n s(X^i) - s(X^i)\|^\Sigma = \|L_n\, {}_i\sigma - {}_i\sigma\|^\Sigma,$$

so that the latter converges to 0, fulfilling the hypotheses of Proposition 1.

Thus, to prove our result we only have to see that for each $X \in \mathcal{C}(\mathcal{F}_c(\mathbb{R}^p))$, $\gamma(s(X), L_n)$ goes to 0. We have that

$$\gamma(s(X), L_n) = \sup_{x\in[0,1]} \sup_{(r,\alpha)\in P} |(L_{n\,(x)} s(X))_{(r,\alpha)}(x) - s(X)_{(r,\alpha)}(x)|$$

$$= \sup_{x\in[0,1]} \sup_{(r,\alpha)\in P} |(L_n s(X(x)))_{(r,\alpha)}(x) - s(X)_{(r,\alpha)}(x)|$$

$$\leq \sup_{t\in[0,1]} \sup_{x\in[0,1]} \sup_{(r,\alpha)\in P} |(L_n s(X(x)))_{(r,\alpha)}(t) - s(X)_{(r,\alpha)}(x)|$$

$$= \sup_{t\in[0,1]} \sup_{x\in[0,1]} \sup_{(r,\alpha)\in P} |s(T_n(X(x)))_{(r,\alpha)}(t) - s(X)_{(r,\alpha)}(x)|$$

$$= \sup_{t\in[0,1]} \sup_{x\in[0,1]} \sup_{(r,\alpha)\in P} |s(r,\alpha,T_n(X(x)))(t) - s(r,\alpha,X(x))|$$

$$= \sup_{x\in[0,1]} \sup_{t\in[0,1]} d_\infty(T_n(X(x))(t), X(x)) = \sup_{x\in[0,1]} D_\infty(T_n(X(x)), X(x));$$

if we suppose that $\sup_{x\in[0,1]} D_\infty(T_n(X(x)), X(x))$ does not converge to 0, then there exists $\varepsilon > 0$ and $\{T_{n_k}\}_k \subset \{T_n\}_n$ such that for all $k \in \mathbb{N}$

$$\sup_{x\in[0,1]} D_\infty(T_{n_k}(X(x)), X(x)) > \varepsilon,$$

whence given T_{n_k}, there exists $x_{n_k} \in [0,1]$ such that

$$D_\infty(T_{n_k}(X(x_{n_k})), X(x_{n_k})) > \varepsilon.$$

The sequence $\{x_{n_k}\}_k$ is included in $[0,1]$, so there is a convergent subsequence (which will be denoted in the same way) whose limit is a value $x \in [0,1]$.

Since X is continuous, we have that

$$X(x_{n_k}) \xrightarrow{d_\infty} X(x) \quad \text{or, equivalently,} \quad X(x_{n_k}) \xrightarrow{D_\infty} X(x)$$

(recall that the latter means convergence of constant functions).

If we set $\eta_k = D_\infty(X(x_{n_k}), X(x))$, then for all $\varepsilon > \eta_k$ we obtain that $X(x) \subset X(x_{n_k}) + \varepsilon I_B$, and by using $\mathcal{F}_c(\mathbb{R}^p)$-linearity and $\mathcal{F}_c(\mathbb{R}^p)$-positivity of T_{n_k} we have that

$$T_{n_k}(X(x)) \subset T_{n_k}(X(x_{n_k})) + \varepsilon T_{n_k}(I_B) \subset T_{n_k}(X(x_{n_k}))$$
$$+\varepsilon \sup_{t\in[0,1]} \|(T_{n_k}(I_B))_0(t)\| I_B.$$

In the same way,

$$T_{n_k}(X(x_{n_k})) \subset T_{n_k}(X(x)) + \varepsilon \sup_{t\in[0,1]} \|(T_{n_k}(I_B))_0(t)\| I_B$$

and, hence,

$$D_\infty(T_{n_k}(X(x_{n_k})), T_{n_k}(X(x))) \leq \eta_k \sup_{t\in[0,1]} \|(T_{n_k}(I_B))_0(t)\|.$$

In virtue of the hypothesis, $D_\infty(T_{n_k} I_B, I_B) \xrightarrow{k} 0$, whence there exists $r \in \mathbb{N}$ such that for all $k \geq r$ we have that

$$1 \geq D_\infty(T_{n_k} I_B, I_B) \geq D_\infty(T_{n_k} I_B, I_{\{0\}}) - D_\infty(I_{\{0\}}, I_B) = D_\infty(T_{n_k} I_B, I_{\{0\}}) - 1$$
$$= \sup_{t\in[0,1]} d_\infty((T_{n_k} I_B)(t), I_{\{0\}}) - 1 = \sup_{t\in[0,1]} \|(T_{n_k} I_B)_0(t)\| - 1$$

so, for all $k \geq r$ we conclude that

$$\sup_{t\in[0,1]} \|(T_{n_k} I_B)_0(t)\| \leq 2,$$

and, therefore,

$$\lim_{k\to\infty} D_\infty(T_{n_k}(X(x_{n_k})), T_{n_k}(X(x))) \leq \lim_{k\to\infty} 2\eta_k = 0.$$

But, for all $k \geq r$

$$\varepsilon < D_\infty(T_{n_k}(X(x_{n_k})), X(x_{n_k}))$$

$$\leq D_\infty(T_{n_k}(X(x_{n_k})), T_{n_k}(X(x))) + D_\infty(T_{n_k}(X(x)), X(x))$$

$$+ D_\infty(X(x), X(x_{n_k}))$$

and all terms in the sum vanish as k tends to ∞, so that we reach a contradiction.

Therefore,

$$\sup_{x\in[0,1]} D_\infty(T_n(X(x)), X(x)) \longrightarrow 0,$$

whence

$$\gamma(s(X), L_n) \longrightarrow 0 \quad \text{for all} \quad X \in \mathcal{C}(\mathcal{F}_c(\mathbb{R}^p)).$$

Now, by applying Proposition 1, we have that for all $X \in \mathcal{C}(\mathcal{F}_c(\mathbb{R}^p))$

$$L_n s(X) \xrightarrow{\|\cdot\|^\Sigma} s(X), \quad \text{that is,} \quad s(T_n(X)) \xrightarrow{\|\cdot\|^\Sigma} s(X)$$

or, equivalently,

$$\lim_{n\to\infty} D_\infty(T_n(X), X) = 0$$

and the proof is complete. ■

We should remark that in the condition *ii)* of Theorem 1, the case $i = 0$ is redundant. Moreover, it suffices to demand the last part of that condition only for elements A in $\mathcal{F}_c(\mathbb{R}^p)$ such that $\|A_0\| = 1$. Both claims are straightforward to check.

On the other hand, following the same ideas as in Vitale (1979) we can give the following quantitative result which generalizes the classical estimate of Shisha and Mond (1968):

$$D_\infty(TX, X) \leq \omega_X(\mu)(\|(TI_B)_0\| + 1) + \sup_{x\in[0,1]} D_\infty(T(X(x)), X(x))$$

where $\mu^2 = \sup_{x\in[0,1]} \|T((t-x)^2 I_B)_0\|(x)$.

4 Bernstein approximants of fuzzy random variables

In this section, and on the basis of Theorem 1, some aspects of the Bernstein approximation of continuous fuzzy random variables $X : [0,1] \longrightarrow \mathcal{F}(\mathbb{R}^p)$ are studied. We should indicate that these results can also be obtained from the direct study of the Bernstein approximant as in Terán and López-Díaz (2001a).

Given a mapping $X : [0,1] \to \mathcal{F}(\mathbb{R}^p)$, its nth *Bernstein approximant* is defined as the mapping $B_n(X,\cdot) : [0,1] \to \mathcal{F}(\mathbb{R}^p)$ such that

$$B_n(X,\cdot) = \sum_{j=0}^{n} f_{nj}(\cdot) X\left(\frac{j}{n}\right), \qquad \text{where} \qquad f_{nj}(t) = \binom{n}{j} t^j (1-t)^{n-j}.$$

Clearly, if X is $\mathcal{F}_c(\mathbb{R}^p)$-valued then $B_n(X,\cdot)$ is also $\mathcal{F}_c(\mathbb{R}^p)$-valued. In the same way, we have that for any fuzzy random variable $X : [0,1] \longrightarrow \mathcal{F}(\mathbb{R}^p)$, $B_n(X,\cdot)_\alpha = B_n(X_\alpha,\cdot)$ for all $\alpha \in [0,1]$, and therefore $B_n(X,\cdot)$ is also a fuzzy random variable, where

$$B_n(X_\alpha,\cdot) = \sum_{j=0}^{n} f_{nj}(\cdot) X_\alpha\left(\frac{j}{n}\right).$$

Bernstein approximants of an $\mathcal{F}_c(\mathbb{R}^p)$-valued mapping satisfy the following property:

Proposition 2. *Let $X : [0,1] \to \mathcal{F}_c(\mathbb{R}^p)$ be a continuous fuzzy random variable. Then,*

$$B_n(X,\cdot) \xrightarrow{d_\infty} X \quad \textit{uniformly.}$$

Proof. In order to apply Theorem 1, we are first going to see that $B_n(X,\cdot) \in C(\mathcal{F}_c(\mathbb{R}^p))$.

Since f_{nj} is continuous, then $\omega_{f_{nj}} \longrightarrow 0$ as $\delta \to 0^+$, where $\omega_{f_{nj}}$ denotes the modulus of continuity of f_{nj}. On the other hand,

$$\omega_{B_n(X,\cdot)}(\delta) = \sup_{|t-x|\le\delta} d_\infty(B_n(X,t), B_n(X,x))$$

$$= \sup_{|t-x|\le\delta} d_\infty\left(\sum_{j=0}^{n} f_{nj}(t) X\left(\frac{j}{n}\right), \sum_{j=0}^{n} f_{nj}(x) X\left(\frac{j}{n}\right)\right)$$

$$\le \sup_{|t-x|\le\delta} \sum_{j=0}^{n} d_\infty\left(f_{nj}(t) X\left(\frac{j}{n}\right), f_{nj}(x) X\left(\frac{j}{n}\right)\right)$$

$$\le \sup_{|t-x|\le\delta} \sum_{j=0}^{n} |f_{nj}(t) - f_{nj}(x)| \|X_0\| \left(\frac{j}{n}\right)$$

$$= \left(\sup_{y \in [0,1]} \|X_0\|(y) \right) \sum_{j=0}^{n} \omega_{f_{n_j}}(\delta) \stackrel{\delta \to 0^+}{\longrightarrow} 0,$$

that is, $B_n(X, \cdot)$ is continuous and hence $B_n(X, \cdot) \in \mathcal{C}(\mathcal{F}_c(\mathbb{R}^p))$.

Then we can define the sequence $\{B_n\}_n$, $B_n : \mathcal{C}(\mathcal{F}_c(\mathbb{R}^p)) \to \mathcal{C}(\mathcal{F}_c(\mathbb{R}^p))$ with $B_n(X) = B_n(X, \cdot)$. Clearly, B_n is $\mathcal{F}_c(\mathbb{R}^p)$-linear and $\mathcal{F}_c(\mathbb{R}^p)$-positive.

To see that $\lim_{n\to\infty} D_\infty(B_n(X, \cdot), X) = 0$ we will apply Theorem 1.

We have to see that $\lim_{n\to\infty} D_\infty(B_n(X^i, \cdot), X^i) = 0$ for $i = 0, 1, 2$. Let $g^i : [0,1] \to \mathbb{R}$ with $g^i(t) = t^i$. Then,

$$B_n(X^i, t) = \sum_{j=0}^{n} f_{nj}(t) X^i \left(\frac{j}{n} \right) = \sum_{j=0}^{n} f_{nj}(t) g^i \left(\frac{j}{n} \right) I_B$$

$$= \left(\sum_{j=0}^{n} f_{nj}(t) g^i \left(\frac{j}{n} \right) \right) I_B = B_n(g^i, t) I_B$$

where $B_n(g^i, \cdot)$ denotes the nth Bernstein approximant of the real function g^i.

By the classical results, $B_n(g^i, \cdot) \to g^i$ uniformly, and thus $B_n(g^i, \cdot) I_B \to g^i I_B$, that is, $D_\infty(B_n(X^i, \cdot), X^i)$ vanishes as n tends to ∞ for $i = 0, 1, 2$.

On the other hand, if $A \in \mathcal{F}_c(\mathbb{R}^p)$, then

$$B_n(A, t) = \sum_{j=0}^{n} f_{nj}(t) A = \left(\sum_{j=0}^{n} f_{nj}(t) \right) A = A.$$

As a consequence of this, and in accordance with Theorem 1, we have that

$$B_n(X, \cdot) \stackrel{d_\infty}{\longrightarrow} X \quad \text{uniformly}$$

for all $X \in \mathcal{C}(\mathcal{F}_c(\mathbb{R}^p))$. ∎

Corollary 1. *Let $X : [0,1] \to \mathcal{F}_c(\mathbb{R}^p)$ be a continuous fuzzy random variable. Then,*

$$B_n(X, \cdot) \stackrel{d_q}{\longrightarrow} X \quad \textit{uniformly for all } q \in [1, \infty).$$

Proof. The result follows directly from the fact that $d_q(U, V) \leq d_\infty(U, V)$ for all $U, V \in \mathcal{F}_c(\mathbb{R}^p)$. ∎

The following proposition shows that if we drop the assumption that the values of the fuzzy random variable are convex, then the approximation result will no longer hold. Instead, we have convergence to the convex hull of the fuzzy random variable.

Proposition 3. *Let $X : [0,1] \to \mathcal{F}(\mathbb{R}^p)$ be a continuous fuzzy random variable. Then,*

$$B_n(X,\cdot) \xrightarrow{d_\infty} \operatorname{co} X$$

uniformly in $[\varepsilon, 1-\varepsilon]$ for all $0 < \varepsilon < \frac{1}{2}$.

Proof. Given $X \in \mathcal{C}(\mathcal{F}(\mathbb{R}^p))$, let us consider its convex hull $\operatorname{co} X$. By virtue of the properties of the convex hull, it is straightforward to check that $\operatorname{co} X \in \mathcal{C}(\mathcal{F}_c(\mathbb{R}^p))$ and hence

$$\sup_{t\in[0,1]} d_\infty(B_n(\operatorname{co}X, t), \operatorname{co}X(t)) \to 0.$$

Then we only have to show that

$$\sup_{t\in[\varepsilon,1-\varepsilon]} d_\infty(B_n(X,t), B_n(\operatorname{co}X, t)) \to 0.$$

We have that

$$\sup_{t\in[\varepsilon,1-\varepsilon]} d_\infty(B_n(X,t), B_n(\operatorname{co}X, t))$$

$$= \sup_{t\in[\varepsilon,1-\varepsilon]} \sup_{\alpha\in[0,1]} d_H(B_n(X_\alpha,t), B_n(\operatorname{co}X_\alpha, t))$$

$$= \sup_{t\in[\varepsilon,1-\varepsilon]} \sup_{\alpha\in[0,1]} d_H\left(\sum_{j=0}^{n} f_{nj}(t) X_\alpha\left(\frac{j}{n}\right), \sum_{j=0}^{n} f_{nj}(t) \operatorname{co}X_\alpha\left(\frac{j}{n}\right)\right)$$

$$= \sup_{t\in[\varepsilon,1-\varepsilon]} \sup_{\alpha\in[0,1]} d_H\left(\sum_{j=0}^{n} f_{nj}(t) X_\alpha\left(\frac{j}{n}\right), \sum_{j=0}^{n} \operatorname{co}\left(f_{nj}(t) X_\alpha\left(\frac{j}{n}\right)\right)\right)$$

On the other hand, Shapley-Folkman inequality (see for instance Arrow and Hahn, 1971) states that for all $K_j \in \mathcal{K}(\mathbb{R}^p)$, $0 \le j \le n$,

$$d_H\left(\sum_{j=0}^{n} K_j, \sum_{j=0}^{n} \operatorname{co}K_j\right) \le \max_{0\le j\le n} \|K_j\| \sqrt{p}$$

and hence,

$$\sup_{t\in[\varepsilon,1-\varepsilon]} \sup_{\alpha\in[0,1]} d_H\left(\sum_{j=0}^{n} f_{nj}(t) X_\alpha\left(\frac{j}{n}\right), \sum_{j=0}^{n} \operatorname{co}\left(f_{nj}(t) X_\alpha\left(\frac{j}{n}\right)\right)\right)$$

$$\le \sup_{t\in[\varepsilon,1-\varepsilon]} \sup_{\alpha\in[0,1]} \max_{0\le j\le n} \left\| f_{nj}(t) X_\alpha\left(\frac{j}{n}\right) \right\| \sqrt{p}$$

$$= \max_{0\le j\le n} \sup_{t\in[\varepsilon,1-\varepsilon]} f_{nj}(t) \left\| X_0\left(\frac{j}{n}\right) \right\| \sqrt{p}$$

$$\leq D\sqrt{p} \max_{0\leq j\leq n} \sup_{t\in[\varepsilon,1-\varepsilon]} f_{nj}(t)$$

where $\sup_t \|X_0(t)\| \leq D \in \mathbb{R}$ since $\|X_0\|$ is continuous, then according to classical results we can conclude that

$$\max_{0\leq j\leq n} \sup_{t\in[\varepsilon,1-\varepsilon]} f_{nj}(t) = o(n^{-1/2}),$$

so that the proof is complete. ■

With respect to the behavior of B_n at the endpoints of $[0,1]$ it is straightforward to check that $B_n(X,0) = X(0)$ and $B_n(X,1) = X(1)$.

As a consequence of the preceding result, we obtain immediately the following one for the d_q-metric.

Corollary 2. *Let $X : [0,1] \to \mathcal{F}(\mathbb{R}^p)$ be a continuous fuzzy random variable. Then,*

$$B_n(X,\cdot) \xrightarrow{d_q} \operatorname{co} X$$

uniformly in $[\varepsilon, 1-\varepsilon]$ for all $0 < \varepsilon < \frac{1}{2}$ and $q \in [1,+\infty)$.

5 Concluding remarks

Bernstein approximants show a 'convexifying' asympthotical behavior, as seen in Proposition 3. Roughly speaking, the process of repeatedly adding (using Minkowski sum or its extension as defined in Section 2) tends to blur the distinction between convex and non-convex elements. Further comments on this respect can be found Vitale (1979, 1983).

Another suggestive example of this behavior appears in laws of large numbers for random sets and fuzzy random variables (for instance Artstein and Vitale, 1975, Colubi *et al.*, 1999). These are descriptions of the asympthotical behavior of an averaging process. Given a fuzzy random variable X and a sequence $\{X_n\}_n$ of independent fuzzy random variables with the same distribution as X, we have that the sequence of arithmetical means $\frac{1}{n}\sum_{i=1}^{n} X_i$ does converge almost surely to the convex hull of the integral of X, not to the integral of X itself (see Puri and Ralescu, 1986, for the definition of the integral of a fuzzy random variable). That is to say, even a mean-value scheme may not provide proper approximations in the non-convex case.

We should indicate that Theorem 1 can also be derived from those results in Keimel and Roth (1992) and in Terán and López-Díaz (2001b).

Acknowledgements

The research in this paper has been partially supported by the Spanish FPI Grant No. 98-71701353, and DGESIC Grants No. DGE-99-PB98-1534 and No.

DGE-98-PB97-1286. We would like to thank the editors for the possibility to contribute to this volume.

References

1. Altomare, F. and Campiti, M. (1994). *Korovkin-type approximation theory and its applications.* de Gruyter Studies in Mathematics, 17. Walter de Gruyter and Co., Berlin.
2. Arrow, K.J. and Hahn, F.H. (1971). *General Competitive Analysis.* Holden-Day, San Francisco.
3. Artstein, Z. and Vitale R.A. (1975). A strong law of large numbers for random compact sets, *Ann. Probab.* **3**, 879-882.
4. Colubi, A., López-Díaz, M., Domínguez-Menchero, J.S. and Gil, M.A. (1999). A generalized strong law of large numbers, *Probab. Th. Relat. Fields* **114**, 401-417.
5. Debreu, G. (1966). *Integration of correspondences,* in Proc. Fifth Berkeley Symp. Math. Statist. Prob., 351-372, Univ. of California Press, Berkeley.
6. Diamond, P. and Kloeden, P.E. (1990). Metric spaces of fuzzy sets. *Fuzzy Sets and Systems* **35**, 241-249; Corrigendum **45**, (1992) 123.
7. Diamond, P. and Kloeden, P.E. (1994). *Metric spaces of fuzzy sets. Theory and Applications.* World Scientific, Singapur.
8. Hiai, F. and Umegaki, H. (1977). Integrals, conditional expectations, and martingales of multivalued functions, *J. Multivar. Anal.* **7**, 149-182.
9. Keimel, K. and Roth, W. (1992). *Ordered cones and approximation.* Lecture Notes in Mathematics, **1517**, Springer-Verlag, Berlin.
10. Klement, E.P., Puri, M.L. and Ralescu, D.A. (1986). Limit theorems for fuzzy random variables, *Proc. R. Soc, Lond.* **A 407**, 171-182.
11. Korovkin, P.P. (1953). On convergence of linear positive operators in the space of continuous functions (in Russian), *Doklady Akad. Nauk SSSR (N.S.)* **90**, 961–964.
12. Korovkin, P. P. (1960). *Linear operators and approximation theory.* Hindustan Publishing Corp., Delhi 1960.
13. Matheron, G. (1975). *Random Sets and Integral Geometry.* J. Wiley & Sons, New York.
14. Puri, M.L. and Ralescu, D. (1981). Différentielle d'une fonction floue, *C.R. Acad. Sci. Paris Sér. A* **293**, 237-239.
15. Puri, M.L. and Ralescu, D. (1986). Fuzzy random variables, *J. Math. Anal. Appl.* **114** 409-422.
16. Shisha, O. and Mond, B. (1968). The degree of convergence of linear positive operators, *Proc. Nat. Acad. Sci. U.S.A.* **60**, 1196-1200.
17. Terán, P. and López-Díaz, M. (2001a). On the Bernstein approximants and the φ-variation of a fuzzy random variable, *Inform. Sci.* (accepted, in press).
18. Terán, P. and López-Díaz, M. (2001b). Approximation of mappings with values which are upper semicontinuous functions (submitted for publication).
19. Vitale, R.A. (1979). Approximation of convex set-valued functions, *J. Approx. Theory* **26** 301-316.
20. Vitale, R.A. (1983). Some developments in the theory of random sets, *Bull. Int. Statist. Inst.* **50**, 863-871.

Several notions of differentiability for fuzzy set-valued mappings

Luis J. Rodríguez-Muñiz

Departamento de Estadística e I.O. y D.M., Universidad de Oviedo, Facultad de Ciencias, C/ Calvo Sotelo s/n, 33007 Oviedo, Spain

Abstract. This paper presents a survey on several definitions of differentiability and the relationships among them. A new definition is also introduced, and its properties and connections with the previous ones are analysed.

1 Introduction

The problem of how to define the differential or derivative of a fuzzy set-valued mapping arises in the first eighties, as a natural step into the evolution of the Fuzzy Set Theory. Some of the former definitions of differential of a fuzzy set-valued mapping were based on those for set-valued mappings. Since then, several notions of differentiability/derivability for fuzzy mappings has been proposed. Clearly, such notions of differential can be applied in wide class of problems, and that is why, frequently, authors have proposed definitions *ad hoc* in order to solve concrete problems and needs to work in a specific framework.

Among the more relevant definitions of differential or derivative for a set-valued mapping we can quote Hukuhara (1967), Banks and Jacobs (1970), Bridgland (1970), De Blasi (1976), Aubin and Frankowska (1990) or Artstein (1995). Some of them are basically *analytical* and some other ones are mainly *geometrical.*

For the fuzzy set-valued case, first definitions were given by Dubois and Prade (1982, 1987), Puri and Ralescu (1983), Goetschel and Voxman (1986) or Kaleva (1987(. Recently, more approaches have been developed by Kandel *et al.* (1996), Buckley and Feuring (2000) and Rodríguez-Muñiz *et al.* (2001).

A group of these definitions are based on the use of an isometrical embedding in order to define a Fréchet-type differentiability between normed spaces. On the other hand, some authors prefer to weaken, in some sense, the usual Fréchet differential conditions to be able to work within a simpler space. In the first sense, the use of the support function is a tool which allows us to work in a normed space.

In this article, we recall some of these definitions and study the relationships among them. The paper is organised as follows: in Section 2 some preliminary concepts and results are presented, in Section 3 a first group of definitions of differentials are presented and their relationships are analized, in Section 4 the definitions based on the support function – especially the s-differentiability– are introduced and studied.

2 Preliminaries

Let $\mathcal{K}(\mathbb{R}^p)$ be the class of nonempty compact subsets of $\mathbb{R}^p$, and let $\mathcal{K}_c(\mathbb{R}^p)$ denote the class of the convex elements in $\mathcal{K}(\mathbb{R}^p)$. On $\mathcal{K}(\mathbb{R}^p)$ the *Hausdorff distance* is defined as follows:

$$d_H(A,B) = \max\left\{ \sup_{a\in A}\inf_{b\in B} \|a-b\|, \sup_{b\in B}\inf_{a\in A} \|a-b\| \right\},$$

where $A, B \in \mathcal{K}(\mathbb{R}^p)$, and $\|\cdot\|$ is the Euclidean norm in $\mathbb{R}^p$. It can be proved that $(\mathcal{K}(\mathbb{R}^p), d_H)$ is a complete and separable metric space (see, for instance, Debreu, 1967). On this metric space, a semilinear structure can be defined (it is semilinear because of the absence of the opposite element for the addition) by means of Minkowski's addition and the product by a scalar, which are given by

$$A + B = \{a + b \,|\, a \in A, b \in B\},$$

$$\lambda A = \{\lambda a \,|\, a \in A\} \text{ for all } A, B \in \mathcal{K}(\mathbb{R}^p),\ \lambda \in \mathbb{R}.$$

The *support function* of an element $A \in \mathcal{K}_c(\mathbb{R}^p)$ is defined as the function $s_A : S^{p-1} \to \mathbb{R}$ given by

$$s_A(r) = \sup_{a\in A} \langle r, a\rangle \qquad \text{for all } r \in S^{p-1},$$

where S^{p-1} is the unit sphere in $\mathbb{R}^p$ and $\langle\cdot,\cdot\rangle$ denotes the scalar product on $\mathbb{R}^p$.

A fuzzy subset of $\mathbb{R}^p$ is a function $A : \mathbb{R}^p \to [0,1]$. The α-level sets or α-cuts of a fuzzy subset are the sets $A_\alpha = \{u \in \mathbb{R}^p : A(u) \geq \alpha\}$ when $\alpha \in (0,1]$, while the 0-cut A_0 will be defined as the convex hull of the support of A (where $\text{supp}A = \{u \in \mathbb{R}^p : A(u) > 0\}$). Let $\mathcal{F}(\mathbb{R}^p)$ denote the class of fuzzy subsets of $\mathbb{R}^p$ whose α-level sets lie in in the class $\mathcal{K}(\mathbb{R}^p)$ for all $\alpha \in [0,1]$, whereas $\mathcal{F}_c(\mathbb{R}^p)$ is the subclass of fuzzy subsets in $\mathcal{F}(\mathbb{R}^p)$ such that they are convex, that is, their α-level sets belong to $\mathcal{K}_c(\mathbb{R}^p)$ for all $\alpha \in [0,1]$. The Hausdorff metric can be extended to the class of fuzzy sets (see Puri and Ralescu, 1985) by means of the following sup-norm:

$$d_\infty(A,B) = \sup_{\alpha\in[0,1]} d_H(A_\alpha, B_\alpha) \qquad \text{for all } A, B \in \mathcal{F}(\mathbb{R}^p).$$

Klement *et al.* (1986) and Puri and Ralescu (1986) have proved that $(\mathcal{F}(\mathbb{R}^p), d_\infty)$ is a complete and nonseparable metric space. This metric space can be endowed with a semilinear structure, by means of the sum and the product by a scalar based on Zadeh's extension principle, in accordance with which these operations are inherited from those defined on $\mathcal{K}(\mathbb{R}^p)$ on the level sets. More precisely, for all $\alpha \in [0,1]$ (see Nguyen, 1978):

$$(A+B)_\alpha = A_\alpha + B_\alpha\,,$$

$$(\lambda A)_\alpha = \lambda A_\alpha \qquad \text{for all } A, B \in \mathcal{F}(\mathbb{R}^p),\ \lambda \in \mathbb{R}.$$

Based on the notion of support function of a compact convex subset (see, for instance, Hörmander, 1955, or Artstein, 1974), the *support function* of an element $A \in \mathcal{F}_c(\mathbb{R}^p)$ is defined (see Puri and Ralescu, 1985) as the function $s_A : [0,1] \times S^{p-1} \to \mathbb{R}$ such that

$$s_A(\alpha, r) = \sup_{a \in A_\alpha} \langle r, a \rangle \qquad \text{for all } (\alpha, r) \in [0,1] \times S^{p-1}\,.$$

We will denote by $\mathcal{F}_{cc}(\mathbb{R}^p)$ the class of subsets $A \in \mathcal{F}_c(\mathbb{R}^p)$ satisfying that the mapping $\alpha \mapsto A_\alpha$ is d_H-continuous, and by $\mathcal{F}_{cl}(\mathbb{R}^p)$ we will denote the class of subsets $A \in \mathcal{F}_c(\mathbb{R}^p)$ satisfying the *Lipschitz condition* given by

$$d_H(A_\alpha, A_\beta) \le M|\alpha - \beta|$$

for all $\alpha, \beta \in [0,1]$ and a positive real number M.

By using support functions of its elements, Puri and Ralescu (1985) have proved that the class $\mathcal{F}_{cl}(\mathbb{R}^p)$ can be isometrically embedded into the Banach space $(C([0,1] \times S^{p-1}), \|\cdot\|_\infty)$, where $C([0,1] \times S^{p-1})$ is the space of all continuous functions from $[0,1] \times S^{n-1}$ to $\mathbb{R}$, and the mapping $j : \mathcal{F}_{cl}(\mathbb{R}^p) \to C([0,1] \times S^{p-1})$ such that $j(A) = s_A$ for all $A \in \mathcal{F}_{cl}(\mathbb{R}^p)$ is the corresponding isometrical embedding. Recently, Rojas-Medar *et al.* (1999) extended this embedding to the class $\mathcal{F}_{cc}(\mathbb{R}^p)$ and proved that this is the maximal class which can be embedded into $(C([0,1] \times S^{p-1}), \|\cdot\|_\infty)$.

3 Former definitions of differentials

In this section we are introducing the main definitions of differential which can be found in the literature and we are going to deal with. First one is De Blasi's definition. It was introduced for set-valued mappings by De Blasi (1976), and later extended for fuzzy set-valued mappings (see Diamond and Kloeden, 1994).

Definition 1. Let T be a nonempty open subset of $\mathbb{R}^k$ and let $t_0 \in T$. The mapping $F : T \to \mathcal{F}_c(\mathbb{R}^p)$ is said to be *De Blasi-differentiable* at t_0 if there exists a positively homogeneous and upper semicontinuous mapping $DF_{t_0} : \mathbb{R}^k \to \mathcal{F}_c(\mathbb{R}^p)$ (which is referred to as the *De Blasi differential* of F at t_0) such that

$$d_\infty(F(t_0 + \Delta t), F(t_0) + DF_{t_0}(\Delta t)) = o(\|\Delta t\|)\,.$$

Main properties of this definition are the following:

- if it exists then it is unique,
- if a mapping is De Blasi differentiable then it is continuous,
- the differential vanishes if and only if the mapping is constant,
- the differential is linear with respect to the addition of mappings and the product of a mapping by a scalar (see referred papers).

De Blasi's definition is a usual tool to avoid the problem of the linearity, and it is on the way of the latest definitions of differential, requiring fewer properties than in Fréchet's definition in order to get better results. De Blasi's interests were mainly focussed on differential equations, so the definition was *conceived* to be useful for this purpose.

Example 1. Let $F : \mathbb{R} \to \mathcal{F}_c(\mathbb{R})$ be a fuzzy set-valued mapping given by $F(t) = |t|\mathbf{1}_{\{a\}}$, for certain $a \in \mathbb{R}$. Clearly, F is De Blasi differentiable at $t_0 = 0$, since it is positively homogeneous and continuous, it coincides with its own De Blasi differential, and hence $DF_0 = F$.

Based on Banks and Jacobs' definition (1970) for set-valued mappings, the concept of π-differentiability was extended to fuzzy set-valued mappings by Puri and Ralescu (1983). To establish this concept, Rådström's embedding Theorem is needed to transform the differentiability problem into one concerning differentials of functions between normed spaces. So, we are defining the normed (and therefore, linear) space structure we need for this target.

On $\mathcal{F}_c(\mathbb{R}^p) \times \mathcal{F}_c(\mathbb{R}^p)$, we define the equivalence relation $\sim$ as follows:

$$(A, B) \sim (C, D) \text{ if, and only if, } A + D = B + C$$

for every $A, B, C, D \in \mathcal{F}_c(\mathbb{R}^p)$. We will denote by $\langle A, B\rangle$ the equivalence class containing (A, B), and by $\mathbf{F}^p$ the quotient space $\mathcal{F}_c(\mathbb{R}^p) \times \mathcal{F}_c(\mathbb{R}^p)/\sim$.

On $\mathbf{F}^p$ the addition and the product by a real number are defined by means of the following expressions:

$$\langle A, B\rangle + \langle C, D\rangle = \langle A + C, B + D\rangle,$$

$$\lambda \langle A, B\rangle = \begin{cases} \langle \lambda A, \lambda B\rangle & \text{if } \lambda \geq 0, \\ \langle |\lambda| B, |\lambda| A\rangle & \text{if } \lambda < 0, \end{cases}$$

The space $(\mathbf{F}^p, +, \cdot)$ defined in this way is a linear space, and a norm on $\mathbf{F}^p$ is given by

$$\|\langle A, B\rangle\|_{\mathbf{F}^p} = d_\infty(A, B)$$

for all $A, B \in \mathcal{F}_c(\mathbb{R}^p)$. The mapping $\pi : \mathcal{F}_c(\mathbb{R}^p) \to \mathbf{F}^p$ defined by $\pi(A) = \langle A, \mathbf{1}_{\{0\}}\rangle$ is an isometrical embedding (Rådström, 1952).

Once we have defined the required structure, the *π-differentiability* (or differentiability in the sense of Puri and Ralescu) is defined as follows:

Definition 2. Let T be a non empty open subset of $\mathbb{R}^k$ and let $t_0 \in T$. The mapping $F : T \to \mathcal{F}_c(\mathbb{R}^p)$ is said to be π-*differentiable* at t_0 if the composed mapping $\pi \circ F = \hat{F} : T \to \mathbf{F}^p$ is differentiable in Fréchet's sense at t_0, that is, there exists a linear continuous mapping $D\hat{F}_{t_0} : \mathbb{R}^k \to \mathbf{F}^p$ (which is referred to as the π-*differential* of F at t_0) satisfying that

$$\|\hat{F}(t_0 + \Delta t) - \hat{F}(t_0) - D\hat{F}_{t_0}(\Delta t)\|_{\mathbf{F}^p} = o(\|\Delta t\|).$$

By means of the embedding, the above expression can be rewritten as follows:

$$d_\infty(F(t_0 + \Delta t) + B_{t_0}(\Delta t), F(t_0) + A_{t_0}(\Delta t)) = o(\|\Delta t\|),$$

where $D\hat{F}_{t_0}(\Delta t) = \langle A_{t_0}(\Delta t), B_{t_0}(\Delta t)\rangle \in \mathbf{F}^p$.

Banks and Jacobs' interests were centered in control theory when they developed the π-differentiability for set-valued mappings. They also performed applications of this concept to integral calculus. In the fuzzy set-valued case, Puri and Ralescu (1983) gave the definition and the main properties of the π-differentiability, but they did not examine further developments of the theory. In fact, many difficulties arise in handling the involved structures. Recently, Buckley and Feuring (2000) used the π-differentiability (among many other definitions, like Kaleva or Goetschel and Voxman ones) to solve fuzzy-valued differential equations.

Example 2. Let us consider a real-valued mapping $f : \mathbb{R} \to \mathbb{R}$, which is Fréchet differentiable at $t_0 \in \mathbb{R}$ and such that $f(x) < 0$ for all $x \in \mathbb{R}$. Let $F : \mathbb{R} \to \mathcal{F}_c(\mathbb{R}^2)$ be the fuzzy set-valued mapping given by $F(t) = f(t) \cdot A$, where $A \in \mathcal{F}_c(\mathbb{R}^2)$. For the definition of the operations in $\mathbf{F}^2$, we have that

$$\langle f(t) \cdot A, \mathbf{1}_{\{0\}}\rangle = f(t)\langle \mathbf{1}_{\{0\}}, -A\rangle.$$

Therefore, F is π-differentiable at t_0 and

$$D\hat{F}_{t_0}(\Delta t) = f'(t_0)(\Delta t)\langle \mathbf{1}_{\{0\}}, -A\rangle$$

for all $\Delta t \in \mathbb{R}$.

A special type of π-differentiability can be formalized as follows (see Puri and Ralescu, 1983) when the differential yields in the embedded cone.

Definition 3. Let $u_1, \ldots, u_k$ be a basis in $\mathbb{R}^k$, T be a nonempty open subset of $\mathbb{R}^k$, and F be a π-differentiable mapping at $t_0 \in T$. F is said to be *conically differentiable* at t_0 if $D\hat{F}_{t_0}(u_i) = \pi\big(A_{t_0}(u_i)\big)$ for $i = 1, 2, \ldots, k$. Obviously, if F is conically differentiable at t_0, then

$$D\hat{F}_{t_0}(\Delta t) = \sum_{i=1}^{k} \Delta t^i \pi\big(A_{t_0}(u_i)\big).$$

We will denote the *conical differential* by $\langle DF_{t_0}(\Delta t), \mathbf{1}_{\{0\}}\rangle$.

This could be seen as the best situation in the case of the π-differential. Both conical and π-differentials are Fréchet differentials, so they have their classical properties.

Example 3. Let $g : \mathbb{R} \to \mathbb{R}$ be a Fréchet differentiable at $t_0 \in \mathbb{R}$ real-valued mapping, and let $F : \mathbb{R} \to \mathcal{F}_c(\mathbb{R})$ be a fuzzy set-valued mapping given by $F(t) = \mathbf{1}_{\{g(t)\}} + A$, where $A \in \mathcal{F}_c(\mathbb{R})$. Clearly, F is conically differentiable at t_0, and its conical differential is given by $\langle \mathbf{1}_{\{g'(t_0)(\Delta t)\}}, \mathbf{1}_{\{0\}} \rangle$ for each $\Delta t \in \mathbb{R}$.

Hukuhara (1967), proposed the following definition to represent the difference between two sets in $\mathcal{K}_c(\mathbb{R}^p)$: $A -_h B$ is defined as the set $C \in \mathcal{K}_c(\mathbb{R}^p)$ such that verifies $B + C = A$. The difference $A -_h B$ does not always exists and it does not coincide, in general, with the set $A + (-B) = \{a - b : a \in A, b \in B\}$. Next condition is necessary and sufficient for $A -_h B$ to exist (Hukuhara, 1967):

Proposition 1. *Let $A, B \in \mathcal{K}_c(\mathbb{R}^p)$, $A -_h B$ exists if, and only if, for every a in the boundary of A there exists at least one c such that $a \in B + \{c\} \subseteq A$.*

For the class $\mathcal{F}_c(\mathbb{R}^p)$, there exists and extension of the Hukuhara difference. It was introduced by Puri and Ralescu (1983) as follows: $A -_h B$ is the fuzzy subset $C \in \mathcal{F}_c(\mathbb{R}^p)$ such that $B + C = A$. It is obvious that $[A -_h B]_\alpha = A_\alpha -_h B_\alpha$ for every $\alpha \in [0, 1]$, when $A -_h B$ exists.

We will also use the concept of derivative introduced by Hukuhara (1967) for set-valued mappings and extended by Puri and Ralescu (1983) for fuzzy set-valued mappings.

Definition 4. Let T a non-empty open interval of $\mathbb{R}$. The mapping $F : T \to \mathcal{F}_c(\mathbb{R}^p)$ is said to be *Hukuhara derivable* at t_0 if it exists $F'(t_0) \in \mathcal{F}_c(\mathbb{R}^p)$ such that the limits (taken in the d_∞ sense) and Hukuhara diferences

$$\lim_{\Delta t \to 0+} \frac{F(t_0 + \Delta t) -_h F(t_0)}{\Delta t} \quad \text{and} \quad \lim_{\Delta t \to 0+} \frac{F(t_0) -_h F(t_0 - \Delta t)}{\Delta t}$$

exist, and they are equal to $F'(t_0)$, which is called the *Hukuhara derivative* of F at t_0.

Hukuhara's set-valued definition was introduced to develope a differential and integral calculus for multifunctions. Hukuhara himself derived some extensions of fundamentals theorems of integral and differential calculus to the set-valued case. Some of these results can be also extended to the fuzzy set-valued case (see, for instance, Diamond and Kloeden, 1994). So, despite of the restrictive definition, it is often useful when working with fuzzy-valued integrals.

Hukuhara derivative is unique –if it exists– , and a Hukuhara derivable mapping is also continuous, the derivative vanishes if, and only if, the mapping is constant, and the derivative is linear with respect to the addition of mappings and the product of a mapping by a scalar (see referred papers).

Example 4. Let $F : [-1,1] \to \mathcal{F}_c(\mathbb{R})$ be a fuzzy set-valued mapping whose α-leves are given by

$$[F(t)]_\alpha = [1 - |t| - .5(1-\alpha), 1 - |t| + .5(1-\alpha)]$$

for every $\alpha \in [0,1]$. We can easily check that F is Hukuhara derivable at $t_0 = 0$ and its derivative is $\mathbf{1}_{\{1\}}$.

A shared property of all these definitions of differential and derivative is the relationship with the α-level mapping. Let $F : T \to \mathcal{F}_c(\mathbb{R}^p)$ be a fuzzy set-valued mapping, their associated α-level mappings are defined as $F_\alpha : T \to \mathcal{K}_c(\mathbb{R}^p)$ given by $F_\alpha(t) = (F(t))_\alpha$ for every $\alpha \in [0,1]$ and $t \in T$. Consequently, the property says that if F is differentiable (in any sense), then F_α is differentiable in the set-valued correspondant of the used fuzzy set-valued criterium.

The converse of this result does not hold in general. The reason is that is not sufficient the existence of the differentials of F_α to guarantee the existence of the differential of F. In virtue of the Representation Lemma by Negoita and Ralescu (1975), the differentials of F_α need to be nested in α to configure a fuzzy set. When the differentials of F_α satisfy this Representation Lemma, then the fuzzy set-valued mapping F is differentiable –in the correspondant sense (see, for instance, Diamond and Kloeden, 1994).

As regards the relationships among these definitions of differentials, it is easy to check that the conical differentiability is not only an special case of the π-differentiality but also of the De Blasi differentiability. In fact, conical differential can be viewed as a linear and continuous De Blasi differential.

Proposition 2. *Let $F : T \to \mathcal{F}_c(\mathbb{R}^p)$ be a fuzzy set-valued mapping. If F is conically differentiable at t_0, with conical differential $\langle DF_{t_0}(\Delta t), \mathbf{1}_{\{0\}}\rangle$, then F is De Blasi differentiable at t_0 and its De Blasi differential is DF_{t_0}.*

As a consequence, the converse holds obviously when the De Blasi differential is linear and continuous. On the other hand, there is no direct relationship between De Blasi differential and π-differential. There are counterexamples of both implications (for instance, Example 1 is a De Blasi differentiable mapping which is not π-differentiable, while Example 2 is valid as a π-differentiable mapping which is not De Blasi differentiable).

Concerning the Hukuhara derivate, Puri and Ralescu (1983) proved that it can be viewed as a particular case of the conical differentiability.

Proposition 3. *Let $F : T \to \mathcal{F}_c(\mathbb{R}^p)$ be a fuzzy set-valued mapping. If F is Hukuhara derivable at t_0, with Hukuhara derivative $F'(t_0)$, then F is conically differentiable at t_0 and its conical differential is $\Delta t\langle F'(t_0), \mathbf{1}_{\{0\}}\rangle$.*

The converse result is not true, since the Hukuhara derivative implies the existence of the Hukuhara differences between certain values of the mapping, but this assertion is not necessarily true in the case of conically differentiable mappings (Example 5 will be a counterexample of the falsity of this result).

As we will need this result later, we can now remark that when a De Blasi diferentiable mapping takes on values in $\mathcal{F}_{cc}(\mathbb{R}^p)$, then the De Blasi differential DF_{t_0} also yields in this class. And the same property holds obviously for the conical differential. In the case of the π-differentiability we can only ensure that the two of the components of the π-differential $(A_{t_0}(\Delta t), B_{t_0}(\Delta t))$ are both in or out of the class $\mathcal{F}_{cc}(\mathbb{R}^p)$.

4 Differentials using the support function

Another way to define the differential of a fuzzy set-valued mapping is by using the natural embedding by means of the support function, wich allows us to consider the class $\mathcal{F}_{cc}(\mathbb{R}^p)$ as a cone in the compact Banach space $(C([0,1] \times S^{p-1}), \|\cdot\|_\infty)$, which has a simpler structure than the metric space constructed for the π-differential (and even is complete, which is not true in the case of the space $\mathbf{F}^p$– see, for instance, Debreu, 1967). This tool was used by Bobylev (1981) to define a new concept of differentiabiliy.

Definition 5. Let $F : T \to \mathcal{F}_{cc}(\mathbb{R}^p)$ be a fuzzy set-valued mapping. It is said to be *Bobylev differentiable* at $t_0 \in T$ if the real-valued mapping $s_F(\alpha, r) : T \to \mathbb{R}$ given by $s_F(\alpha, r)(t) = s_{F(t)}(\alpha, r)$ is Fréchet differentiable, uniformly in $(\alpha, r) \in [0,1] \times S^{p-1}$, and there exists a fuzzy set $A \in \mathcal{F}_{cc}(\mathbb{R}^p)$ such that:

$$s_A(\alpha, r) = s'_{F(t_0)}(\alpha, r),$$

where $s'_{F(t_0)}(\alpha, r)$ represents the differential of $s_F(\alpha, r)$ in t_0 for every $(\alpha, r) \in [0,1] \times S^{p-1}$. The fuzzy set A defined in this way is called the *Bobyleb differential* of F.

Actually, Bobylev definition was introduced for fuzzy set-valued mappings taking on values in $\mathcal{F}_c(\mathbb{R}^p)$, but Bobylev used another slightly different definition of support function (introduced by himself), se we have decide to rename the definition in order to be able to use Puri and Ralescu's definition of support function.

Example 5. Let $F : (0, 2\pi) \to \mathcal{F}_{cc}(\mathbb{R})$ be a fuzzy set-valued mappings whose α-levels are given by

$$[F(t)]_\alpha = (1-\alpha)(2+\sin t)[-1,1]$$

for each $\alpha \in [0,1]$. Its support function is given by

$$s_{F(t)}(\alpha, r) = (1-\alpha)(2+\sin t)$$

for all $(\alpha, r) \in [0,1] \times \{-1,1\}$ and $t \in (0, 2\pi)$. If F is defined in this way, it is Bobylev differentiable at all $t_0 \in (0, 2\pi)$ and its differential is given, levelwise, by

$$[F'_{t_0}]_\alpha = (1-\alpha) \cos t_0 [-1,1]$$

(see Diamond and Kloeden, 1994).

From our point of view, Bobylev's definition seems to be very restrictive because of the condition of existence of such a fuzzy set A whose support function coincides with the differential of the support function. So, in order to weaken this definition, we can introduce the Fréchet differentiability of the support function, that is, of the mapping $s_F : T \to C([0,1] \times S^{p-1})$ given by $s_F(t) = s_{F(t)}$, and without requiring for the existence of a fuzzy set having the Fréchet differential as its own support function.

Remark 1. The mapping which will be given in Counterexample 1 is itself an example of Fréchet differentiable support function but not Bobylev differentiable mapping (the Fréchet differential of the support function is not support function of any fuzzy subset).

Appart from the Fréchet differentiability of the support function (which, as we well see, extends in some sense the π-differentiability) we can introduce another concept which will be very related to the De Blasi differentiability:

Definition 6. Let $F : T \to \mathcal{F}_{cc}(\mathbb{R}^p)$ be a fuzzy set-valued mapping. F is said to be *s-differentiable* at $t_0 \in T$ if there exists a positively homogeneous mapping $F'_{t_0} : \mathbb{R}^k \to C([0,1] \times S^{p-1})$ (which will be referred as the *s-differential* of F at t_0) such that

$$\|s_{F(t_0+\Delta t)} - s_{F(t_0)} - F'_{t_0}(\Delta t)\|_\infty = o(\|\Delta t\|)\,, \tag{1}$$

and the mappings $F'_{t_0}(\cdot)(\alpha, r) : \mathbb{R}^k \to \mathbb{R}$ given by $F'_{t_0}(u)(\alpha, r)$ are upper semicontinuous, uniformly in $(\alpha, r) \in [0,1] \times S^{p-1}$.

The study of the s-differential is motivated by the aim of obtain results in this analytical way, which could be applied to Statistics (Point Estimation, Information Theory, etc)

Among the properties of the s-differential we can quote:

- if it exists, it is unique,
- the s-differentiable mapping is also continuous,
- the s-differential vanishes if, and only if, the mapping is constant,
- it is linear with respect to the addition of mappings and the product of a mapping by a scalar.

Therefore, as we can see, the s-differential *inheritates* the desiderable properties of the De Blasi differential.

Example 6. Let us consider the fuzzy set-valued mapping $G : (-1,1) \to \mathcal{F}_{cc}(\mathrm{I\!R})$ given by

$$[G(t)]_\alpha = \left[-\frac{1}{(1+\alpha)(1+|t|)}, \frac{1}{(1+\alpha)(1+|t|)}\right]$$

for every $\alpha \in [0,1]$, whose support function is

$$s_{G(t)}(\alpha, r) = \sup_{x \in [G(t)]_\alpha} \langle x, r \rangle = \frac{1}{(1+\alpha)(1+|t|)}$$

for every $(\alpha, r) \in [0,1] \times \{-1,1\}$. We can easily deduce that the s-differential of G at $t_0 = 0$ is given by the mapping $G_0' : \mathrm{I\!R} \to C([0,1] \times S^0)$ defined as $G_0'(\Delta t) : [0,1] \times S^0 \to \mathrm{I\!R}$, with

$$G_0'(\Delta t)(\alpha, r) = -\frac{|\Delta t|}{1+\alpha}$$

for every $(\alpha, r) \in [0,1] \times S^0$ and every $\Delta t \in R$.

Let us see how the s-differentiability is related to the previous concepts we have seen. Obviuosly, the s-differentiability is a weaker concept than the Fréchet differentiability of the support function. With respect to the De Blasi differential we have that

Theorem 1. *Let $F : T \to \mathcal{F}_{cc}(\mathrm{I\!R}^p)$ be a fuzzy set-valued mapping. If F is De Blasi differentiable at t_0, then F is s-differentiable at t_0. Moreover,*

$$F_{t_0}'(\Delta t) = s_{DF_{t_0}(\Delta t)}$$

for all $\Delta t \in \mathrm{I\!R}^k$, where DF_{t_0} is the De Blasi differential of F at t_0 and F_{t_0}' is the s-differential of F at t_0.

Hence, the s-differential is a more general concept than the De Blasi differential in the case of $\mathcal{F}_{cc}(\mathrm{I\!R}^p)$-valued mappings. Of course, they are note similar concepts, as we can see in the following:

Counterexample 1. Let us consider the mapping $F : (.5, 2) \to \mathcal{F}_{cc}(\mathrm{I\!R})$ whose α-level sets are

$$\big(F(t)\big)_\alpha = \left[-\frac{1}{(1+\alpha)t}, \frac{1}{(1+\alpha)t}\right],$$

for every $\alpha \in [0,1]$ and $t \in (.5, 2)$. We can easily prove that this mapping is s-differentiable at 1 (actually, its support function is Fréchet differentiable at 1) but it is not De Blasi differentiable at this point (there does not exist a fuzzy set in $\mathcal{F}_{cc}(\mathrm{I\!R})$ whose support function is F_1').

On the other hand, we can ask for sufficient conditions to verify the converse result of that in Theorem 1. These conditions guarantee the existence of a fuzzy set whose support function is the s-differential of the mapping, and hence, the s-differentiability implies the De Blasi differentiability (see Rodríguez-Muñiz *et al.*, 2001, for details).

As regards the π-differentiability and its relationship with the s-differentiability, we have the following result:

Theorem 2. *Let $F : T \to \mathcal{F}_{cc}(\mathbb{R}^p)$ be a fuzzy set-valued mapping. If F is π-differentiable at t_0 and its π-differential at t_0, written as $D\hat{F}_{t_0}(\Delta t) = \langle A_{t_0}(\Delta t), B_{t_0}(\Delta t)\rangle \in \mathbf{F}$, is such that $A_{t_0}(\Delta t) \in \mathcal{F}_{cc}(\mathbb{R}^p)$ for all $\Delta t \in \mathbb{R}^k$, then F is s-differentiable at t_0 (moreover, its support function is Fréchet differentiable at t_0). And the s-differential satisfies that*

$$F'_{t_0}(\Delta t) = s_{A_{t_0}(\Delta t)} - s_{B_{t_0}(\Delta t)}$$

for all $\Delta t \in \mathbb{R}^k$.

The converse result is not true, in general, as we can see in the following:

Counterexample 2. Let us consider the mapping $F : \mathbb{R} \to \mathcal{F}_{cc}(\mathbb{R}^2)$ given by $F(t) = \mathbf{1}_{tB^2}$. This mapping is not π-differentiable at $t_0 = 0$ (see Banks and Jacobs, 1970). But, since it positively homogeneous and continuous, F is De Blasi differentiable at $t_0 = 0$ (in fact, $DF_{t_0} = F$), and due to Theorem 1, F is s-differentiable at $t_0 = 0$.

In order to obtain the converse result of that in Theorem 2 we have to require the Fréchet differentiability of the support function of the mapping. Thus,

Theorem 3. *Let $F : T \to \mathcal{F}_{cc}(\mathbb{R}^p)$ be a fuzzy set-valued mapping. If its support function $s_{F(\cdot)} : T \to C([0,1] \times S^{p-1})$ is Fréchet differentiable at t_0 and its differential $s'_{F(t_0)}(\Delta t)$ satisfies, for every $\Delta t \in \mathbb{R}^k$, that there exist $A_{t_0}(\Delta t), B_{t_0}(\Delta t) \in \mathcal{F}_{cc}(\mathbb{R}^p)$ such that*

$$s'_{F(t_0)}(\Delta t) = s_{A_{t_0}(\Delta t)} - s_{B_{t_0}(\Delta t)},$$

then the mapping F is π-differentiable at t_0, and

$$D\hat{F}_{t_0}(\Delta t) = \langle A_{t_0}(\Delta t), B_{t_0}(\Delta t)\rangle$$

for every $\Delta t \in \mathbb{R}^k$.

Remark 2. Under the hypothesis of Theorem 3, if $B_{t_0}(\Delta t) = \mathbf{1}_{\{0\}}$, for each $\Delta t \in \mathbb{R}^k$, then F is conically differentiable at t_0. Moreover, the hypothesis $A_{t_0}(\Delta t) \in \mathcal{F}_{cc}(\mathbb{R}^p)$ will be unnecessary: since F takes on values in $\mathcal{F}_{cc}(\mathbb{R}^p)$ then $A_{t_0}(\Delta t)$ belongs to in the same class.

5 Conclusions

The use of the differential based on the support function allows us to work in a Banach space, $(C([0,1] \times S^{p-1}, \|\cdot\|)$, which is a much better situation than working in a non complete metric space like $(\mathcal{F}_c(\mathbb{R}^p), d_\infty)$, or a non complete normed space like $\mathbf{F}^p$. Thus, by reducing the class of fuzzy subsets to $\mathcal{F}_{cc}(\mathbb{R}^p)$ we can get the completeness and hence we can use the classical tecniques of integration (instead of Kudō-Aumann integration), and we can try to deduce fundamentals results of integral and differential calculus, related to fuzzy random variables (introduced by Puri and Ralescu, 1986).

Moreover, the use of the support function provides a much easier to handling tool than other differentials taking on values on a normed space –like the π-differential, whose construction gives a much more complicated structure than in the case of the differential using the support function. And, as we have seen in the presented results, in the class $\mathcal{F}_{cc}(\mathbb{R}^p)$ the s-differential and the Fréchet differential of the support function generalise –non trivially– the De Blasi differential and the π-differential.

Acknowledgements

Author wishes to gratefully thank the financial support received from the Spanish Ministery of Education, Culture and Sports (DGESIC Grant DGE-99-PB98-1534), and from the Fundación Banco Herrero.

References

1. Artstein, Z. (1974). On the calculus of closed set-valued functions, *Indiana Univ. Math. J.* **24**, 433-441.
2. Artstein, Z. (1995). A calculus for set-valued maps and set-valued evolution equations, *Set-Valued Anal.* **3**, 213-261.
3. Aumann, R.J. (1965). Integrals of set-valued functions, *J. Math. Anal. Appl.* **12**, 1-12.
4. Banks, H.T. and Jacobs, M.Q. (1970). A differential calculus for multifunctions, *J. Math. Anal. Appl.* **29**, 246-272.
5. Bridgland, T.F. (1970). Trajectory integrals of set valued functions, *Pacific J. Math.* **33**, 43-67.
6. Bobylev, V.N. (1985a). Support function for a fuzzy set and its characteristic properties, *Mat. Zametki* **37**, 507-513.
7. Bobylev, V.N. (1985b). Cauchy problem under fuzzy control, *BUSEFAL* **21**, 117-126.
8. Buckley, J.J. and Feuring, T. (1999). Introduction to fuzzy partial differential equations. Fuzzy analysis and related topics (Prague, 1997), *Fuzzy Sets and Systems***105**, 241-248.
9. Buckley, J.J. and Feuring, T. (2000). Fuzzy differential equations, *Fuzzy Sets and Systems* **110**, 43-54.
10. De Blasi, F.S. (1976). On the differentiability of multifunctions, *Pacific J. Math.* **66**, 67-81.

11. De Blasi, F.S. and Myjak, J. (1986). Weak convergence of convex sets in Banach spaces, *Arch. Math. (Basel)* **47**, 448-456.
12. Debreu, G. (1967). Integration of correspondences, *Proc. Fifth Berkeley Sympos. Math. Statist. and Probability (Berkeley, Calif., 1965/66)*, Vol. II: Contributions to Probability Theory, Part 1, pp.351–372. Univ. California Press, Berkeley, Calif.
13. Diamond, P. and Kloeden, P. (1994). *Metric Spaces of Fuzzy Sets. Theory and applications*, World Scientific Publishing Co., Inc., River Edge, NJ.
14. Dubois, D. and Prade, H. (1982). Towards fuzzy differential calculus. Part 3: Differentiation, *Fuzzy Sets and Systems* **8**, 225-233.
15. Dubois, D. and Prade, H. (1987). On several definitions of the differential of a fuzzy mapping, *Fuzzy Sets and Systems* **24**, 117-120.
16. Goetschel, R. and Voxman, W. (1986). Elementary fuzzy calculus, *Fuzzy Sets and Systems* **18**, 31-43.
17. Hörmander, L. (1955). Sur la fonction d'appui des ensembles convexes dans un espace localement convexe, *Ark. Mat.* **3**, 181-186.
18. Hukuhara, M. (1967). Intégration des applications mesurables dont la valeur est un compact convexe, *Funkcial. Ekvac.* **10**, 205-223.
19. Kaleva, O. (1987). Fuzzy differential equations, *Fuzzy Sets and Systems* **24**, 301-317.
20. Kandel, A., Friedman, M. and Ming, M. (1996). On fuzzy dynamical processes, *Proc. FUZZ-IEEE'96*, 1813-1818.
21. Klement, E.P., Puri, M.L. and Ralescu, D.A. (1986). Limit theorems for fuzzy random variables, *Proc. Roy. Soc. London Ser. A* **407**, 171-182.
22. Kudō, H. (1954). Dependent experiments and sufficient statistics, *Nat. Sci. Rep. Ochanomizu Univ.* **4**, 151-163.
23. Lowen, R. (1980). Convex fuzzy sets, *Fuzzy Sets and Systems* **3**, 291-310.
24. Negoita, C.V. and Ralescu, D.A. (1975). *Applications of fuzzy sets to system analysis*, John Wiley & Sons, New York.
25. Nguyen, H.T. (1978). A note on the extension principle for fuzzy sets, *J. Math. Anal. Appl.* **64**, 369-380.
26. Puri, M.L. and Ralescu, D.A. (1983). Differentials of fuzzy functions, *J. Math. Anal. Appl.* **91**, 552-558.
27. Puri, M.L. and Ralescu, D.A. (1985). The concept of normality for fuzzy random variables, *Ann. Probab.* **13**, 1373-1379.
28. Puri, M.L. and Ralescu, D.A. (1986). Fuzzy random variables, *J. Math. Anal. Appl.* **114**, 409-422.
29. Rådström, H. (1952). An embedding theorem for spaces of convex sets, *Proc. Amer. Math. Soc.* **3**, 165-169.
30. Rojas-Medar, M., Bassanezi, R.C. and Román-Flores, H. (1999). A generalization of the Minkowski embedding theorem and applications, *Fuzzy Sets and Systems* **102**, 263-269.
31. Rodríguez-Muñiz, L.J., López-Díaz, M., Gil, M.A. and Ralescu, D.A. (2001). The s-diferentiability of fuzzy mappings (submitted for publication).
32. Seikkala, S. (1987). On the fuzzy initial value problem, *Fuzzy Sets and Systems* **24**, 319-330.
33. Zadeh, L.A. (1965). Fuzzy sets, *Information and Control* **8**, 338-353.

Part 3

POSSIBILITY, PROBABILITY AND FUZZY MEASURES

Average level of a fuzzy set

Dan A. Ralescu[1]

Department of Matematical Sciences
University of Cincinnati, Cincinnati, OH 45221-0025, USA

Abstract. In this paper the average level set of a fuzzy set is defined on the basis of the Kudo-Aumann integral of a set-valued mapping. Some properties of the average level as well as its particularization for some special cases are analyzed.

1 Introduction

Given a fuzzy set $u : \mathbb{X} \to [0,1]$ the α-level of u is defined as the set $L_\alpha u = \{x \in \mathbb{X} \,|\, u(x) \geq \alpha\}$.

The mapping $Lu : [0,1] \to \mathcal{P}(\mathbb{X})$ such that $Lu(\alpha) = L_\alpha u$ is said to be the *level map* of the fuzzy set u of $\mathbb{X}$ (Negoita and Ralescu, 1974).

To average a fuzzy set we are going to follow an interpretation different from that in Dubois and Prade (1987) and Heilpern (1992). For this purpose they have considered a probabilistic framework which does not seem to be natural in this setting. Also, they considered only the case $\mathbb{X} = \mathbb{R}$.

In this paper we make use of the Kudo-Aumann integral to average levels of the considered fuzzy set.

The notion introduced in this paper makes easier to prove properties of the average, like the additivity, and we will consider the more general case $\mathbb{X} = \mathbb{R}^n$. The results in this paper can be extended to the case in which $\mathbb{X}$ is a Banach space, and even to the case in which $\mathbb{X}$ is a Polish space.

The present paper has the following structure: in Section 2 we describe our concept of average level set; in Section 3 we compare it with the Dubois and Prade (1987) mean value of a fuzzy number, and we will give later different properties of this new concept.

2 Average level set

Let $\mathcal{K}(\mathbb{R}^n)$ denote the class of all compact, nonempty subsets of $\mathbb{R}^n$. On $\mathcal{K}(\mathbb{R}^n)$ we will consider the well-known *Hausdorff metric* given by

$$d_H(A,B)$$

$$= \max\left\{\sup_{a\in A}\inf_{b\in B}\|a-b\|, \sup_{b\in B}\inf_{a\in A}\|a-b\|\right\},$$

where $\|A\| = d_H(A,\{0\}) = \sup_{a\in A}\|a\|$.

Let $(\Omega, \mathcal{A}, \mu)$ be a finite, nonatomic measure space. A *set-valued function* $F : \Omega \to \mathcal{P}(\mathbb{R}^n)$ is always assumed such that $F(\omega) \neq \emptyset$ for all $\omega \in \Omega$.

The mapping F is called *measurable* if its graph $\{(\omega, x) \,|\, x \in F(\omega)\} \in \mathcal{A} \times \mathcal{B}$, where $\mathcal{B}$ is the class of the Borel sets in $\mathbb{R}^n$.

The space of integrable *selectors* of F is given by

$$S(F) = \{f \,|\, f(\omega) \in F(\omega)\, a.e., f \text{ integrable}\}.$$

Our main object of interest is the *integral* of F (first defined by Kudo; then developed by Aumann and Debreu):

$$\int_\Omega F\, d\mu = \left\{ \int_\Omega f\, d\mu \,|\, f \in S(F) \right\}.$$

Note that the integral of a set-valued function is a set.

Some important properties of this concept are collected in the next results.

First, we are recalling some properties of the integral which are due to Aumann (1965) and Debreu (1967).

Theorem 1.
(a) *$\int_\Omega F\, d\mu$ is a convex set (possibly empty);*
(b) *If F is measurable and $\int_\Omega \|F\|\, d\mu < \infty$, then $\int_\Omega F\, d\mu \neq \emptyset$;*
(c) *If F has closed values and $\int_\Omega \|F\|\, d\mu < \infty$, then $\int_\Omega F\, d\mu$ is compact.*

Next, as we have mentioned, we will consider fuzzy sets $u : \mathbb{R}^n \to [0, 1]$. The levels are defined by

$$L_\alpha u = \begin{cases} \{x \in \mathbb{R}^n \,|\, u(x) \geq \alpha\} & \text{if } \alpha \in (0, 1] \\ \operatorname{supp} u = \operatorname{cl}\{x \in \mathbb{R}^n \,|\, u(x) > 0\} & \text{if } \alpha = 0. \end{cases}$$

Each level set $L_\alpha u$ is an approximation of fuzzy set u (containing only those elements of $\mathbb{R}^n$ whose membership degrees exceed α). But a fuzzy set u is "many" ordinary sets; actually, the totality of levels $\{L_\alpha u \,|\, 0 \leq \alpha \leq 1\}$ represents u. Can we average all these level sets to obtain a single approximation of a fuzzy set? The answer is YES and is provided by the set-valued integral of the level set map. More exactly:

Definition 1. The*average level of a fuzzy set* u is given by

$$Au = \int_{[0,1]} Lu(\alpha) d\alpha = \int_{[0,1]} L_\alpha u\, d\alpha.$$

In the above formula we have considered the Kudo-Aumann integral of the level set map Lu, with respect to the Lebesgue measure in $[0, 1]$. The quantity Au is a set.

By adapting Theorem 1 to the present case we get:

Proposition 1. *If the membership function u is upper semicontinuous and the support of u is compact, then Au is a nonempty compact and convex set.*

Proof. Note that $L_\alpha u$ is closed and $\int_{[0,1]} \|L_\alpha u\| d\alpha < \infty$, giving the result. What is surprising in this framework is the convexity of Au: it is not necessary to assume that the levels of u are convex! This is a well-known property of the set-valued integral (cf. Theorem 1 (a) above). ∎

Remark 1. The hypothesis that $\operatorname{supp} u$ is compact can be replaced by the weaker one that $d_1(u, \{0\}) = \int_{[0,1]} \|L_\alpha u\| d\alpha$ is finite. Here d_1 is the distance introduced by Klement *et al.* (1986).

The average level set is now particularized to some special situations.

Examples.

(1) Let u be an ordinary nonfuzzy set A, i.e., $u = I_A$. Assume A is compact, but not necessarily convex. Then $Au = \operatorname{co} A$ (with co the convex hull of A, that is, the smallest convex set which contains A).

If A is compact, convex, then $Au = A$. This result is obvious: all level sets of I_A are equal to A, so their "average" is A. Note, however, that the "average" is convexifying.

(2) An extension of the above example is the case of a fuzzy set u assuming only finitely many values, $1 = a_1 > a_2 > \ldots > a_r \geq 0 = a_{r+1}$. Assume convex levels. Then, the level set map is a simple multivalued function:

$$Lu = \sum_{j=1}^{r} L_{a_j} u \times I_{(a_{j+1}, a_j]}.$$

In this case,

$$Au = \sum_{j=1}^{r} L_{a_j} u \times (a_j - a_{j+1})$$

(this formula was obtained by Dubois and Prade [1987, p. 281]).

The next result covers the case of (not necessarily convex) fuzzy sets in $\mathbb{R}$:

Proposition 2. *Let* $u : \mathbb{R} \to [0,1]$ *be upper semicontinuous, with compact support, and such that* $L_1 u \neq \emptyset$. *Then,*

$$Au = \left[\int_{[0,1]} \inf L_\alpha u \, d\alpha, \int_{[0,1]} \sup L_\alpha u \, d\alpha \right].$$

Proof. Let $y \in Au = \int_{[0,1]} L_\alpha u \, d\alpha$; then, $y = \int_{[0,1]} f(\alpha) \, d\alpha$, where $f(\alpha) \in L_\alpha u$ a.e.. But then

$$\inf L_\alpha u \leq f(\alpha) \leq \sup L_\alpha u$$

which, after integration, gives

$$\int_{[0,1]} \inf L_\alpha u \, d\alpha \leq y \leq \int_{[0,1]} \sup L_\alpha u \, d\alpha.$$

To prove the other inclusion, note that $\inf L_\alpha u, \sup L_\alpha u \in L_\alpha u$ since $L_\alpha u$ is compact. Thus, $\int_{[0,1]} \inf L_\alpha u \, d\alpha \in Au$, $\int_{[0,1]} \sup L_\alpha u \, d\alpha \in Au$. By Proposition 1, Au is a convex set, thus it contains the entire interval, and the proof ends. ■

Remark 2. The functions $\varphi(\alpha) = \inf L_\alpha u$, $\phi(\alpha) = \sup L_\alpha u$ are selectors of Lu.

In Figure 1, the horizontal segments represent level sets:

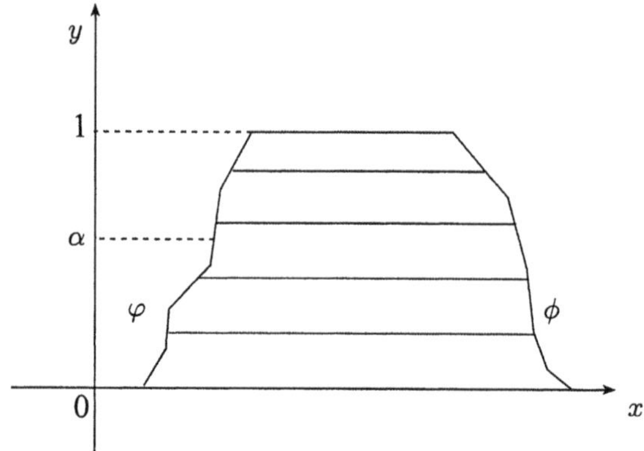

Fig. 1. Level sets of a fuzzy set

Example. Consider a triangular fuzzy number u with support $[a - \delta, a + \delta]$, $u(a) = 1$ (Figure 2).

Then,

$$Au = \left[a - \frac{\delta}{2}, a + \frac{\delta}{2}\right]$$

3 Properties

Let us first show that for fuzzy sets in $\mathbb{R}$, Au coincides with the mean value defined by Dubois and Prade (1987):

Proposition 3. *Let $u : \mathbb{R} \to [0,1]$ be upper semicontinuous and with compact support, and let $M(u)$ be the mean value of u as in Dubois and Prade (1987). Then, $Au = M(u)$.*

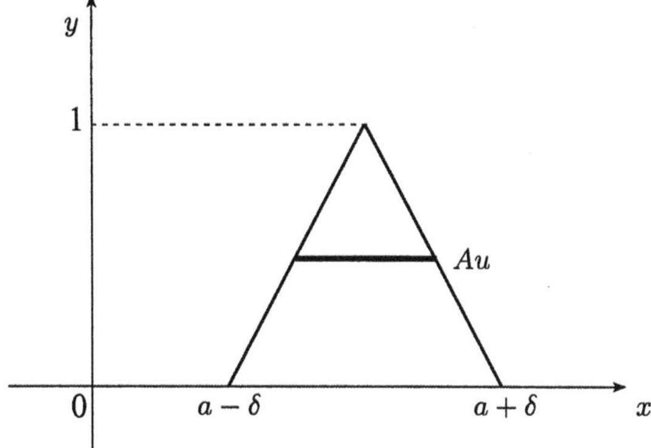

Fig. 2. Triangular fuzzy number

Proof. Recall how $M(u)$ is calculated: the level set map $Lu : [0,1] \to \mathcal{K}(\mathbb{R})$ generates an upper probability U given by

$$U(A) = \sup_{x \in A} u(x),$$

then

$$M(u) = \{E_P\varphi \,|\, P \leq U\}.$$

In the above set, P is a probability measure (on $\mathbb{R}$) bounded by U, $\varphi : \mathbb{R} \to \mathbb{R}$ is the identity function $\varphi(x) = x$, and $E_P\varphi = \int_{\mathbb{R}} \varphi(x)\, dP(x)$ is the ordinary expected value with respect to P. We can also write

$$M(u) = \left\{ \int_{\mathbb{R}} x\, dF(x) \,|\, P \leq U \right\},$$

where $F(x) = P(-\infty, x]$ is the distribution function (under P) of the random variable φ.

Clearly $M(u)$ is a convex set, thus an interval in $\mathbb{R}$; its endpoints are (see Wasserman, 1990):

$$\inf_P E_P\varphi = \int_{[0,1]} \left(\inf_{x \in L_\alpha u} x \right) d\alpha,$$

$$\sup_P E_P\varphi = \int_{[0,1]} \left(\sup_{x \in L_\alpha u} x \right) d\alpha.$$

Clearly, these are, respectively, $\int_{[0,1]} \inf L_\alpha u\, d\alpha$, $\int_{[0,1]} \sup L_\alpha u\, d\alpha$. From Proposition 2 it follows that $M(u) = A(u)$. We now prove the linearity of the average level for fuzzy sets in $\mathbb{R}^n$ (in Dubois and Prade, 1987, this was proved for $\mathbb{R}$ only and the proofs there are much longer than ours). ■

Proposition 4. *Let $u, v : \mathbb{R}^n \to [0,1]$ be upper semicontinuous, with compact support, and $L_1 u \neq \emptyset$, $L_1 v \neq \emptyset$. Let $\lambda \in \mathbb{R}$, and $u+v$, λu be the operations of addition and scalar multiplication of fuzzy sets. Then,*

(*i*) $A(u+v) = Au + Av$,
(*ii*) $A(\lambda u) = \lambda\, Au$.

Proof. We show (*i*) only; (*ii*) is proved similarly.
Thus:

$$A(u+v) = \int_{[0,1]} L_\alpha(u+v)\, d\alpha = \int_{[0,1]} (L_\alpha u + L_\alpha v)\, d\alpha$$

$$= \int_{[0,1]} L_\alpha u\, d\alpha + \int_{[0,1]} L_\alpha v\, d\alpha = Au + Av. \qquad \blacksquare$$

We have used the linearity of the set-valued integral, as well as properties of fuzzy sets. A simple approximation to Au in $\mathbb{R}$ is given in

Proposition 5. *If $u : \mathbb{R} \to [0,1]$, then*

$$Au \approx .5(L_0 u + L_1 u).$$

Proof. From Figure 1 we see that

$$\int_{[0,1]} \inf L_\alpha u\, d\alpha = \int_{[0,1]} \varphi(a) d\alpha \approx \frac{\inf L_0 u + \inf L_1 u}{2},$$

$$\int_{[0,1]} \sup L_\alpha u\, d\alpha = \int_{[0,1]} \phi(a) d\alpha \approx \frac{\sup L_0 u + \sup L_1 u}{2},$$

from which the result follows. $\blacksquare$

The next property shows that for trapezoidal or triangular fuzzy numbers (in $\mathbb{R}$), the average level set is $L_{.5} u$:

Proposition 6. *Let u be a trapezoidal or a triangular fuzzy number. Then,*

$$Au = L_{.5} u = \{x \in \mathbb{R} \,|\, u(x) \geq .5\}.$$

Proof. Consider u triangular (the proof in the other case is similar); let $\operatorname{supp} u = [a,b]$, and $u(c) = 1$, $a \leq c \leq b$. It is very easy to see that

$$Au = \left[\frac{a+c}{2}, \frac{b+c}{2}\right] = L_{.5} u.$$

Is it possible in other cases that Au is not equal to the .5-level? The following example provides an affirmative answer:

Example. Let $u(x) = 4(x - x^2)$ for $0 \le x \le 1$, $u(x) = 0$ elsewhere. Then, simple calculations show that

$$Au = \left[\frac{1}{6}, \frac{5}{6}\right] = L_{5/9}u.$$

Actually, in many cases Au is not a level set at all, i.e., $Au \neq L_\alpha u$, for any $\alpha \in [0,1]$:

Example. Let $u(x) = 2x$ for $0 \le x \le .5$, $u(x) = 4(x - x^2)$ for $.5 \le x \le 1$, $u(x) = 0$ elsewhere.

Then,

$$Au = \left[\frac{1}{4}, \frac{5}{6}\right],$$

and $u(1/4) \neq u(5/6)$.

However, in symmetric cases, Au is always equal to one of the levels:

Proposition 7. *Let $u : \mathbb{R} \to [0,1]$ have support $[a-\delta, a+\delta]$ and be symmetric with respect to a (i.e., $u(a - x) = u(a + x)$ for every x). Then, $Au = L_{\alpha_0}u$, where $\alpha_0 = u(a + .5\int_{-\delta}^{\delta} u(a + x)\,dx)$.*

Proof. Note that $L_\alpha u = [\varphi(\alpha), \phi(\alpha)]$ and that $L_\alpha u$ is symmetric with respect to a. Thus, $\varphi(\alpha) + \phi(\alpha) = 2a$,which gives

$$\int_{[0,1]} \varphi(\alpha)\,d\alpha + \int_{[0,1]} \phi(\alpha)\,d\alpha = 2a.$$

Let

$$Au = [k, l] = \left[\int_{[0,1]} \varphi(\alpha)\,d\alpha, \int_{[0,1]} \phi(\alpha)\,d\alpha\right].$$

We also have

$$\int_{[0,1]} \phi(\alpha)\,d\alpha - \int_{[0,1]} \varphi(\alpha)\,d\alpha = \int_{a-\delta}^{a+\delta} u(x)\,dx = \int_{-\delta}^{\delta} u(a + x)\,dx.$$

Solving these simple equations, we get:

$$k = \int_{[0,1]} \varphi(\alpha)\,d\alpha = a - .5\int_{-\delta}^{\delta} u(a + x)\,dx,$$

$$l = \int_{[0,1]} \phi(\alpha)\,d\alpha = a + .5\int_{-\delta}^{\delta} u(a + x)\,dx.$$

Thus, Au is a symmetric interval, thus a level $L_{\alpha_0}u$, with $\alpha_0 = u(a + .5\int_{-\delta}^{\delta} u(a + x)\,dx)$. ■

Acknowledgements

The research in this paper has been supported in part by NSF MRI Grant 9871345 and by a JSPS Fellowship.

References

1. Aumann, R.J. (1965). Integrals of set-valued functions, *J. Math. Anal. Appl.* **12**, 1-12.
2. Debreu, A. (1967). Integration of correspondences, *Proc. Fifth Berkeley Symp. Math. Stat. Prob.*, 351-372.
3. Dubois, D. and Prade, H. (1987). The mean value of a fuzzy number, *Fuzzy Sets and Systems* **24**, 279-300.
4. Heilpern, S. (1992). The expected value of a fuzzy number, *Fuzzy Sets and Systems* **47**, 81-86.
5. Klement, E.P., Puri, M.L. and Ralescu, D.A (1986). Limit theorems for fuzzy random variables, *Proc. R. Soc. Lond. A* **407**, 171-182.
6. Negoita, C.V. and Ralescu, D.A. (1974), *Fuzzy Sets and Their Applications*, Wiley, New York.
7. Wasserman, L.A. (1990), Prior envelopes based on belief functions, *Annals of Statistics* **18**, 454-464.

Second order possibility measure induced by a fuzzy random variable

Inés Couso, Susana Montes, and Pedro Gil

University of Oviedo
C/ Calvo Sotelo, s/n 33007 Oviedo, Spain

Abstract. Random sets and fuzzy random variables are commonly used to model situations where two different types of uncertainty (imprecision/vagueness and randomness) appear simultaneously. In this context, the meaning of random sets is clear. The same does not happen for the case of fuzzy random variables. The meaning depends on the particular interpretation of fuzzy sets chosen.

In this paper, we consider the possibilistic interpretation introduced by Zadeh (1978) and show some situations where the imprecise information about some characteristic of the individuals of the population may be represented by a fuzzy random variable. We deal with the concept of induced probability measure in this more general context. We examine different ways to extend this definition to the case of random sets, and show the advantages and disadvantages of each one. As a generalization of these studies, we propose a new way to describe the available information about the "original" probability measure when fuzzy random variables are used. The model introduced is closely related to second order possibility measures, recently studied by several authors.

1 Introduction

We will consider situations in which two different types of uncertainty (vagueness and randomness) appear simultaneously. Let consider, for instance, that we choose at random some piece of fruit from a box and then we weigh it on a scales. The scales registers some value, imagine 315 grams. But this is not necessarily the real weight. The error margin of the measurements of the scales is usually shown by the manufacturer. If, for example, we can not guarantee an error lower than the 5%, the true weight could be any quantity between 300 and 331 grams. Furthermore, it is possible that, for a new measurement of the same piece of fruit, we obtain a different value, say 305 grams. So, it is clear that we can not model this experiment by a random variable. In that case, which would be the image of the chosen piece, 315 or 305 grams?.

We will describe a random experiment by a probability space, $(\Omega, \mathcal{A}, P)$, where Ω is the set of all possibility outcomes of the experiment, $\mathcal{A}$ is a σ-algebra of subsets of Ω and the set function P, defined on $\mathcal{A}$, is a probability measure. We will represent by a random variable, $U_0 : \Omega \to \Omega'$, certain characteristic of the elements of the referential set, Ω, where Ω' is the set of all possible values of the attribute. When our measurement is not totally precise, we don't know the exact value, $U_0(\omega)$, of the characteristic for the individual ω. In the case of the example above mentioned, we only know that the true weight of the piece ω, $U_0(\omega)$, belongs to some subset of Ω' that we denote $\Gamma(\omega)$. Hence, we can define a multi-valued mapping, $\Gamma : \Omega \longrightarrow \mathcal{P}(\Omega')$, that represents the imprecise perception of the random variable $U_0 : \Omega \longrightarrow \Omega'$. Following the notation established by Kruse and Meyer (1987), and Meyer and Kruse (1990) we will call U_0 the *original random variable.*

Now consider another example. Suppose that we have different information about the precision of the scales. Let imagine that 90% times, measurements are in a 5% error margin, but, in general, we cannot guarantee an error lower than 10%. If w is the number that we observe in the scales, we can state that, with probability 0.9, the true weight, w_0, is included in the interval from $0.95\,w$ to $1.05\,w$. We also know that w_0 is between $0.91\,w$ and $1.11\,w$. We can consider the experiment that consists on selecting at random a number of the unit interval, $[0,1]$. If the selected value is less than or equal to 0.9 then the measure is precise up to a 5%, but, when it is greater than 0.9, then we only get a 10% precision. We can represent this information by the random set $\Gamma : [0,1] \to \mathcal{P}(\mathbb{R})$ given by

$$\Gamma(\alpha) = \begin{cases} [0.95\,w, 1.05\,w] & \text{if } \alpha \leq 9/10 \\ \\ [0.91\,w, 1.11\,w] & \text{if } \alpha > 9/10 \end{cases}$$

In this example, we see that $P_*([0.95\,w, 1.05\,w]) = 0.9$ and $P_*([0.91\,w, 1.11\,w])$ $= 1$. This means that we are sure that w_0 belongs to the interval $[0.91\,w, 1.11\,w]$ and that the probability that it is in the interval $[0.95\,w, 1.05\,w]$ is, at least, 0.9. Obviously, this probability value can be greater than 0.9, but we can not assure it.

In general, we can obtain, from an imprecise observation of $U_0(\omega)$, a family of nested sets, $A_i(\omega) \supseteq A_{i+1}(\omega)$, $i = 1, \ldots, n-1$, with fixed lower probabilities, $P_{\omega *}(A_i) = 1 - \alpha_i$, such that $\alpha_i \leq \alpha_{i+1}$, $i = 1, \ldots, n-1$. The set function $P_{\omega *}$ defined on the family of subsets of Ω', $\{A_1(\omega), \ldots, A_n(\omega)\}$ may be extended by *natural extension* (Walley, 1991) to the power set, $\mathcal{P}(\Omega')$, as a *necessity measure* (Dubois and Prade, 1992), (Shafer, 1976), as we prove in Couso *et al.* (2000). So, its *dual*, De Campos and Bolaños (1989), Delgado and Moral (1989), P^*_ω, defined as $P^*_\omega(A) = 1 - P_{\omega *}(A^c)$, $\forall A \subseteq \Omega'$ is a *possibility measure* (De Cooman and Aeyels, 1999), (Dubois and Prade, 1992), (Shafer, 1976). We may associate this possibility measure to a fuzzy set, $\widetilde{X}(\omega)$, by the rule: $\widetilde{X}(\omega)(\omega') := P^*_\omega(\{\omega'\})$, $\forall \omega' \in \Omega'$. The fuzzy set $\widetilde{X}(\omega)$ contains the same information as the set function P^*_ω, since the last can be obtained from the former by the formulas $P^*_\omega(A) = \sup_{\omega' \in A} \widetilde{X}(\omega)(\omega')$, $\forall A \subseteq \Omega'$. In this case, $\widetilde{X}(\omega)(\omega')$ represents the degree of possibility that the true value $U_0(\omega)$ coincides with ω'. This interpretation agrees with the possibilistic semantic of fuzzy sets introduced by Zadeh (1978). This last is not the unique interpretation of fuzzy sets. For a detailed discussion, see, for instance, Dubois and Prade (1997).

Thus, we have introduced two different models to represent the imprecise observation of the values of a random variable $U_0 : \Omega \to \Omega'$, depending on the kind of this imprecise information: multi-valued and fuzzy set mappings. The last of the two models generalizes the former one.

The aim of this paper is to extend the concept of probability induced by a random variable to these more general cases of multi-valued and fuzzy set mappings. A first approach to the solution of this problem for the case of random sets (measurable multi-valued mappings) could start on defining a σ-algebra on the set $\mathcal{P}(\Omega')$ or on a proper subset of it, $\mathcal{C} \subseteq \mathcal{P}(\Omega')$, that we will denote by $\mathcal{A}_\mathcal{C}$. Then we could consider the probability measure induced by the $\mathcal{A}$-$\mathcal{A}_\mathcal{C}$ measurable mapping $\Gamma : \Omega \to \mathcal{C}$, $P \circ \Gamma^{-1} : \mathcal{A}_\mathcal{C} \to [0,1]$. In random set literature, we can observe that several authors identify a random set with its induced probability measure (see, for instance, Matheron, 1975). However, as we will show later, this generalization of the concept of induced probability measure may be unsuitable when the random set represents the imprecise observation of a random variable. We will show that the set function $P \circ \Gamma^{-1}$ does not contain all the information determined by Γ about the probability measure induced by the original random variable, $P_{U_0} = P \circ U_0^{-1} : \sigma(\Omega') \to [0,1]$.

2 Random sets

The theory of random sets has been extensively developed by several authors such that Aumann (1965), Debreu (1965), Kendall (1974) and Matheron (1975). Even though the concept is essentially the same in all the cases, different authors develop their ideas in different contexts. For example, Kendall (1974) and Matheron (1975) relate random set theory with Stochastic Geometry and the Theory of Capacities of Choquet (Aumann, 1965). In our work, we will make use of the concept of random set to represent the imprecise observation of a random variable, $U_0 : \Omega \to \Omega'$, according to other works by Aumann (1965) and Kruse and Meyer (1987). In any case, a *random set* is a mapping defined on Ω with values on $\mathcal{P}(\Omega')$, $\Gamma : \Omega \to \mathcal{P}(\Omega')$, which is measurable for some σ-algebra defined on some subset of $\mathcal{P}(\Omega')$. In this paper, we consider the σ-algebra generated by the class $\boldsymbol{C}(\sigma(\Omega')) = \{\mathcal{C}_B : B \in \sigma(\Omega')\}$, where $\sigma(\Omega')$ is some σ-algebra defined on Ω' and $\mathcal{C}_B = \{C \subseteq \Omega' : C \cap B \neq \emptyset\}$, for all $B \in \sigma(\Omega')$. For any measurable subset of Ω', $B \in \sigma(\Omega')$, the set $\Gamma^{-1}(\mathcal{C}_B)$ will be called the *upper inverse* (Nguyen, 1977) of B.

What we want here is to find a suitable way to summarize the information that the random set Γ contains about the probability measure $P \circ U_0^{-1}$. To this purpose, Dempster (1967) defines the *upper* and *lower probabilities* of any measurable set B by the formulas:

$$\mathrm{Pl}(B) = P\{\omega \in \Omega : \Gamma(\omega) \cap B \neq \emptyset\} / P\{\omega \in \Omega : \Gamma(\omega) \neq \emptyset\}$$

$$= P\left[\Gamma^{-1}(\mathcal{C}_B)\,|\,\Gamma^{-1}(\mathcal{C}_{\Omega'})\right],\ \forall B \in \sigma(\Omega')$$

$$\mathrm{Bel}(B) = P\{\omega \in \Omega : \Gamma(\omega) \subseteq B,\, \Gamma(\omega) \neq \emptyset\}/P\{\omega \in \Omega : \Gamma(\omega) \neq \emptyset\}$$

$$= P\left[[\Gamma^{-1}(\mathcal{C}_{B^c})]^c\,|\,\Gamma^{-1}(\mathcal{C}_{\Omega'})\right],\ \forall B \in \sigma(\Omega').$$

Notice that they are well defined for the measurability condition above mentioned. Throughout this paper, we will not condition to the event $\Gamma^{-1}(\mathcal{C}_{\Omega'})$. So, we will define the *upper* (resp. the *lower probability*) of B as the probability of the upper (resp. the lower) inverse of B, i.e.:

$$P^*(B) := P[\Gamma^{-1}(\mathcal{C}_B)], \quad P_*(B) := P\left[[\Gamma^{-1}(\mathcal{C}_{B^c})]^c\right],\ \forall B \in \sigma(\Omega').$$

When $\Omega \backslash \Gamma^{-1}(\mathcal{C}_{\Omega'})$ is a null subset of Ω, these set functions coincide with those defined by Dempster. We can suppose that this last condition is true when Γ represents the imprecise observation of some U_0, since, in that case, we have that $\Gamma(\omega) \ni U_0(\omega), \forall \omega \in \Omega$, so that $\Omega \setminus \Gamma^{-1}(\mathcal{C}_{\Omega'}) = \emptyset$. It is easy to prove that the true probability value of some event $B \in \sigma(\Omega')$, $P_{U_0}(B)$, belongs to the interval of values $[P_*(B), P^*(B)]$. But, this interval may be not sufficiently precise to represent our knowledge about $P_{U_0}(B)$. All we know about it is that it belongs to the set:

$$P_\Gamma(B) = \{P_U(B) : U : \Omega \to \Omega' \text{ measurable, } U(\omega) \in \Gamma(\omega),\ \forall \omega \in \Omega\}.$$

The values $P_*(B)$ and $P^*(B)$ are bounds of this last set of possible values but, as far as we know, it has not been proved that they coincide with its supremum and infimum. The fact is already known for the particular case of finite referential sets (see, for example, Grabisch *et al.*, 1995), but not for the case of general referential sets. Although L. Wasserman affirmed in 1987 that, for a measurable multi-valued mapping, $\Gamma : \Omega \to \mathcal{P}(\mathbb{R})$, and a Borel set, $A \in \mathrm{Borel}(\mathbb{R})$, the values $P^*(A)$ and $P_*(A)$ respectively coincide with the supremum and the infimum of $P_\Gamma(A)$, in that proof, cited in Grabisch *et al.* (1995), it was supposed that, for each $\epsilon > 0$ and each measurable set $A \in \mathrm{Borel}(\mathbb{R})$ there exists some measurable selection, φ_ϵ, so that $\inf I_A(\Gamma(\omega)) \leq (I_A \circ \varphi_\epsilon)(\omega) \leq \inf I_A(\Gamma(\omega)) + \epsilon$, $\forall \omega \in \Omega$. Since set characteristic functions can only take the values 0 and 1, this assumption makes φ_ϵ satisfy the condition $I_A \circ \varphi_\epsilon \equiv I_{A_*}$. According to this, the equality $P_*(A) = P_{\varphi_\epsilon}(A)$ would be fulfilled, and therefore $P_*(A) \in P_\Gamma(A)$, and this last condition is stronger than the one we wanted to prove and, furthermore, it is not necessarily true.

On the other hand, the set of possible values, $P_\Gamma(B)$, is not convex in general. The fact that all what we need in most cases is to find upper and lower bounds for $P_{U_0}(B)$ could leads us to replace the original set $P_\Gamma(B)$ by its convex hull $\mathrm{co}\,[P_\Gamma(B)]$. Several authors, as Kyburg and Pitarelli (1996) and Couso *et al.* (1999) have verified that convex sets of probability measures are not sufficient to treat some kinds of information. We will later examine these two problems of reaching bounds and convexity in a more detail.

When the multi-valued mapping Γ represents our knowledge about the original random variable U_0, all what we know about the probability measure $P_{U_0} = P \circ U_0^{-1}$ is that it belongs to the set:

$$\mathcal{P}(\Gamma) := \{P_U : U : \Omega \to \Omega' \text{ measurable, } U(\omega) \in \Gamma(\omega),\ \forall \omega \in \Omega\}.$$

Now we will show the relationships among the three sets of probability measures $\mathcal{P}(\Gamma)$, $\Delta(\Gamma) := \{Q \in \mathcal{P} : Q(A) \in P_\Gamma(A),\ \forall A \in \sigma(\Omega')\}$ and $\mathcal{M}(P^*) = \{Q \text{ prob. measure} : Q(A) \leq P^*(A)\ \forall A \subseteq \Omega'\}$. It is easy to check that the inclusions $\mathcal{P}(\Gamma) \subseteq \Delta(\Gamma) \subseteq \mathcal{M}(P^*)$ are always satisfied. On the other hand, the three sets only coincide in particular cases, as we will show below. The first of them $\mathcal{P}(\Gamma)$, is the most precise set we can manage to represent the available information about $P \circ U_0^{-1}$, but its calculation may be large and difficult in many cases. On the other hand, the calculation of the set $\mathcal{M}(P^*)$ is simpler. The information given by this last set is determined by the probability measure $P \circ \Gamma^{-1}$, since the set function $P^* : \sigma(\Omega') \to [0,1]$ may be considered as the "restriction" of $P \circ \Gamma^{-1}$ to the set $\{\mathcal{C}_B : B \in \sigma(\Omega')\}$. The converse is also true, since the class $\{(\mathcal{C}_B)^c : B \in \sigma(\Omega')\}$ is closed for finite intersection. The problem of replacing $\mathcal{P}(\Gamma)$ by the wider set $\mathcal{M}(P^*)$ is that we can lose important information with this change, as we will show below. The information contained in the set $\mathcal{M}(P^*)$, that is equivalent to the information given by $P \circ \Gamma^{-1}$, is not sufficient in the general case.

Example 1. Let consider the initial space $\Omega = \{\omega_1, \omega_2, \omega_3\}$ with the σ–algebra $\mathcal{P}(\Omega)$ and the discrete uniform probability measure. Consider the final set, $\Omega' = \{0,1\}$ and the multi-valued mapping $\Gamma : \Omega \longrightarrow \mathcal{P}(\Omega')$ given by $\Gamma(\omega_1) = \{0\}$, $\Gamma(\omega_2) = \{1\}$, $\Gamma(\omega_3) = \{0,1\}$. Suppose that Γ represents the imprecise information about the random variable U_0. The set of possible probability measures for $U_0 : \Omega \to \Omega'$ is:

$$\mathcal{P}(\Gamma) = \{(1/3, 2/3), (2/3, 1/3)\}.$$

Let now consider the initial probability space $([0,1], \beta_{[0,1]}, \lambda_{[0,1]})$ (Lebesgue measure space) and the multi-valued mapping $\Gamma' : [0,1] \to \mathcal{P}(\Omega')$ such that $\Gamma'(\omega) = \{0\}$, if $\omega \leq 1/3$ and $\Gamma'(\omega) = \{1\}$, if $\omega \in (1/3, 2/3]$. Imagine that Γ' represents the imprecise observation of another random variable $U_0' : [0,1] \to \Omega'$. The set of all possible probability distributions for U_0' is the following:

$$\mathcal{P}(\Gamma') = \{(p, 1-p) : 1/3 \leq p \leq 2/3\}$$

As we can see, it coincides with the convex hull of $\mathcal{P}(\Gamma)$. We can also verify that their Dempster upper probabilities coincide. So, we get the equality:

$$\mathcal{M}(P_\Gamma^*) = \mathcal{M}(P_{\Gamma'}^*).$$

If we now consider, for each case, the set of all possible values of Shannon's entropy, we obtain different results:

$$\{H(U) : U \in C(\Gamma)\} = \{H(1/3, 2/3)\} = \{\log_2 3 - 2/3\}$$

$$\{H(U) : U \in C(\Gamma')\} = [H(1/3, 2/3), H(1/2, 1/2)] = [\log_2 3 - 2/3, 1]$$

So, in the first case, we know exactly the value of the entropy of the probability measure $P \circ U_0^{-1}$. However, in the second case, we only know that $H(\lambda \circ U_0^{-1})$ belongs to the interval $[\log_2 3 - 2/3, 1]$.

With this last example, we have shown that the upper probability of a random set or, equivalently, its induced probability measure is not sufficient to determine the available information about the entropy value of the original random variable. Now we show another example.

Example 2. Consider the initial space ($\Omega = \{\omega_1, \omega_2, \omega_3\}$, with the σ–algebra $\mathcal{P}(\Omega)$ and the probability measure P, where $P(\{o_1\}) = 0.2$, $P(\{o_2\}) = 0.2$ and $P(\{o_3\}) = 0.6$. Suppose that the random set $\Gamma : \{\omega_1, \omega_2, \omega_3\} \to \mathcal{P}(\{0,1\})$, defined as $\Gamma(\omega_1) = \{0\}$, $\Gamma(\omega_2) = \{1\}$ and $\Gamma(\omega_3) = \{0,1\}$, represents the imprecise observation of some U_0. Now consider a simple random sample, (U_0^1, U_0^2), (U_0^1 and U_0^2 are independent and identically distributed as U_0). The set of possible joint probability measures for (U_0^1, U_0^2) under the available information is:

$$\mathcal{P} = \{(0'04, 0'16, 0'16, 0'64), (0'64, 0'16, 0'16, 0'04)\}.$$

Now suppose that we replace our information about each marginal probability measure by information conveyed by the upper probability of the random set, P^*. The set of all products of pairs of identical probability measures, each one of them belong to $\mathcal{M}(P^*)$ is given by

$$\mathcal{P}' = \{(p^2, p(1-p), p(1-p), (1-p)^2) : p \in [0'2, 0'8]\}.$$

We can observe that the probability measure $(0'25, 0'25, 0'25, 0'25)$ belongs to this last set but, on the other hand, it is not contained in the convex hull of the set of probabilities obtained in the first case, $\mathrm{co}(\mathcal{P}) = \{(p^2, 0'16, 0'16, (1-p)^2) : p \in [0'2, 0'8]\}$.

Thus, if we make use of all the initial information, we know that the pair of values $(0,1)$ has probability 0.16. However, in the second case, we only know that this value belongs to the interval $[0.16, 0.25]$.

These last problems appear when the set of probabilities $\mathcal{P}(\Gamma)$ is not convex. We don't have available results about sufficient conditions to guarantee the convexity of $\mathcal{P}(\Gamma)$, but for its images.

Proposition 1. *(Couso, 1997, 1999) Let $(\Omega, \mathcal{A}, P)$ be non-atomic and let consider another measurable space $(\Omega', \mathcal{A}')$. Let $\Gamma : \Omega \longrightarrow \mathcal{P}(\Omega')$ be a multi-valued mapping. Then, for all $A \in \sigma(\Omega')$, the set $P_\Gamma(A)$ is convex.*

This proposition recalls the result given by J. Aumann (1965) about the convexity of random sets expectation for non-atomic initial spaces.

As we have explained before, we can not find in the literature a result that guarantees that Dempster bounds are attained in the general case. Next we offer some particular results which proofs can be found in Couso (1997, 1999) and Couso *et al.* (1998).

Proposition 2. *Consider the probability space* $(\Omega, \mathcal{A}, P)$ *and a Polish space* (E, τ). *Let* $\Gamma : \Omega \rightarrow \mathcal{P}(E)$ *be* $\mathcal{A}$-$\sigma\langle\{\mathcal{C}_B : B \in \text{Borel}(E)\}\rangle$ *measurable* (Borel(E) *denotes the Borel* σ*-algebra on* (E, τ)), *closed* ($\Gamma(\omega)$ *closed,* $\forall \omega \in \Omega$) *and non-empty* ($\Gamma(\omega) \neq \emptyset$, $\forall \omega \in E$). *Then:*

(a) $P^*(A) = \max P_\Gamma(A)$, $\forall A \in \mathcal{F}(E) = \{F \subseteq E : F \text{ is closed}\}$.
(b) $P_*(A) = \min P_\Gamma(A)$, *if* A *may be expressed as* $A = \cap_{n=1}^{\infty} A_n$, $\{A_n\}_{n \in \mathbb{N}} \subseteq \mathcal{G}(E) = \{G \subseteq E : G \text{ is open}\}$.

Any closed set of a Polish space may be written as the countable union of open sets, so, as a corollary, we obtain the following result.

Corollary 1. *Consider the probability space* $(\Omega, \mathcal{A}, P)$*and a Polish space* (E, τ). *Let* $\Gamma : \Omega \rightarrow \mathcal{P}(E)$ *be* $\mathcal{A}$-$\sigma\langle\{\mathcal{C}_B : B \in \text{Borel}(E)\}\rangle$ *measurable, closed* ($\Gamma(\omega) \in \mathcal{F}(E)$, $\forall \omega \in \Omega$) *and non-empty* ($\Gamma(\omega) \neq \emptyset$, $\forall \omega \in E$). *Then:*

$$P^*(A) = \max P_\Gamma(A) \text{ and } P_*(A) = \min P_\Gamma(A), \ \forall A \in \mathcal{F}(E) \cup \mathcal{G}(E).$$

We have also proved (Couso, 1999) the following result for measurable sets in a Polish space.

Proposition 3. *Consider the probability space* $(\Omega, \mathcal{A}, P)$*and a Polish space* (E, τ). *Let* $\Gamma : \Omega \rightarrow \mathcal{P}(E)$ *be* $\mathcal{A}$-$\sigma\langle\{\mathcal{C}_B : B \in \text{Borel}(E)\}\rangle$ *measurable, compact* ($\Gamma(\omega)$ compact, $\forall \omega \in \Omega$) *and non-empty* ($\Gamma(\omega) \neq \emptyset$, $\forall \omega \in E$). *Then:*

$$P^*(A) = \sup P_\Gamma(A), \text{ and } P_*(A) = \inf P_\Gamma(A), \ \forall A \in \text{Borel}(E),$$

It is easy to prove the following result.

Proposition 4. *Let consider a probability space* $(\Omega, \mathcal{A}, P)$, *and a measurable space* $(\Omega', \sigma(\Omega'))$. *Let* $\Gamma : \Omega \longrightarrow \mathcal{P}(\Omega')$ *be* $\mathcal{A}$-$\sigma\langle\{\mathcal{C}_B : B \in \sigma(\Omega')\rangle$ *measurable. If* Γ *is simple (it has a finite quantity of different images), then we have the equalities*

$$P^*(A) = \max P_\Gamma(A) \text{ and } P_*(A) = \min P_\Gamma(A), \ \forall A \in \sigma(\Omega').$$

Proposition 5. *Let consider a probability space* $(\Omega, \mathcal{A}, P)$, *and a normed space* $(E, \|\ \|)$. *Let also consider an* $\mathcal{A}$-$\sigma\langle\{\mathcal{C}_B : B \in \sigma(\Omega')\rangle$ *measurable multi-valued mapping,* $\Gamma : \Omega \longrightarrow \mathcal{P}(\Omega')$. *Suppose that* $\Gamma(\omega)$ *is open and bounded,* $\forall \omega \in \Omega$. *Then:*

$$P^*(A) = \max P_\Gamma(A) \text{ and } P_*(A) = \min P_\Gamma(A), \ \forall A \in \text{Borel}(E).$$

When the initial space is non-atomic and Γ fulfills the conditions of some of the above results, the information given by P^* coincides with the information described by $\Delta(\Gamma)$. On the other hand, in examples 2 and 2 given before, the sets of probability measures $\Delta(\Gamma)$ and $\mathcal{P}(\Gamma)$ are the same and they represent all available information about P_{U_0}. But they do not coincide in general, as we show below.

Example 3. Consider the probability space $(\Omega, \mathcal{A}, P)$, where $\Omega = \{\omega_1, \omega_2\}$, $\mathcal{A} = \mathcal{P}(\Omega)$ and $P(\{\omega_1\}) = 1/3$, $P(\{\omega_2\}) = 2/3$. Let also consider the random set $\Gamma : \Omega \to \mathcal{P}(\{1,2,3\})$ defined as follows:

$$\Gamma(\omega) = \begin{cases} \{1,2,3\} & \text{if } \omega = \omega_1 \\ \{1,2\} & \text{if } \omega = \omega_2 \end{cases}$$

The sets of possible probability values for each event are given by

$$\begin{aligned} P_\Gamma(\emptyset) &= 0 \\ P_\Gamma(\{1\}) &= \{0, 1/3, 2/3, 1\} = P_\Gamma(\{2\}) \\ P_\Gamma(\{3\}) &= \{0, 1/3\} \\ P_\Gamma(\{1,2\}) &= \{2/3, 1\} \\ P_\Gamma(\{1,3\}) &= \{0, 1/3, 2/3, 1\} = P_\Gamma(\{2,3\}) \\ P_\Gamma(\{1,2,3\}) &= 1 \end{aligned}$$

On the other hand, the set of possible probability measures, under the available information is the following:
$\mathcal{P}(\Gamma) = \{(1,0,0), (1/3,2/3,0), (2/3,1/3,0), (0,1,0), (2/3,0,1/3), (0,2/3,1/3)\}$.

We see that the probability measure $Q_0 \equiv (1/3, 1/3, 1/3)$, does not belong to this last set. However, its images are contained in the images of the mapping P_Γ. So, in this example, the set of probability measures that are compatible with P_Γ, $\Delta(\Gamma) = \{Q \in \mathcal{P} : Q(A) \in P_\Gamma(A),\ \forall A \in \mathcal{P}(\{1,2,3\})\}$, does not coincide with the set of probability measures that are compatible with Γ, $\mathcal{P}(\Gamma)$.

The mapping $P_\Gamma : \sigma(\Omega') \to \mathcal{P}([0,1])$ contains all the available information about the probability value of each event. However, this is not all the available information about the original probability measure, P_{U_0}.

Most authors identify a random set with its induced probability or, equivalently, with its Dempster upper probability. But these mappings do not contain in general all the information given by the random set about the original probability distribution, as we have shown in this section with several examples. The set of probability measures $\mathcal{P}(\Gamma)$ represents all the available information. The wider set $\Delta(\Gamma)$ contains all the information about the probability of each event. On the other hand, the convex set $\mathcal{M}(P^*)$ contains all the probability measures dominated by P^*. In each particular problem, we should analyze whether the information conveyed by $\Delta(\Gamma)$ or else by $\mathcal{M}(P^*)$ seems to be sufficient or not.

3 Fuzzy random variables

With the example of the scales given in the introduction we have shown that a different kind of imprecise information about U_0 may be described by a fuzzy random variable, $\widetilde{X} : \Omega \to \mathcal{P}(\Omega')$, instead of a random set $\Gamma : \Omega \to \mathcal{P}(\Omega')$.

The term fuzzy random variable has been used by Zadeh (1975), Nguyen (1977), Nahmias (1978, 1979), Hirota (1981) and Stein and Talati (1981) in different contexts as here. Our definition is related to those ones give by Féron (1976), Kwakernaak (1989), Puri and Ralescu (1986) and Stojaković (1992). In all these cases, a *fuzzy random variable* is defined as a mapping which values are fuzzy subsets of the final space that satisfies some measurability condition. Different definitions defer on the kind of the final space and the particular measurability condition considered. We will impose the strong measurability introduced by Nguyen (1978) and follow the scheme proposed by by Kruse and Meyer (1987). According to them, the fuzzy random variable $\widetilde{X}$ represents the imprecise observation of the original random variable. This way, we will interpret the quantity $\widetilde{X}(\omega)(\omega')$ as the degree of acceptability that the unknown random variable takes the value ω' for the individual ω. A similar reasoning leads as to define a fuzzy set on the class of all random variables from Ω to Ω', $\mu_{\widetilde{X}}$, that assigns, to each measurable mapping, $U : \Omega \to \Omega'$ the value:

$$\mu_{\widetilde{X}}(U) := \inf\{[\widetilde{X}(\omega)](U(\omega)) : \omega \in \Omega\}.$$

Following the interpretation given by Kruse and Meyer, this value may be interpreted as the degree of certainty of the proposition "$U = U_0$", since it is true when, for each element of the initial space, $\omega \in \Omega$, the proposition "$U_0(\omega) = U(\omega)$" is true. This way, the degree of acceptability of the initial proposition will be given by

$$\text{acc}(U = U_0) = \text{acc}(\{\forall\, \omega \in \Omega \text{ "the value of the original r.v. on}\, \omega \text{ is } U(\omega)\text{"}\})$$

$$= \inf\{\text{acc}(U_0(\omega) = U(\omega)) : \omega \in \Omega\} = \inf\{\widetilde{X}(\omega)(U(\omega)) : \omega \in \Omega\}.$$

For two measurable spaces $(\Omega, \mathcal{A})$ and $(\Omega', \mathcal{A}')$, we define a fuzzy random variable as a mapping $X : \Omega \to \mathcal{P}(\Omega')$ such that, for each $\alpha \in [0,1]$, the multi-valued mapping $\widetilde{X}_\alpha$ defined as:

$$\widetilde{X}_\alpha(\omega) := [\widetilde{X}(\omega)]^{[\alpha]} = \{\omega' \in \Omega' : \widetilde{X}(\omega) \geq \alpha\},\ \forall \omega \in \Omega$$

is strong measurable. Notice that we don't need to impose any additional condition to the final set.

We can obtain this type of information from an imprecise measurement of the attribute studied as we describe in the introduction of this paper. That way we obtain a family of nested sets, $\{A_1(\omega), \ldots, A_n(\omega)\}$, each one of them is associated to a lower probability value $P_*(A_i(\omega)) = 1 - \alpha_i$. We can observe that, in this case, for each $\alpha \in (0,1)$, the set $G_\omega(\alpha) := \bigcap_{\alpha_i \leq \alpha} A_i(\omega)$ represents the most precise set that contains the unknown value $U_0(\omega)$ with lower probability greater than or equal to $1-\alpha$. As we have proved in Couso *et al.* (2000), the family of nested sets $\{G_\omega(\alpha)\}_{\alpha \in [0,1)}$ coincides with the family of strong α-cuts of the associated fuzzy set, $\widetilde{X}(\omega)$.

Our objective in this section consist on finding a suitable way to describe the available information about the original probability measure, P_{U_0}. We could solve the problem of extending the concept of induced probability by considering the probability distribution $P \circ \widetilde{X}^{-1}$ when $\widetilde{X}^{-1}$ is measurable for some σ-algebra defined on a subset of $\widetilde{\mathcal{P}}(\Omega')$. This would not be a generalization of the concept of "induced probability measure", but only a particular case where the values of the random measurable mapping $\widetilde{X}$ are fuzzy subsets of a certain set Ω. We can observe that the probability measure $P \circ \widetilde{X}^{-1}$ does not contain in general all the available information about P_{U_0}, as we have shown in the example 2 for the particular case of random sets.

Since our imprecise knowledge about U_0 has been represented by a fuzzy set defined on the set of all random variables from $(\Omega, \mathcal{A})$ to $(\Omega', \sigma(\Omega'))$, it seems reasonable that our imprecise information about the induced probability measure, $P \circ U_0^{-1}$, is represented by a fuzzy set defined on the class of all probability measures on $(\Omega', \sigma(\Omega'))$. This fuzzy set will be defined by the membership function:

$$\mathcal{P}(\widetilde{X})(Q) := \sup\{\mu_{\widetilde{X}}(U) : P_U \equiv Q\}, \ \forall Q \in \mathcal{P}_{\sigma(\Omega')},$$

where $\mathcal{P}_{\sigma(\Omega')}$ represents the set of all probability measures that may be defined on $\sigma(\Omega')$. In the case that $\widetilde{X}$ has crisp images, this definition coincides with the definition of $\mathcal{P}(\Gamma)$ given in Section 2. We can interpret the membership value $\mathcal{P}(\widetilde{X})(Q)$ as the degree of acceptability of the proposition

"Q is the original probability measure,"

as we show in Couso (1999).

On the other hand, for an event $B \in \sigma(\Omega')$ of the final space, we can summarize the information we have about the probability value $P_{U_0}(B)$ with the fuzzy set $P_{\widetilde{X}}(B) \in \widetilde{\mathcal{P}}([0,1])$ given by

$$P_{\widetilde{X}}(B)(p) := \sup\{\mu_{\widetilde{X}}(U) : U : \Omega \to \Omega' \text{ v.a.}, \ P_U(B) = p\}.$$

We interpret the membership value $P_{\widetilde{X}}(B)(p)$ as the degree of acceptability of the proposition "p is the true value of the probability of B".

Example 4. Let $(\Omega, \mathcal{A}, P)$ be non-atomic. Let consider a random variable $U : \Omega \to I\!R$ with uniform distribution on the interval $(0,1)$. Let also consider the fuzzy mapping $\widetilde{X} : \Omega \to \widetilde{\mathcal{P}}(I\!R)$, where the membership function of the fuzzy set $\widetilde{X}(\omega)$ is given (see Figure 1) by

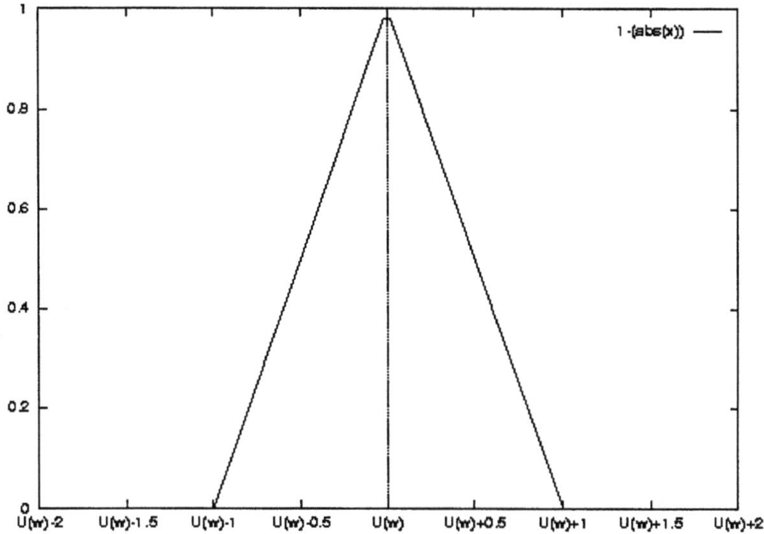

Fig. 1. Membership function of $\widetilde{X}(\omega)$

$$\widetilde{X}(\omega)(x) = \begin{cases} 1 - |U(\omega) - x| & \text{if } x \in [U(\omega) - 1, U(\omega) + 1] \\ 0 & \text{in other case} \end{cases}$$

The weak α–cuts of $\widetilde{X}(\omega)$ are given by the formulas:

$$\widetilde{X}_\alpha(\omega) = [U(\omega) - (1 - \alpha), U(\omega) + (1 - \alpha)], \ \forall \alpha \in [0, 1].$$

Let now calculate, for example, the fuzzy set $P_{\widetilde{X}}(-\infty, 1/4]$. Since the initial space is non-atomic, its α–cuts are convex. They are given by

$$[P_{\widetilde{X}}(-\infty, 1/4]]^{[\alpha]} = \begin{cases} [0, 1] & \text{if } \alpha \leq 1/4 \\ [0, 5/4 - \alpha] & \text{if } \alpha \in (1/4, 3/4) \\ [\alpha - 3/4, 5/4 - \alpha] & \text{if } \alpha \geq 3/4 \end{cases}$$

The graphical representation of the fuzzy set $P_{\widetilde{X}}(-\infty, 1/4]$ is that in Figure 2.

Now, we will give the expression of the fuzzy set $P_{\widetilde{X}}(-\infty, 1/2]$. Its α–cuts are given by

$$[P_{\widetilde{X}}(-\infty, 1/2]]^{[\alpha]} = \begin{cases} [0, 1] & \text{if } \alpha \leq 1/2 \\ [\alpha - 1/2, 3/2 - \alpha] & \text{if } \alpha > 1/2 \end{cases}$$

We can also represent it graphically (see Figure 3).

We observe that, in the last two cases, the membership value 1 is attained for $P(U \leq 1/4)$ and $P(U \leq 1/2)$, respectively. On the other hand, the supports of the fuzzy sets $P_{\widetilde{X}}(-\infty, 1/4)$ and $P_{\widetilde{X}}(-\infty, 1/2)$ coincide with the in-

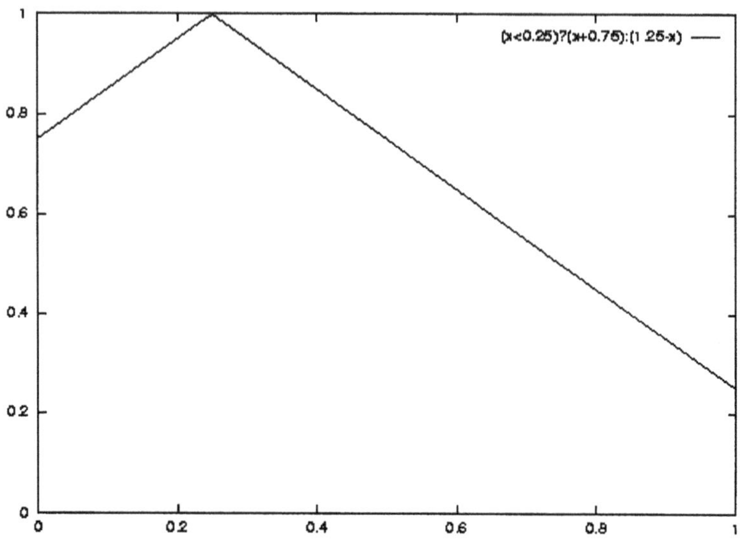

Fig. 2. Representation of $P_{\widetilde{X}}(-\infty, 1/4]$

tervals $[P(U+1 \leq 1/4), P(U-1 \leq 1/4)]$ and $[P[U+1 \leq 1/2], P[U-1 \leq 1/2]]$, respectively.

Proposition 6. *Let $(\Omega, \mathcal{A}, P)$ be a probability space and let $(\Omega', \mathcal{A}')$ be a measurable space. Let consider, for a fuzzy function $\widetilde{X} : \Omega \longrightarrow \widetilde{\mathcal{P}}(\Omega')$, the family of multivalued mappings $\{\widetilde{X}_\alpha\}_{\alpha \in [0,1]}$ where the images of $\widetilde{X}_\alpha : \Omega \to \mathcal{P}(\Omega')$ coincide with the α-cuts of the images of $\widetilde{X}$. Then, the following conditions hold:*

1. $[\mathcal{P}(\widetilde{X})]^{(\alpha)} \subseteq \mathcal{P}(\widetilde{X}_\alpha) \subseteq [\mathcal{P}(\widetilde{X})]^{[\alpha]}, \ \forall \alpha \in [0,1]$
2. $[P_{\widetilde{X}}(B)]^{(\alpha)} \subseteq P_{\widetilde{X}_\alpha}(B) \subseteq [P_{\widetilde{X}}(B)]^{[\alpha]}, \ \forall B \in \mathcal{A}, \ \alpha \in [0,1],$

where $[\mathcal{P}(\widetilde{X})]^{(\alpha)}$ and $[P_{\widetilde{X}}(B)]^{(\alpha)}$ respectively denote the strong α-cuts of $\mathcal{P}(\widetilde{X})$ and $P_{\widetilde{X}}(B)$, while $[\mathcal{P}(\widetilde{X})]^{[\alpha]}$ and $[P_{\widetilde{X}}(B)]^{[\alpha]}$ represent the weak α-cuts.

From this result, we can deduce that there exists a close relationship between this extension of the concept of induced probability and the concept of expectation of a fuzzy random variable introduced by Puri and Ralescu (1986).

The probabilistic envelope induced by a fuzzy random variable, $P_{\widetilde{X}} : \sigma(\Omega') \to \widetilde{\mathcal{P}}(\mathbb{R})$, is a particular case of *possibilistic probability* (De Cooman, 1998). For any class, $\mathcal{K}$, of real bounded functions defined on a set Ω', De Cooman (1998) defines a *possibilistic prevision* as a function, p, defined on $\mathcal{K}$ which values are fuzzy subsets of the unit interval $[0,1]$. In particular,

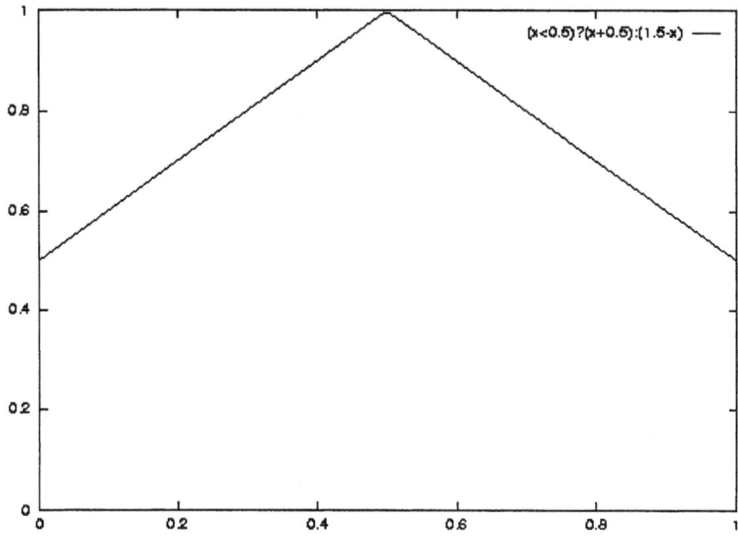

Fig. 3. Representation of $P_{\tilde{X}}(-\infty, 1/2]$

when all the elements of $\mathcal{K}$ are indicator functions of events of the final space, the possibilistic prevision is called a possibilistic probability. Thus, the probabilistic envelope induced by a fuzzy random variable is a possibilistic probability defined on the set of indicator functions of the elements of $\sigma(\Omega')$. According to the author (De Cooman, 1998), we could obtain a possibilistic prevision from a *second order possibility distribution.* This type of model consists on a possibility measure defined on the set of all linear previsions (Walley, 1991) on some set of bounded real mappings, $\mathcal{K}$. A *linear prevision,* $P : \mathcal{K} \to [0,1]$ is a linear operator that takes positive values for positive functions and has a fixed point at 1. In particular, when $\mathcal{K}$ coincides with the class of indicator functions of the events of a σ–algebra $\sigma(\Omega')$, the concept coincides with the definition of finitely additive measure. According to De Cooman (1998), we can construct a possibilistic prevision from some second order possibility, $\Pi : \mathcal{P}_{\mathcal{K}} \to [0,1]$. It will be defined as $p_\Pi : \mathcal{K} \to \mathcal{P}(\mathbb{R})$, where

$$p_\Pi(X)(x) = \sup\{\Pi(P) : P(X) = x\},\ \forall X \in \mathcal{K},\ \forall x \in \mathbb{R}.$$

De Cooman also analyzes the converse problem. The author studies the conditions that a possibilistic prevision must satisfy to be able to be represented by a second order possibility that way. In that case, the possibilistic prevision is called *representable,* and the respective second order possibility is called a *representation* of it. This concept of representability is related to the concept of *coherence* (Walley, 1991) for imprecise probabilities models. In De Cooman (1998), the author defines, for each possibilistic prevision, p, the function $\mathcal{M}(p) : \mathcal{P}_{\mathcal{L}(\Omega')} \to [0,1]$:

$$\mathcal{M}(p)(Q) := \inf\{p(X)(Q(X)) : X \in \mathcal{K}\}, \ \forall Q \in \mathcal{P}_{\mathcal{L}(\Omega')}.$$

He also defines, from $\mathcal{M}(p)$, another possibilistic prevision, $e : \mathcal{L}(\Omega) \to \mathcal{P}(\mathbb{R})$, by *natural extension* techniques (Walley, 1991):

$$e(X)(x) := \sup\{\mathcal{M}(p)(Q) : Q \in \mathcal{P}_{\mathcal{L}(\Omega')}, \ Q(X) = x\}, \ \forall x \in \mathbb{R}, \forall X \in \mathcal{L}(\Omega).$$

According to the author, this is the (pointwise) greatest representable possibilistic prevision on $\mathcal{L}(\Omega')$ that is dominated by p. He also proves that the coincidence between e and p on its domain $\mathcal{K}$ is a necessary and sufficient condition for a *normal* possibilistic prevision p to be representable.

For our particular case, we have proved (Couso, 1999) that, whenever its images are normal fuzzy sets, the probabilistic envelope of a fuzzy random variable, $P_{\widetilde{X}} : \sigma(\Omega') \to \widetilde{\mathcal{P}}(\mathbb{R})$ is representable. Two possible second order models associated to it are $\Delta(\widetilde{X})$ and $\mathcal{P}(\widetilde{X}) : \mathcal{P}_{\mathcal{L}(\Omega')} \to [0,1]$, where $\Delta(\widetilde{X})(Q) = \inf\{P_{\widetilde{X}}(A)(Q(A)) : A \in \sigma(\Omega')\}$. We can observe that these two models do not coincide in general, as we show in the example 2 for the particular case of random sets. $\Delta(\widetilde{X})$ is the least precise second order possibility that represents $\mathcal{P}(\widetilde{X})$. So it provides the same information as $P_{\widetilde{X}}$. On the other hand, $\mathcal{P}(\widetilde{X})$ represents al the available information about the original probability measure, $P \circ U_0^{-1}$.

De Cooman (1998) and Walley (1996, 1997) find second order possibilities very useful to model some usual situations. Models of imprecise probabilities (Huber, 1981, Walley, 1991, etc.) are often used to represent the imprecise information about the probability measure of a random experiment. In this case, the information is described by a class of probability measures of which the true unknown probability measure is a member. In many cases, however, it seems difficult to draw the line between the probability measures that will be included in the class, and the ones that have to be excluded. At the same time, not all members of the class will deserve the same status. Furthermore, as the authors indicate in De Cooman and Walley (1999) with several examples, second order possibility models are realistic in problems where the expert has little information about the correct first (imprecise probability) model. Another important reason to assume this type of model is that possibility measures are particularly easy to asses and to characterize through their possibility distribution, which is a point function. Moreover, possibility measures also enable us to model weak states of information, e.g., through the vacuous upper probability, which is the possibility measure whose distribution is equal to one everywhere. Besides, possibility measures seem to be good models for some types of vague information expressed in natural language (see Walley and De Cooman, 1998, for a detailed discussion).

According to Walley (1996, 1997), second order possibility measures are compatible with first order models. He asserts that any second order possibility model can be reduced into a first order model of imprecise probabilities by using natural extension techniques. This reduction will be convenient for the decision making stage. To adapt it to our particular case, we will replace linear previsions by probability measures. On the other hand, the sets of bounded real functions considered will coincide, in our case, with $\sigma-$ algebras of subsets of the referential. We will consider a class of probability measures $\mathbb{P}$ defined on a set of probability measures, $\mathcal{P}_1$, each of them is defined on the σ-algebra $\sigma(\Omega')$. We will suppose, for simplicity, that the set $\mathcal{P}_1$ is finite. A similar reasoning remains valid for the general case. The upper probability associated to $\mathbb{P}$, $\Pi_{\mathbb{P}} : \mathcal{P}(\mathcal{P}_1) \to [0,1]$, is supposed to be a possibility measure, i.e., $\Pi_{\mathbb{P}}(\{Q\}) = \sup\{\mathbb{P}(\{Q\}) : \mathbb{P} \in \mathbb{P}\}$, $\forall Q \in \mathcal{P}_1$ and $\Pi_{\mathbb{P}}(\mathcal{Q}) = \sup\{\Pi_{\mathbb{P}}(\{Q\}) : Q \in \mathcal{Q}\}$, for all $\mathcal{Q} \subseteq \mathcal{P}_1$. We can observe that this type of model contains, as particular cases, second order probability models (in this case, $\mathbb{P}$ is a singleton) and imprecise probability models (where $\mathbb{P}$ is a set of ... probability measures and $\Pi_{\mathbb{P}}$ takes the values 0 and 1 for each probability measure of $\mathcal{P}_1$. For each probability measure, $\mathbb{P} \in \mathbb{P}$, we will construct a new probability measure (also denoted $\mathbb{P}$) defined on the product space $\sigma(\Omega') \otimes \mathcal{P}(\mathcal{P}_1)$ by the formulas: $\mathbb{P}(A|\{Q\}) = Q(A)$, $\forall Q \in \mathcal{P}_1$ and $\mathbb{P}(\Omega' \times \{Q\}) = \mathbb{P}(\{Q\})$. For each two-dimensional probability measure, $\mathbb{P}$, constructed this way, we calculate its marginal on $\sigma(\Omega')$, so that, when $\mathcal{P}_1 = \{Q_1, \dots, Q_n\}$ and $\mathbb{P} \in \mathbb{P}$, then: $\mathbb{P}(A \times \mathcal{P}_1) = \sum_{i=1}^n \mathbb{P}(A|\{Q_i\})\,\mathbb{P}(Q_i) = \sum_{i=1}^n Q_i(A)\,\mathbb{P}(Q_i)$. (The marginal probability measure of $\mathbb{P}$ in the first set is the linear convex combination of $\{Q_1, \dots, Q_n\}$ weighed by $\mathbb{P}$). The upper and lower probabilities of an event $A \in \sigma(\Omega')$ will be given by

$\overline{P}(A) = \sup\{\mathbb{P}(A \times \mathcal{P}_1) : \mathbb{P} \in \mathbb{P}\}$, $\underline{P}(A) = \inf\{\mathbb{P}(A \times \mathcal{P}_1) : \mathbb{P} \in \mathbb{P}\}$. This way, we obtain a first order model on $\sigma(\Omega')$, form the initial second order one. Walley (1997) proves that, under these conditions, $\overline{P}$ and $\underline{P}$ may be expressed by

$$\overline{P}(A) = \int_0^1 \overline{P}_\alpha(A)\,d\alpha, \quad \underline{P}(A) = \int_0^1 \underline{P}_\alpha(A)\,d\alpha,$$

where, for each index, $\alpha \in [0,1]$, $\overline{P}_\alpha$ and $\underline{P}_\alpha$ are defined as follows:

$$\overline{P}_\alpha(A) = \sup\{Q(A) : Q \in \mathcal{P}_1, \Pi_{\mathbb{P}}(\{Q\}) \geq \alpha\} \text{ and}$$

$$\underline{P}_\alpha(A) = \inf\{Q(A) : Q \in \mathcal{P}_1, \Pi_{\mathbb{P}}(\{Q\}) \geq \alpha\}.$$

Example 5. Suppose that we have an urn with 10 balls, which are coloured either red or white. We know that it has 3 white and 3 red. We do not know the colour of the remaining 4 ones, but we know that, with probability 1/2, two of them are white. One ball is chosen at random from the urn. To model this random experiment, we can construct the random sets $\Gamma_1, \Gamma_2 : \{1, \ldots, 10\} \to \mathcal{P}([0,1])$ as follows:

$$\Gamma_1(i) = \begin{cases} \{W\} & \text{if } i = 1,2,3 \\ \{R\} & \text{if } i = 4,5,6 \\ \{W,R\} & \text{if } i = 7,8,9,10 \end{cases}$$

$$\Gamma_2(i) = \begin{cases} \{W\} & \text{if } i = 1,2,3,7,8 \\ \{R\} & \text{if } i = 4,5,6 \\ \{W,R\} & \text{if } i = 9,10. \end{cases}$$

With probability 1/2, the set of possible probability measures is $\mathcal{P}_2 = \{(5/10, 5/10), (6/10, 4/10), (7/10, 3/10)\}$. Furthermore, we are sure that the true probability measure belongs to the set $\mathcal{P}_1 = \{(3/10, 7/10), (4/10, 6/10), (5/10, 5/10),$
$(6/10, 4/10), (7/10, 3/10)\}$. The associated set of probability measures defined on $\mathcal{P}(\mathcal{P}_1)$ is $\mathbb{P}$, given by
$I\!P \in \mathbb{P}$ if:

- $I\!P(\{Q\}) = 1$ and $I\!P(\{Q'\}) = 0, \ \forall Q' \neq Q$, if $Q \in \mathcal{P}_2$, or else
- $I\!P(\{Q_1\}) = 1/2 = I\!P(\{Q_2\})$, where $Q_1 \in \mathcal{P}_1$, $Q_2 \in \mathcal{P}_2$, $Q_1 \neq Q_2$ and $I\!P(\{Q'\}) = 0, \ \forall Q' \notin \{Q_1, Q_2\}$

The possibility measure associated to $\mathbb{P}$ is $\Pi_{\mathbb{P}} : \mathcal{P}(\mathcal{P}_1) \to [0,1]$ such that
$$\Pi_{\mathbb{P}}(\{Q\}) = \begin{cases} 1 & \text{if } Q \in \mathcal{P}_2 \\ 1/2 & \text{if } Q \in \mathcal{P}_1 \cap \mathcal{P}_2^c \end{cases}$$
If we define the fuzzy random variable $\widetilde{X} : \{1, \ldots, 10\} \to \widetilde{\mathcal{P}}(\{W, R\})$ by

$$\widetilde{X}(i)^{[\alpha]} := \begin{cases} \Gamma_1(i) & \text{if } \alpha \leq 1/2 \\ \Gamma_2(i) & \text{if } \alpha > 1/2 \end{cases}$$

we can observe that the fuzzy set $\mathcal{P}(\widetilde{X})$ coincides with the last second order possibility distribution. Now, we will obtain the associated first order model, by making use of Walley's procedure (1997). For each probability measure $I\!P \in \mathbb{P}$, we calculate the value $I\!P(\{W\}) = \sum_{Q \in \mathcal{P}_1} I\!P(Q)\, Q(\{W\})$. The upper and lower probability values of the event $\{W\}$ are:

$\overline{P}(\{W\}) = \sup\{I\!P(\{W\}) : I\!P \in \mathbb{P}\} = \int_0^1 \sup(P_{\widetilde{X}}(\{W\})^{[\alpha]}\, d\alpha =$
$7/10 \cdot 1/2 + 7/10 \cdot 1/2 = 7/10.$
$\underline{P}(\{W\}) = \inf\{I\!P(\{W\}) : I\!P \in \mathbb{P}\} = \int_0^1 \inf(P_{\widetilde{X}}(\{W\})^{[\alpha]}\, d\alpha =$
$3/10 \cdot 1/2 + 5/10 \cdot 1/2 = 4/10.$

Acknowledgements

The research in this paper has been partially supported by the Spanish DGESIC Grant No. DGE-98-PB97-1286. This financial support is gratefully acknowledged.

References

1. Aumann, J. (1965). Integral of set valued functions, *J. Math. Anal. Appl.* **12,** 1–12.
2. Bolaños, M.J., Lamata, M.T. and Moral, S. (1988). Decision making problems in a general environment, *Fuzzy Sets and Systems***25,** 135–144.
3. Cano, A., Moral, S. and Verdegay-López, J.F. (1992). Partial inconsistency of probability envelopes, *Fuzzy Sets and Systems* **52,** 201–21.
4. Choquet, G. (1954). Theory of Capacities, *Ann. Inst. Fourier* **V,** 131–295.
5. Couso, I. (1997). La envolvente probabilística: definición y propiedades. Research Report. Departament of Statistics, O.R. and M.D. University of Oviedo, Spain.
6. Couso, I. (1999) . Teoría de la probabilidad para datos imprecisos. Algunos aspectos. Departament of Statistics, O.R. and M.D. Thesis, University of Oviedo, Spain.
7. Couso, I., Gil, P. (1997). La envolvente probabilística de variables aleatorias difusas, *Proc. ESTYLF'97*, Tarragona, Spain.
8. Couso, I., Montes, S. and Gil, P. (1998). Función de distribución y mediana de variables aleatorias difusas, *Proc. ESTYLF'98*, Pamplona, Spain.
9. Couso, I., Montes, S. and Gil, P. (2000). Some remarks about the possibilistic interpretation of fuzzy sets, *Proc. FSTA'2000*, Liptowski Mikulash, Slovakia.
10. Couso, I., Moral, S. and Walley, P. (1999). Examples of Independence for Imprecise Probabilities, *Proc. 1st Intern. Symposium on Imprecise Probabilities and Their Applications*, Ghent, Belgium.
11. Debreu, G. (1965). Integration of correspondences, *Proc. Fifth Berkeley Symp. Math. Statist. and Probability*, Berkeley, USA **2**, 351-372.
12. De Campos, L.M. and Bolaños, M.J. (1989). Representation of Fuzzy Measures through Probabilities, *Fuzzy Sets and Systems* **31,** 23–36.
13. De Cooman, G. (1998). Possibilistic previsions, *Proc. IPMU 98*, Paris, France **1** 2-9.
14. De Cooman, G. and Aeyels, D. (1999). A random set description of a possibility measure and its natural extension, *IEEE Trans. Syst., Man Cyber.* (accepted, in press).
15. De Cooman, G. and Walley, P. (1999). An imprecise hierarchical model for behaviour under uncertainty, (submitted for publication).
16. Delgado, M. and Moral, S. (1989). Upper and lower fuzzy measures, *Fuzzy Sets and Systems* **33,** 191–200.
17. Dempster, A.P. (1967). Upper and Lower Probabilities Induced by a Multivalued Mapping, *Ann. Math. Statistics* **38,** 325–339.
18. Dubois, D., Prade, H. (1992). When upper probabilities are possibility measures, *Fuzzy Sets and Systems* **49,** 65–74.
19. Dubois, D. and Prade, H. (1997). The three semantics of fuzzy sets, *Fuzzy Sets and Systems* **90,** 141–150.
20. Féron, R. (1976). Ensembles aléatoires flous, *C.R. Acad. Sci. Paris Ser. A* **282,** 903–906.
21. Goodman, I.R., and Nguyen, H.T. (1985). *Uncertainty Models for Knowledge-Based Systems*, Elsevier Science Publishers.
22. Goutsias, J., Mahler, R.P.S. and Nguyen, H.T. (1997). *Random Sets. Theory and Applications.* The IMA Volumes in Mathematics and its Applications. Vol. 97, Springer-Verlag, Heidelberg.

23. Grabisch, M., Nguyen, H.T. and Walker, E.A. (1995). *Fundamentals of Uncertainty Calculi with Applications to Fuzzy Inference.* Kluwer Academic Publishers, New York.
24. Hirota, K. (1981). Concepts of probabilistic sets, *Fuzzy Sets and Systems* **5,** 31–46.
25. Huber, P.J. (1981). *Robust Statistics*, John Wiley & Sons, New York.
26. Kendall, D.G. (1974). Foundations of a theory of random sets, In *Stochastic Geometry* (Harding, E.F., Kendall, D.G. eds.), pp. 322–376. John Wiley & Sons, New York.
27. Kruse, R. and Meyer, K.D. (1987). *Statistics with vague data*, D. Reidel Publishing Company, Dordrecht.
28. Kyburg, H.E. and Pittarelli, M. (1996). Set-Based Bayesianism, *IEEE Transactions on Systems, Man and Cybernetics* **26,** 324–339.
29. Kwakernaak (1989). Fuzzy random variables. Definition and theorems, *Inform. Sci.* **15,** 1–29.
30. Matheron, G. (1975). *Random Sets and Integral Geometry*, John Wiley & Sons, New York.
31. Meyer, K.D. and Kruse, R. (1990). On calculating the covariance in the presence of vague data, In *Progress in Fuzzy Sets and Systems* (Janko, W.H. *et al.*, eds.), pp. 125–133. Kluwer Academic Publishers, New York.
32. Nahmias, S. (1978). Fuzzy variables, *Fuzzy Sets and Systems*, **1,** 97–110.
33. Nahmias, S. (1979). Fuzzy variables in a random envirnment, In *Advances in Fuzzy Set Theory and Applications* (Gupta, M.M., Ragade, R.K. and Yager, R.R., eds.), pp. 165–180. North Holland, Amsterdam.
34. Nguyen, H.T.(1977). On fuzziness and linguistic probabilities, *J. Math. Anal. Appl.* **61,** 658–671.
35. Nguyen, H.T. (1978). On random sets and belief functions, *J. Math. Anal. Appl.* **63,** 531–542.
36. Puri, M.L. and Ralescu, D. (1986). Fuzzy Random Variables, *J. Math. Anal. Appl.* **114,** 409–422.
37. Shafer, G. (1976). *A mathematical theory of evidence*, Princeton University Press, Princeton.
38. Stein, W.E. and Talati, K. (1981). Convex fuzzy random variables, *Fuzzy Sets and Systems* **6,** 271–283.
39. Stojaković, M. (1992). Fuzzy conditional expectation, *Fuzzy Sets and Systems* **52,** 53–60.
40. Walley, P. (1991). *Statistical Reasoning with Imprecise Probabilities*, Chapman and Hall, London.
41. Walley, P. (1996). Measures of Uncertainty in Expert Systems, *Artificial Intelligence* **83,** 1–58.
42. Walley, P. (1997). Statistical inferences based on a second-order possibility distribution, *International Journal of General Systems* **26,** 337–384.
43. Walley, P., De Cooman, G. (1998). A behavioural model for linguistic uncertainty, In *Computing with words*, (Paul P. Wang, Ed.), (accepted, in press).
44. L.A. Zadeh (1975). The concept of a linguistic variable and its applicaction in approximate reasoning, Part 2, *Inform. Sci.* **8,** 301–357.
45. L.A. Zadeh (1978). Fuzzy sets as a basis for a theory of possibility, *Fuzzy Sets and Systems* **1,** 3–28.

Measure extension from meet-systems and falling measures representation

P.Z. Wang[1], Y.C. Chen[2], B.T. Low[3]

[1] AEI, West Texas A & M University, Canyon, TX79016, USA
[2] Ming Chuan University, Taipei 111, Taiwan, Republic of China
[3] Department of SEEM, Chinese University of Hong Kong, Shatin N.T., Hong Kong

Abstract. A renewed measure extension theorem is presented in this paper which extends measure from a meet-system $\mathcal{M}$ to $\mathcal{B}$, the σ-algebra generated from $\mathcal{M}$. And based on it, the falling measures representation theorems are given.

1 Introduction

Through random sets to deal with fuzzy theory and applications has been developed several years. Among this flow there is the Falling Shadow Representation theory presented by the authors (see papers by Wang and Wang *et al.* in References), which aims to treat fuzzy information and fuzzy data by means of random sets and set-valued statistics. It does not mean from the theory that the fuzziness can be totally transferred into and covered by the randomness; The fuzziness is much different from the randomness. But we also believe that there are some relationships between them. Usually, "a phenomena with fuzziness in the 'ground' U can be treaded as a phenomena with randomness in the 'sky', the power of U" (see Wang, 1985). Fuzzy theory and probability can benefit each other. From this aspect, many interested results have been found. Based on the author's work (Wang, 1985), in this paper we will introduce one of the foundations of the Falling Shadow Representation theory. That is the measures' extension theorem extending measure starting from meet-systems. The result is not only important for falling measures' representation, it can be also shared in the researches of probability and measure theory.

In Section 2, a renewed measure extension theorem which extending measures from meet-systems will be given. In Section 3, the representation theorems for the four kinds of falling measures will be introduced. Finally, a simple conclusion can be found in Section 4.

2 Measure extension from meet-systems

A non-negative, σ-additive set-function m with $m(\emptyset) = 0$ is called a measure. The classical measure-extension theorem told us that a measure can be extended from $\mathcal{F}$, an algebra (or field) to $\mathcal{B}$, the σ-algebra generated from $\mathcal{F}$. $\mathcal{F}$ is the start-point of extension. Most of the non-additive set functions used in the researches of subjective measurements are always concerned with some additive measures on the "sky", the power of the universe of discussion $\mathcal{P}(X)$. In these cases, the classical measure extension theory is not competent enough to fit the practical needs excepted we can move the start-point to the simple structure: meet-systems. In this section we will give a renewed measure extension theorem which will extend a finitely additive measure from $\mathcal{M}$ to $\mathcal{F}$, and extend a "measure" from $\mathcal{M}$ to $\mathcal{B}$, the σ-algebra generated from $\mathcal{M}$.

A *meet-system* on X is a non-empty group of subsets $\mathcal{M} \subseteq \mathcal{P}(X)$ satisfies:

1. $X \in \mathcal{M}$;
2. If $A, B \in \mathcal{M}$, then $A \cap B \in \mathcal{M}$.

From a meet-system $\mathcal{M}$ going to an algebra $\mathcal{F}$ needs to pass through the following steps:

1. $\mathcal{M}_\cup = [\mathcal{M}]_\cup = \{\cup_{i \in I} A_i : |I| \geq 0, A_i \in \mathcal{M}\}$,
2. $\mathcal{D} = \mathcal{M}(\setminus)[\mathcal{M}]_\cup = \{A \setminus \cup_{i \in I} B_i : |I| \geq 0, A, B_i \in \mathcal{M}\}$,
3. $\mathcal{F} = [\mathcal{D}]_\cup = \{\cup_{i \in I} D_i : |I| \geq 1, D_i \in \mathcal{D}\}$.

$\mathcal{F}$ is the algebra generated from $\mathcal{M}$.

To simplify, we only talk about the extension of probability, which is a kind of measures with $p(X) = 1$.

For easy to typing, we write $m[n], m[n+1]$ to denote the double index m_n, m_{n+1}, and so on. We promise that $I_n = \{1, \ldots, n\}$, $J_m = \{1, \ldots, m\}$ in this paper.

The *2-ary additive property* of a probability on $\mathcal{F}$ is represented as:

$$p(B_1 \cup B_2) = p(B_1) + p(B_2) - p(B_1 \cap B_2). \tag{1}$$

When $A \supseteq B_1 \cup B_2$, $p(A) \geq p(B_1 \cup B_2)$, so that

$$p(A) - (p(B_1) + p(B_2)) + p(B_1 \cap B_2) \geq 0. \tag{2}$$

The formula (2) reflects the additive property and the monotony of probability together. For any $A, B_1, B_2 \in \mathcal{F}$, we have that $A \supseteq (B_1 \cup B_2) \cap A = (B_1 \cap A) \cup (B_2 \cap A)$, then (2) becomes

$$p(A) - p(B_1 \cap A) + p(B_2 \cap A) + p(B_1 \cap B_2 \cap A) \geq 0. \tag{2'}$$

The generalized formulae of (1) and (2') can be written as follows:

$$p\left(\cup_{i \in I_n} B_i\right) = \sum_{\emptyset \neq I \subseteq I_n} (-1)^{|I|-1} p\left(\cap_{i \in I_n} B_i\right) \tag{3}$$

which is the generalization of (1) and is called the *recurrent additive property* of probability. For any $A, B_1, \ldots, B_n \in \mathcal{F}$,

$$\sum_{I \subseteq I_n} (-1)^{|I|} p(\cap_{i \in I} B_i \cap A) \geq 0 \tag{4}$$

which is the generalization of (2') where $\cap i \in IB_I = X$ whenever $I = \emptyset$. (4) is the same as

$$p(A) - \sum_{\emptyset \neq I \subseteq I_n} (-1)^{|I|-1} p(\cap_{i \in I} B_i \cap A) \geq 0. \tag{4'}$$

On the algebra, the 2-ary additive property (1) implies the recurrent additive property (3). Unfortunately, on a meet-system, the later cannot be deduced from the former. So we need some definition for describing the character of probability when it is just defined on a meet-system.

Definition 1. Let $\mathcal{M}$ be a meet-system on X. The mapping $P^0 : \mathcal{M} \to [0, 1]$ is called a *proba* on $\mathcal{M}$ if $P^0(X) = 1$, and (3) and (4) hold, i.e., for any $A, B_1, \ldots, B_n \in \mathcal{F}$,

$$\sum_{I \subseteq I_n} (-1)^{|I|} P^0(\cap_{i \in I} B_i \cap A) \geq 0,$$

and "=" is taken whenever $A = B_1 \cup \ldots \cup B_n$. A proba is called a *probabi* on $\mathcal{M}$ if it is continuous, i.e., for any monotone sequence $B_n \subseteq \mathcal{M}$, $\lim_{n \to \infty} B_n = A \in \mathcal{M}$, we have that

$$P^0(B_n) \to P^0(A) \quad (\text{as } n \to \infty).$$

To get the renewed measure extension theorem, at first, we need to prove a lemma which just consider a finite meet-system on X.

Suppose that $\mathcal{C} = \{C_1, \ldots, C_n\} \subseteq \mathcal{P}(X)$, and $X \in \mathcal{C}$. Set

$$d(\mathcal{C}) = \{\alpha_1 \cap \ldots \cap \alpha_n : \alpha_i = C_i \text{ or } C_i^c, i = 1, \ldots, n\}$$

Which is called the (X)*division* generated by $\mathcal{C}$. Set

$$a(\mathcal{C}) = \left\{ \sum_{i \in I} D_i : |I| \geq 0; D_i \in d(\mathcal{C}) \right\},$$

which is the algebra generated from $\mathcal{C}$, and where $\sum_{i \in I} D_i = \emptyset$ when $|I| = 0$.

A mapping $P^* : d(\mathcal{C}) \to [0, 1]$ satisfying that

1. $P^*(\emptyset) = 0$ if $\emptyset \in C$,
2. $\sum_{D \in d(\mathcal{C})} P^*(D) = 1$,

is called a *basic distribution* on $d(\mathcal{C})$, by which a probability p can be uniquely determined on $a(\mathcal{C})$ as follows:

$$p\left(\sum_{i\in I} D_i\right) = \sum_{i\in I} P^*(D_i) \quad (D_i \in d(\mathcal{C})).$$

Lemma 1. *For any $n \geq 1$, let $\mathcal{C} = \{C_1, \ldots, C_n\} \subseteq \mathcal{P}(X)$ be a meet-system consists of n subsets of X. Let P^0 be a proba on $\mathcal{C}$. Then, P^0 can be uniquely extended to a probability defined on $\mathcal{C}$, the algebra generated from $\mathcal{C}$.*

Proof. Use the method of mathematical induction. If $n = 1$: $\mathcal{C} = \{X\}$, $d(\mathcal{C}) = \{X, \emptyset\}$, $a(\mathcal{C}) = \{X, \emptyset\}$. There is one, and only one, way to define a probability p on $a(\mathcal{C})$:

$$p(X) = P^0(X) = 1; \; p(\emptyset) = 0,$$

whence Lemma 1 is true.

If $n = 2$: $\mathcal{C} = \{X, A\}$, $d(\mathcal{C}) = \{A, A^c, \emptyset\}$, $\mathcal{F} = \{A, A^c, X, \emptyset\}$. There is one, and only one, way to define a probability p on $a(\mathcal{C})$:

$$p(A) = P^0(A), \; p(A^c) = 1 - P^0(A), \; p(X) = P^0(X) = 1, \; p(\emptyset) = 0,$$

whence Lemma 1 is true.

Suppose that Lemma is true for n, and consider $n+1$. Let $\mathcal{C} = \{C_1, \ldots, C_n, C_{n+1}\} \subseteq \mathcal{P}(X)$ be a meet-system on X with $n+1$ members. Set $\mathcal{S} = \{C_1, \ldots, C_n\} \subseteq \mathcal{C}$ and make that $X \in \mathcal{S}$. $\mathcal{S}$ is a meet-system on X. Let $d_{.1}$ and $a_{.1}$ be the (X)division and the algebra generated from $\mathcal{S}$, respectively. Set $\mathcal{N} = \{C_1 \cap C_{n+1}, \ldots, C_n \cap C_{n+1}\} \subseteq \mathcal{C}$, which is a meet-systems on $X' = X \cap C_{n+1} = C_{n+1}$. Let $d_{.2}$ and $a_{.2}$ be the (X')division and the algebra generated from $\mathcal{N}$ in the space X', respectively. Let P^0 be a proba defined on $\mathcal{C}$. According to the assumption, Lemma 1 is true for $\mathcal{S}$ and $\mathcal{N}$, then there are determinate probability $p_{.i}$ defined on $a_{.i}$ $(i = 1, 2)$, respectively, which are both extended from P^0.

The (X)division generated from $\mathcal{C}$ is

$$d = d(\mathcal{C}) = \{D \cap \alpha : D \in d_{.1}, \; \alpha = C_{n+1} \text{ or } C_{n+1}^c\}.$$

It is clear that $d_{.2} = \{D \cap C_{n+1} : D \in d_{.1}\}$, so that $D \cap C_{n+1} \in d_{.2}$ when $D \in d_{.1}$.

Define mapping $P^* : d \to [0, 1]$

$$P^*(D \cap C_{n+1}) = p_{.2}(D \cap C_{n+1}) \quad (D \in d_{.1}),$$

$$P^*(D \cap C_{n+1}^c) = p_{.1}(D) - p_{.2}(D \cap C_{n+1}) \quad (D \in d_{.1}).$$

We have that

$$\sum_{D\in d} P^*(D) = \sum_{D\in d_{.1}} P^*(D \cap C_{n+1}) + \sum_{D\in d_{.1}} P^*(D \cap C_{n+1}^c)$$

$$= \sum_{D \in d_{.1}} (P^*(D \cap C_{n+1}) + P^*(D \cap C_{n+1}^c)) = \sum_{D \in d_{.1}} p_{.1}(D) = 1.$$

So that P^* is a basic distribution on d which determines a probability P on $a(\mathcal{C})$. Since $P^*(D) = p_{.i}(D)$, $(D \in d_{.1})$, then $P(C) = p_{.i}(C)$, $(C \in a_{.i})$, $(i = 1, 2)$.

But $p_{.i}$ are extensions of P^0, so that P is an extension of P^0. It is clear that P is the only possible one probability extended from P^0. ∎

Let us go to the first part of the renewed probability extension theorem:

Theorem 1. (Extension Theorem - Part 1) *Suppose that $\mathcal{M}$ is a non-empty meet-system on X. P^0 is a proba on $\mathcal{M}$, then P^0 can be uniquely extended to a finite probability P defined on $\mathcal{F}$, the algebra generated from $\mathcal{M}$.*

Proof.
Step 1. Extending P^0 from $\mathcal{M}$ to $\mathcal{M}_\cup$.

Define $P : \mathcal{M}_\cup \to [0, 1]$

$$P(\cup_{i \in I_n} C_i) = \sum_{\emptyset \neq I \subseteq I_n} (-1)^{|I|-1} P^0(\cap i \in I C_i)(C_i \in \mathcal{M}). \tag{5}$$

It is obvious that $P(C) = P^0(C)$, $(C \in \mathcal{M})$. The definition of P is determinate and P is additive on $\mathcal{M}_\cup$. Indeed, consider any two subsets $B = B_1 \cup \ldots \cup B_n$ and $C = C_1 \cup \ldots \cup C_m$ in $\mathcal{M}_\cup$. Set

$$\mathcal{M}^* = \{(\cap_{i \in I} B_i) \cap (\cap_{j \in J} C_j) : |I| \geq 0, |J| \geq 0, I \subseteq I_n, J \subseteq J_m\} \subseteq \mathcal{M}.$$

$P^0|_{\mathcal{M}^*}$, the constraint of P^0 on $\mathcal{M}^*$, is a proba. By Lemma 1, which uniquely determines a probability P' on $\mathcal{F}^*$, the algebra generated from $\mathcal{M}^*$. Since a probability on $\mathcal{F}^*$ satisfies the recurrent additive property. So that P' satisfies the equation (3), which coincides with the definition of P. So that $P'(C) = P(C)$, $(C \in \mathcal{F}^* \cap \mathcal{M}_\cup)$. Note that $B = B_1 \cup \ldots \cup B_n$, $C = C_1 \cup \ldots \cup C_m \in \mathcal{F}^* \cap \mathcal{M}_\cup$

$$B \cup C = B_1 \cup \ldots \cup B_n \cup C_1 \cup \ldots \cup C_m \in \mathcal{F}^* \cap \mathcal{M}_\cup,$$

$$B \cap C = (B_1 \cap C_1) \cup \ldots \cup (B_n \cap C_m) \in \mathcal{F}^* \cap \mathcal{M}_\cup.$$

We have that $P'(B \cup C) = P'(B) + P'(C) - P'(B \cap C)$, so that $P(B \cup C) = P(B) + P(C) - P(B \cap C)$. It means that P is finitely additive.

From logical consideration, we should take $B = C$ in the beginning. When $B = C$, we have that

$$B_1 \cup \ldots \cup B_n = (B_1 \cap C_1) \cup \ldots \cup (B_n \cap C_m) = C_1 \cup \ldots \cup C_m.$$

Since $P'(B_1 \cup \ldots \cup B_n) = P'((B_1 \cap C_1) \cup \ldots \cup (B_n \cap C_m)) = P'(C_1 \cup \ldots \cup C_m)$.

We have that

$$P(B_1 \cup \ldots \cup B_n) = P((B_1 \cap C_1) \cup \ldots \cup (B_n \cap C_m)) = P(C_1 \cup \ldots \cup C_m).$$

It means that the definition of P is determinate.

Step 2. Generating a proba $P\hat{}$ on $\mathcal{D}$.

Note that $\mathcal{D}$ is a meet-system. Indeed,

$$\begin{aligned}(A \setminus (B_1 \cup \ldots \cup B_n)) \cap (A' \setminus (B_1' \ldots \cup B_m')) \\ = (A \cap B_1^c \cap \ldots \cap B_n^c) \cap (A' \cap B_1'^c \cap \ldots \cap B_m'^c) \\ = (A \cap A') \cap (B_1^c \cap \ldots \cap B_n^c \cap B_1'^c \cap \ldots \cap B_m'^c) \\ = (A \cap A') \setminus (B_1 \cup \ldots \cup B_n \cup B_1' \cup \ldots \cup B_m') \in \mathcal{D}.\end{aligned}$$

Define mapping $P\hat{}: \mathcal{D} \to [0,1]$

$$\begin{aligned}P\hat{}(A \setminus (B_1 \cup \ldots \cup B_n)) = P^0(A) - P(\cup_{i \in I_n}(B_i \cap A)) \\ = P^0(A) - \sum_{\emptyset \neq I \subseteq I_n} (-1)^{|I|-1} P^0(\cap_{i \in I}(B_i \cap A)).\end{aligned}$$

Since P^0 is a proba on $\mathcal{M}$, according to (4') we have that $P\hat{}(A \setminus (B_1 \cup \ldots \cup B_n)) \geq 0$. This definition is determine and $P\hat{}$ is a proba on $\mathcal{D}$. The proof is similar to the proof in Step 1.

Step 3. Extension to $\mathcal{F}$.

Since $P\hat{}$ is a proba on the meet-system $\mathcal{D}$, by the consequence proved in Step 1, $P\hat{}$ can be uniquely extended to a finite probability P defined on $\mathcal{F}$, the algebra generated from $\mathcal{M}$. The proof of the first part of the theorem is completed. ■

To get the second part of the theorem, we need to consider the continuity.

Suppose that $\{A_n\}$, $\{B_n\}$ are two monotone increasing (decreasing) sequences. Write $\{A_n\} \prec \{B_n\}$ ($\{A_n\} \succ \{A_n\}$) if for any $n \geq 1$, there is an $N(n)$ such that $B_k \supseteq A_n$ ($B_k \subseteq A_n$) whenever $k \geq N(n)$. Note that "$\prec$", "$\succ$" are not inverse each other.

We say $\{B_n\}$ is *asymptotically chased* by $\{A_n\}$ if either $\{A_n\} \prec \{B_n\}$ or $\{A_n\} \succ \{B_n\}$.

Definition 2. A monotone decreasing sequence $\{B_n\}$ ($\subseteq \mathcal{M}_\cup$) having a limit $A = A_1 \cup \ldots \cup A_m \in \mathcal{M}_\cup$ is called *regular* if for each $i \in \{1, \ldots, m\}$, there are two monotone decreasing sequences $\{A_n^{(i)}\}, \{V_n^{(i)}\}$ ($\subseteq \mathcal{M}$) with $A_n^{(i)} \downarrow A_i$ and $V_n^{(i)} \setminus A_n^{(i)} \downarrow \emptyset$, such that $\{B_n\}$ is asymptotically chased by $\{A_n^{(1)} \cup \ldots \cup A_n^{(m)} \cup V_n^{(1)} \cup \ldots \cup V_n^{(m)}\}$. A monotone increasing sequence $\{B_n\}$ ($\subseteq \mathcal{M}_\cup$) having a limit $A = A_1 \cup \ldots \cup A_m \in \mathcal{M}_\cup$ is called *regular* if for each $i \in \{1, \ldots, m\}$, there is a monotone increasing sequence $\{A_n^{(i)}\}$ ($\subseteq \mathcal{M}$) converging to A_i such that $\{B_n\}$ is asymptotically chased by $\{A_n^{(1)} \cup \ldots \cup A_n^{(m)}\}$. A meet-system $\mathcal{M}$ is called a regular meet-system on X if for any monotone sequence $\{B_n\}$ ($\subseteq \mathcal{M}_\cup$) having a limit $A = A_1 \cup \ldots \cup A_m \in \mathcal{M}_\cup$ is a regular sequence.

Theorem 2. (Extension Theorem - Part 2) *Suppose that $\mathcal{M}$ is a regular meet-system on X. If P^0 is a probabi on $\mathcal{M}$, then P^0 can be uniquely extended to a probability on B, the σ-algebra generated from $\mathcal{M}$.*

Proof. Since P^0 is a probabi on $\mathcal{M}$, then P^0 is a proba on $\mathcal{M}$, then by the first part of this extension theorem, it can be uniquely extended to a finite probability P defined on $\mathcal{F}$, the algebra generated from $\mathcal{M}$. To prove the second part of the theorem, we only need to prove that P is continuous on $\mathcal{F}$.

We are going to prove that P is continuous from below. Suppose that $C, C_k \in \mathcal{F}$, $C_k \uparrow C$ $(n \to \infty)$ where

$$\begin{aligned} C &= (A_1 \setminus (B_{11} \cup \ldots \cup B_{1n[1]})) \cup \ldots \cup (A_m \setminus (B_{m1} \cup \ldots \cup B_{mn[m]})) \\ &= A_1 B_{11}^c \ldots B_{1n[1]}^c \cup \ldots \cup A_m B_{m1}^c \ldots B_{mn[m]}^c. \end{aligned}$$

Here we omit the symbol "$\cap$" between those subsets which are taking intersection. Set $B_{i0} = A_i^c$ $(i = 1, \ldots, m)$

$$C = (B_{10} \cup \ldots \cup B_{m0}) \cap \left[\cap \left\{\cup_{i=1}^m B_{ij[i]}^c : \text{at least one}\, i : j[i] \neq 0\right\}\right]$$

$$\begin{aligned} &(B_{10} \cup \ldots \cup B_{m0}) \cap \left[\cup \left\{\cap_{i=1}^m B_{ij[i]} : \text{at least one}\, i : j[i] \neq 0\right\}\right]^c \\ &= (A_1 \cup \ldots \cup A_m) \setminus \left[\cup \left\{\cap_{i=1}^m B_{ij[i]} : \text{at least one}\, i : j[i] \neq 0\right\}\right] \\ &= \alpha \setminus (\beta_1 \gamma_1 \cup \ldots \cup \beta_t \gamma_t), \end{aligned}$$

where $\alpha = A_1 \cup \ldots \cup A_m$, β_h are intersections of some members from $\{B_{ij}\}$, and for each s, γ_s is a complementary of a union of some members from $\{A_i\}$.

$$\begin{aligned} C_k &= (A_i^{(k)} \setminus (B_{11}^{(k)} \cup \ldots \cup B_{1n[i]}^{(k)})) \cup \ldots \cup (A_m^{(k)} \setminus (B_{m1}^{(k)} \cup \ldots \cup B_{mn[m]}^{(k)})) \\ &= \alpha^{(k)} \setminus (\beta_1^{(k)} \gamma_1^{(k)} \cup \ldots \cup \beta_t^{(k)} \gamma_t^{(k)}), \end{aligned}$$

where $\alpha^{(k)} = A_1^{(k)} \cup \ldots \cup A_{m[k]}^{(k)}$, $\beta_h^{(k)}$ are intersections of some members from $\{B_{ij}^{(k)}\}$, and for each s, $\gamma_s^{(k)}$ is a complementary of a union of some members from $\{A_i^{(k)}\}$. For any $i \in \{1, \ldots, m\}$, consider that

$$C_k \cap A_i = (\alpha^{(k)} \cap A_i) \setminus ((\beta_1^{(k)} \cap A_i)\gamma_1^{(k)} \cup \ldots \cup (\beta_t^{(k)} \cap A_i)\gamma_t^{(k)}).$$

$C_k \uparrow C$ $(k \to \infty)$ implies that

1. $(\alpha^{(k)} \cap A_i) \uparrow A_i$ $(k \to \infty)$;
2. $((\beta_1^{(k)} \cap A_i)\gamma_1^{(k)} \cup \ldots \cup (\beta_t^{(k)} \cap A_i)\gamma_t^{(k)})(B_{i1} \cup \ldots \cup B_{in[i]})$ $(k \to \infty)$.

Without trivial proof we can see that 2) implies

2'. $(\beta_1^{(k)} \cap A_i) \cup \ldots \cup (\beta_t^{(k)} \cap A_i) \downarrow (B_{i1} \cup \ldots \cup B_{in[i]})$ $(k \to \infty)$.

Since $\mathcal{M}$ is a regular meet-system, and

$$\alpha^{(k)} \cap A_i, (\beta_1^{(k)} \cap A_i) \cup \ldots \cup (\beta_t^{(k)} \cap A_i), B_{i1} \cup \ldots \cup B_{in[i]} \in \mathcal{M}_{\cup}$$

so that there are monotone sequences

$$\{F_i^{(k)}\}, \text{ and } \{G_{i1}^{(k)}\}, \ldots, \{G_{im[i]}^{(k)}\}, \{V_i^{(k)}\}, \ldots, \{V_{im[i]}\}^{(k)}\} \subseteq \mathcal{M},$$

$$F_i^{(k)} \uparrow A_i, \text{ and } G_{i1}^{(k)} \downarrow B_{i1} \cap A_i, \ldots, G_{im[i]}^{(k)} \downarrow G_{im[i]}^{(k)} \cap A_i, \ (k \to \infty).$$

And $\{(\alpha^{(k)} \cap A_i)\}, \{(\beta_1^{(k)} \cap A_i) \cup \ldots \cup (\beta_t^{(k)} \cap A_i)\}$ are asymptotically chased by $\{F_i^{(k)}\}, \{G_{i1}^{(k)} \cup \ldots \cup G_{im[i]}^{(k)} \cup V_{i1}^{(k)} \cup \ldots \cup V_{im[i]}^{(k)}\}$, respectively. Where $V_{ij}^{(k)} \setminus G_{ij}^{(k)} \downarrow \emptyset \, (j = 1, \ldots, m[i])$. Since P^0 is continuous on $\mathcal{M}$, so that

$$\lim_{k\to\infty} P^0(F_i^{(k)}) = P^0(A_i).$$

Denote that $G_{i1}^{(k)} \cup \ldots \cup G_{im}[i]^{(k)} = R_i^{(k)}, V_{i1}^{(k)} \cup \ldots \cup V_{im[i]}^{(k)} = W_i^{(k)}$ and since that $V_{ij}^{(k)} \setminus G_{ij}^{(k)} \downarrow \emptyset$, so that $W_i^{(k)} \setminus R_i^{(k)} \downarrow \emptyset$. Without trivial statement we have that $\lim_{k\to\infty} P(W_i^{(k)} \setminus R_i^{(k)}) = 0$. So that

$$\lim_{k\to\infty} P(R_i^{(k)}) \cup W_i^{(k)}) == \lim_{k\to\infty} P(R_i^{(k)}) + \lim_{k\to\infty} P(W_i^{(k)} \setminus R_i^{(k)})$$

$$= \lim_{k\to\infty} P(R_i^{(k)}) = \lim_{k\to\infty} \sum_{\emptyset \neq J \subseteq J_{m[i]}} (-1)^{|J|-1} P^0(\cap_{j\in J} G_{ij}^{(k)})$$

$$= \sum_{\emptyset \neq J \subseteq J_{m[i]}} (-1)^{|J|-1} \lim_{k\to\infty} P^0(\cap_{j\in J} G_{ij}^{(k)})$$

$$= \sum_{\emptyset \neq J \subseteq J_{m[i]}} (-1)^{|J|-1} P^0(\cap_{j\in J} B_{ij} \cap A_i) = P((B_{i1} \cup \ldots \cup B_{im[i]}) \cap A_i).$$

So that

$$\lim_{k\to\infty} [P^0(F_i^{(k)}) - P(G_{i0}^{(k)} \cup \ldots \cup G_{im[i]}^{(k)} \cup V_{i1}^{(k)} \cup \ldots \cup V_{im[i]}^{(k)})]$$

$$= P^0(A_i) - P((B_{i1} \cup \ldots \cup B_{im[i]}) \cap A_i) = P(A_i \setminus (B_{i1} \cup \ldots \cup B_{im[i]})).$$

But, since

$$P^0(F_i^{(k)}) - P(G_{i0}^{(k)} \cup \ldots \cup G_{im[i]}^{(k)} \cup V_i^{(k)}))$$

$$\leq P(C_k \cap A_i) \leq P(A_i \setminus (B_{i1} \cup \ldots \cup B_{im[i]})).$$

So that $\lim_{k\to\infty} P(C_k \cap A_i) = P(A_i \setminus (B_{i1} \cup \ldots \cup B_{im[i]}))$. So that

$$\lim_{k\to\infty} P(C_k) = \lim_{k\to\infty} P\left(\cup_{i=1}^m (C_k \cap A_i)\right)$$

$$= \lim_{k\to\infty} \sum_{\emptyset\neq I\subseteq I_m} (-1)^{|I|-1} P^0(\cap_{i\in I} A_i \cap C_k)$$

$$= \sum_{\emptyset\neq I\subseteq I_m} (-1)^{|I|-1} \lim_{k\to\infty} P^0(\cap_{i\in I} A_i \cap C_k)$$

$$= \sum_{\emptyset\neq I\subseteq I_m} (-1)^{|I|-1} P(A_i \setminus (B_{i1} \cup \ldots \cup B_{im[i]}) = P(C).$$

It means that P is continuous from below, then P is continuous on $\mathcal{F}$. So that it is a probability on $\mathcal{F}$, according to the classical measure extension theorem, P and then P^0 can be uniquely extended to $\mathcal{B}$, the σ-algebra generated (from $\mathcal{F}$ then) from $\mathcal{M}$. ■

3 Falling measures representation theorems

Choquet gave a famous representation theorem which use the probabilistic distribution in the power of X to represent the Capacity in X. He stated and proved his theorem under some topological assumption. Dempster and Shafer gave a representation theorem which use the probabilistic distribution to represent the non-additive measures such as belief functions etc. on X. They stated and proved the theorems for X being a finite space. In this section we will state and prove the representation theorem in general situation by pure measure theoretical method without topological assumption.

In the first, we need to see two important examples of regular meet-systems on $\mathcal{P}(X)$. For simplicity, we use symbols $\cup$, $\cap$ and ^{c} still to represent the operations of union, intersection, and complementary on $\mathcal{P}(X)$. ·

Example 1. Let $\mathcal{C} = \mathcal{P}(X)$, $\mathcal{F}$ be an algebra on X. Set

$$\underline{A} = \{B \in \mathcal{C} : B \subseteq A\},\ \mathcal{M}_1 = \{\underline{A} : A \in \mathcal{F}\}.$$

We have following a Proposition:

Proposition 1. *$\mathcal{M}_1$ is a regular meet-system on $\mathcal{P}(X)$.*

Proof. Consider a monotone increasing sequence $E_n = \cup_{j=1}^{m[n]} \underline{B}_{nj} \uparrow \underline{A}$ $(B_{nj}, A \in \mathcal{F})$. Since that $E_n \uparrow$, for any B_{nj} there is some $B_{(n+1)j'} \supseteq B_{nj}$. Otherwise, there are $x_{j'} \in B_{nj} \setminus B_{(n+1)j'}$ $(j' = 1, \ldots, m[n+1])$. Then, $\{x_1, \ldots, x_{m[n+1]}\} \in \underline{B}_{nj} \setminus E_{n+1}$. This is a contradiction with $E_n \subseteq E_{n+1}$. So that each B_{nj} belongs to at least an increasing sequence which consists of some sets among $\{B_{nj} : n \geq 1;\ j \in \{1, \ldots, m[n]\}$. Set

$$\mathcal{L} = \left\{L : \exists\{B_{nj[n]}\}\,(n \geq k)(k \geq 1) \text{ such that } L = \lim_{n\to\infty} B_{nj[n]}\right\}.$$

Then, we have that $\cup_{n=1}^{\infty} E_n \subseteq \cup\{\underline{L} : L \in \mathcal{L}\}$. Now, there is an $L^* \in \mathcal{L}$ and $L^* = A$. Otherwise, there are $x_L \in A \setminus L$ $(L \in \mathcal{L})$, then $\{x_L : L \in$

$\mathcal{L}\} \in \underline{A} \setminus \cup\{\underline{L} : L \in \mathcal{L}\} \subseteq \underline{A} \setminus \cup_{n=1}^{\infty} E_n$. This is a contradiction with that $\cup_{n=1}^{\infty} E_n = \underline{A}$.

Track the sequence $\{B_{nj[n]}^*\}$ which converges to L^*, and set

$$H_n = \underline{B}_{nj[n]}^* \text{ when } n \geq k; \; H_n = \emptyset \text{ otherwise.}$$

It is clear that $H_n \uparrow A \; (n \to \infty)$. For any n take $N(n) = n$. Since $B_{nj[n]}^* \in E_n$, so that $E_k \supseteq H_n$ when $k \geq N(n)$. Then $\{E_n\}$ is asymptotically chased by $\{H_n\}$. Therefore, $\{E_n\}$ is regular.

Suppose that $E_n \uparrow \underline{A}_1 \cup \ldots \underline{A}_m \in \mathcal{M}_{1\cup}$. Set $E_n^{(i)} = E_n \cap \underline{A}_i \; (i = 1, \ldots, m)$, we have that $E_n^{(i)} \uparrow \underline{A}_i$. As mentioned above, there are monotone increasing sequences $\{H_n^{(i)}\} \; (i = 1, \ldots, m) \subseteq \mathcal{M}_{1\cup}$. $H_n^{(i)} \uparrow \underline{A}_i \; (n \to \infty)$ and $\{E_n^{(i)}\}$ are asymptotically chased by $\{H_n^{(i)}\}$, i.e., for any $n \geq 1$, there are $N_i(n)$ such that $E_k^{(i)} \supseteq H_n^{(i)}$ whenever $k \geq N_i(n)$. Set

$$N(n) = \max\{N_1(n), \ldots, N_m(n)\}.$$

When $k \geq N(n)$, we have that

$$E_k = E_k \cap (\underline{A}_1 \cup \ldots \cup \underline{A}_m) = E_k^{(1)} \cup \ldots \cup E_k^{(m)} \supseteq H[(1)_n \cup \ldots \cup H_n^{(m)}.$$

It means that $\{E_n\}$ is asymptotically chased by $\{H_n^{(1)} \cup \ldots \cup H_n^{(m)}\}$. Therefore, $\{E_n\}$ is regular.

We have proved the first part, and now we are going to prove the second part of the proposition.

Consider that a monotone decreasing sequence $E_n = \cup_{j=1}^{m[n]} \underline{B}_{nj} \downarrow \underline{A}_1 \cup \ldots \cup \underline{A}_m \in \mathcal{M}_{1\cup} \; (A_i \not\subset A_j \; (i \neq j))$. As mentioned before, since $E_n \downarrow \underline{A}_1 \cup \ldots \cup \underline{A}_m$, for any i and any n, there is $\underline{B}_{nj} \supseteq \underline{A}_i$, then $B_{nj} \supseteq A_i$. For any i, set

$$A_n^{(i)} = \cup\{B_{nj} : B_{nj} \supseteq A_i\} \; (\neq \emptyset) \,|, (n \geq 1).$$

By the same reason mentioned before, we have that for any $B_{nj} \supseteq A_i$, there is some $B_{(n-1)j'} \in E_{n-1} : B_{(n-1)j'} \supseteq B_{nj}$. Obviously, we have that $B_{(n-1)j'} \supseteq A_i$. So that $\{A_n^{(i)}\}$ is a monotone decreasing sequence.

Now we have that $\underline{A}_n^{(i)} \downarrow \underline{A}_i \; (n \to \infty)$. Otherwise, there is $x \in (\cup_{n=1}^{\infty} A_n^{(i)}) \setminus A_i$. Then, $A_n^{(i)} \supseteq (A_i \cup x)$ for any $n \geq 1$. But $A_n^{(i)}$ is the union of some B_{nj} which contains A_i, so that there is $B_{nj[n]} \supseteq (A_i \cup x)$ for any $n \geq 1$. Set

$$G = \cap_{n=1}^{\infty} B_{nj[n]}.$$

$G \notin \underline{A}_i$ since $x \in G$. And $G \notin \underline{A}_j \; (j \neq i)$ since $G \supseteq A_i$ and $A_i \not\subset A_j$. So that $G \notin \underline{A}_1 \cup \ldots \cup \underline{A}_m$. But $G \in E_n$ for any $n \geq 1$, then $G \in \cap_{n=1}^{\infty} E_n$. So that $\cap_{n=1}^{\infty} E_n \neq \underline{A}_1 \cup \ldots \cup \underline{A}_m$. This is a contradiction with that $E_n \downarrow \underline{A}_1 \cup \ldots \cup \underline{A}_m \; (n \to \infty)$.

We hope that $\{E_n\}$ will be chased by $\{\underline{A}_n^{(1)} \cup \ldots \cup \underline{A}_n^{(m)}\}$. Unfortunately, there may have some B_{nj} which do not contain any A_i. Set $U_1 = \cup\{B_{1j} : B_{1j}$ does not contain any $A_i\}$. When U_n has been given, then set $U_{n+1} = \cup\{B_{(n+1)j} : B_{(n+1)j}$ does not contain any A_i, $\exists B_{nj'} \in \underline{U}_n$, $B_{nj'} \supseteq B_{(n+1)j}\}$ $(n \geq 1)$. Then we get a monotone decreasing sequence $\{U_n\}$. (U_n may be the empty set $\emptyset$). Set $U = \lim_{n\to\infty} U_n$. For any $x \in U$, there is a monotone decreasing sequence $s = \{B_{nj[n]}\}$, $B_{nj[n]} \in \underline{U}_n$ such that $\lim_{n\to\infty} B_{nj[n]}$ (denoted as $\underline{s}$) $\ni x$. We can prove that $\underline{s}$ must be contained in some A_i. Otherwise, there are two points $a, b \in \underline{s}$, $\{a, b\} \notin \underline{A}_1 \cup \ldots \cup \underline{A}_m$. Note that for any n, then $\underline{s} \in E_n$. So that $\{a, b\} \in \cap E_n \setminus (\underline{A}_1 \cup \ldots \cup \underline{A}_m)$. This is a contradiction with that $E_n \downarrow \underline{A}_1 \cup \ldots \cup \underline{A}_m$.

Denote $B_{nj0} \in s = \{B_{nj[n]}\}$ if $B_{nj[n]} = B_{nj0}$. Set

$$S = \{s = \{B_{nj[n]}\} : s \text{ is monotone decreasing, } B_{nj[n]} \in U_n \ (n \geq 1)\}.$$

For any i, set

$$V_n^{(i)} = \{B_{nj} \in U_n : \exists s \in S, \underline{s} \subseteq A_i, B_nj \in s\}.$$

$V_n^{(i)}$ are monotone decreasing. It is not difficult to prove that $\lim_{n\to\infty} V_n^{(i)} = V^{(i)}$ must be contained in A_i. Since $A_n^{(i)} \supseteq A_i$, so that

$$V_n^{(i)} \setminus A_n^{(i)} \subseteq V_n^{(i)} \setminus A_i \downarrow V^{(i)} \setminus A_i = \emptyset.$$

Note that $U_n = V_n^{(1)} \cup \ldots \cup V_n^{(m)}$, it is clear that $\{E_n\}$ is chased by $\{\underline{A}_n^{(1)} \cup \ldots \cup \underline{A}_n^{(m)}) \cup V_n^{(1)} \cup \ldots \cup V_n^{(m)}\}$. Therefore, $\{E_n\}$ is a regular decreasing sequence. ∎

Example 1. Let $\mathcal{C} = \mathcal{P}(X)$, $\mathcal{F}$ be an algebra on X. Set

$$\overline{A} = \{B \in \mathcal{C} : B \supseteq A\}, \ \mathcal{M}_2 = \{\overline{A} : A \in \mathcal{F}\}.$$

We have following a Proposition:

Proposition 2. *$\mathcal{M}_2$ is a regular meet-system on $\mathcal{P}(X)$.*

Proof. The proof is similar to that of Proposition 1. ∎

From the two examples, we can get the falling measures' representation theorems.

Definition 3. Let $\mathcal{F}$ be an algebra on X. $(\mathcal{P}(X), \mathcal{B}_i)$ are called the *type-i hyper-measurable space* where $\mathcal{B}_i$ is the algebra generated from $\mathcal{M}_i$ defined in Example i $(i = 1, 2)$.

Definition 4. Given probability p_i on the type-i hyper-measurable space $(\mathcal{P}(X), \mathcal{B}_i)$ $(i = 1, 2)$. Set

$$f_{aa}(C) = p_1(\underline{C}) = p_1\{B \in \mathcal{P}(X) : B \subseteq C\} \ (C \in \mathcal{F}) \tag{6}$$

$$f_{ab}(C) = 1 - p_1(\underline{C^c}) = p_1\{B \in \mathcal{P}(X) : B \cap C \neq \emptyset\} \ (C \in \mathcal{F}) \tag{7}$$

$$f_{ba}(C) = p_2(\overline{C}) = p_2\{B \in \mathcal{P}(X) : B \supseteq C\} \ (C \in \mathcal{F}) \tag{8}$$

$$f_{bb}(C) = 1 - p_2(\overline{C^c}) = p_2\{B \text{ does not } \supseteq C^c\} \ (C \in \mathcal{F}) \tag{9}.$$

They are called four kinds of *falling measures* on $\mathcal{F}$. The related probability is called the *represented probability.*

The four kinds of falling measures are consistent with the definitions of four kinds non-additive measures by Shafer (1976). They are the Belief, Plausibility, Anti-belief, Anti-plausibility measure respectively. Here, we have given a more critical definition to them.

Theorem 3. (Representation Theorem 1) *A set function f defined on an algebra $\mathcal{F}$ is the f_{aa} type falling measure of some probability P in the type-1 hyper-measurable space if, and only if, it satisfies:*

1. $f(X) = 1$, $f(\emptyset) = 0$;
2. $\{B_n\} \subseteq \mathcal{F}$, $B_n \uparrow A \in \mathcal{F}$ *or* $B_n \downarrow A \in \mathcal{F}$ $(n \to \infty)$ *implies that* $f(B_n) \to f(A)$ $(n \to \infty)$;
3. *For any* $n > 0$, $B_1, \ldots, B_n; A \in \mathcal{F}$

$$\sum_{I \subseteq I_n} (-1)^{|I|} f(\cap_{i \in I} B_i \cap A) \geq 0 \tag{10}.$$

Proof. Sufficiency.

Suppose that f satisfies conditions 1), 2) and 3). According to Proposition 1, $\mathcal{M}_1 = \{\underline{A} : A \in \mathcal{F}\}$ is a regular meet-system. Define a mapping on $\mathcal{M}_1$, $P^0 : \mathcal{M}_1 \to [0,1]$ such that $P^0(\underline{C}) = f(C)$ $(\underline{C} \in \mathcal{M}_1)$.

Since f satisfies (10) and for any $C_1, \ldots, C_t \in \mathcal{M}_1$, we have that

$$\underline{C^{(1)} \cap \ldots \cap C^{(t)}} = \underline{C}^{(1)} \cap \ldots \cap \underline{C}^{(t)}.$$

So that, for any $n > 0, B_1, \ldots, B_n; A \in \mathcal{F}$

$$\sum_{I \subseteq I_n} (-1)^{|I|} P^0(\cap_{i \in I} \underline{B_i} \cap \underline{A}) = \sum_{I \subseteq I_n} (-1)^{|I|} P^0(\underline{\cap_{i \in I} B_i \cap A})$$

$$= \sum_{I \subseteq I_n} (-1)^{|I|} f(\cap_{i \in I} B_i \cap A) \geq 0$$

P^0 satisfies (4). And since f satisfies conditions 1) and 2), P^0 is a probabi on $\mathcal{M}_1$. According to the Extension Theorem - Part 1, it uniquely determines a probability P on $\mathcal{B}_1$, and f is the f_{aa}-type falling measure of P. The sufficiency has been proved.

The proof of necessity is similar. ∎

Theorem 4. (Representation Theorem 2) *A set function f defined on an algebra $\mathcal{F}$ is the f_{ab} type falling measure of some probability P in the type-1 hyper-measurable space if, and only if, it satisfies:*

1. $f(X) = 0$, $f(\emptyset) = 1$;
2. $\{B_n\} \subseteq \mathcal{F}$, $B_n \uparrow A \in \mathcal{F}$ *or* $B_n \downarrow A \in \mathcal{F}$ $(n \to \infty)$ *implies that* $f(B_n) \to f(A)$ $(n \to \infty)$;
3. *For any* $n > 0$, $B_1, \ldots, B_n; A \in \mathcal{F}$

$$\sum_{I \subseteq I_n} (-1)^{|I|} f(\cup_{i \in I} B_i \cup A) \leq 0 \qquad (11).$$

Proof. Set $g(A) = 1 - f(A^c)$. For any $n > 0, B_1, \ldots, B_n; A \in \mathcal{F}$ we have that

$$\sum_{I \subseteq I_n} (-1)^{|I|} g(\cap_{i \in I} B_i^c \cap A^c) \geq 0$$

$$\Leftrightarrow \sum_{I \subseteq I_n} (-1)^{|I|} g(\cup_{i \in I} B_i \cup A)c \geq 0$$

$$\Leftrightarrow \sum_{I \subseteq I_n} (-1)^{|I|} [1 - f(\cup_{i \in I} B_i \cup A)] \geq 0$$

$$\Leftrightarrow \sum_{I \subseteq I_n} (-1)^{|I|} f(\cup_{i \in I} B_i \cup A) \leq 0.$$

It means that f satisfies (11) if and only if g satisfies (10). But, according to the definition of falling measures, we know that f is in f_{ab} type if and only if g is in f_{aa} type, so that Representation Theorems 1 and 2 are dual. The rest of the proof is obvious. ■

Without proof, we give the rest two theorems as follows:

Theorem 5. (Representation Theorem 3) *A set function f defined on an algebra $\mathcal{F}$ is the f_{ba} type falling measure of some probability P in the type-1 hyper-measurable space if, and only if, it satisfies:*

1. $f(X) = 0$, $f(\emptyset) = 1$;
2. $\{B_n\} \subseteq \mathcal{F}$, $B_n \uparrow A \in \mathcal{F}$ *or* $B_n \downarrow A \in \mathcal{F}$ $(n \to \infty)$ *implies that* $f(B_n) \to f(A)$ $(n \to \infty)$;
3. *For any* $n > 0$, $B_1, \ldots, B_n; A \in \mathcal{F}$

$$\sum_{I \subseteq I_n} (-1)^{|I|} f(\cap_{i \in I} B_i \cap A) \leq 0 \qquad (12).$$

Theorem 6. (Representation Theorem 4) *A set function f defined on an algebra $\mathcal{F}$ is the f_{ab} type falling measure of some probability P in the type-1 hyper-measurable space if, and only if, it satisfies:*

1. *$f(X) = 1$, $f(\emptyset) = 0$;*
2. *$\{B_n\} \subseteq \mathcal{F}$, $B_n \uparrow A \in \mathcal{F}$ or $B_n \downarrow A \in \mathcal{F}$ $(n \to \infty)$ implies that $f(B_n) \to f(A)$ $(n \to \infty)$;*
3. *For any $n > 0$, $B_1, \ldots, B_n; A \in \mathcal{F}$*

$$\sum_{I \subseteq I_n} (-1)^{|I|} f(\cup_{i \in I} B_i \cup A) \leq 0 \qquad (13).$$

4 Conclusion

As known, the representation theorems was been stated and proved for finite sets or infinite sets with some topological assumptions. Here, we have had a pure measure theoretical statement and a clear proof to the representation theorems in the general situation. It is quite important and useful in theory and applications.

References

1. Choquet, G. (1953-54). Theory of capacities, *Ann. Inst. Fourier*, 131-295
2. Dempster, A.P. (1967). Upper and lower probabilities induced by multivalued mapping, *Ann, Math. Stat.* **38**, 325-339
3. Dubois, D. and Prade, H. (1982). Fuzzy sets and statistical data, *Ensembles Flous-82 Notes, Communications, Articles on 1982.*
4. Goodman, I.R. (1982). Fuzzy sets as equivalance class of random sets, In *Recent Developments in Fuzzy Sets and Possibility Theory*, (R. Yager, ed.).
5. Goodman, I.R. and Nguyen, H.T. (1985). *Uncertainty Models for Knowledge-based Systems*, North-Holland, Amsterdam.
6. Matheron, G. (1975). *Random Sets and Integral Geometry*, John Wiley & Sons, New York
7. Shafer, G. (1976). *A Mathematical theory of Evidence*, Princeton Univ. Press, Princeton.
8. Wang, P.Z. and Sanchez, E. (1982). Treating a fuzzy subset as a projectable random set, In *n Fuzzy Information and decision*, (M.M. Gupta and E. Sanchez eds.), Pergamon Press, 212-219.
9. Wang, P.Z. (1983). From the fuzzy statistics to the falling random sets, In *Advance in Fuzzy Sets Theory and Applications*, (Paul P. Wang ed.), Pergamon Press, 81-95.
10. Wang, P.Z. and Sanchez, E. (1983). Hyper fields and random sets, In *Fuzzy Information, knowledge Representation and Decision Analysis*, (E. Sanchez ed.), Pergamon Press, 335-337.
11. Wang, P.Z. (1985). *Fuzzy Sets and the Falling Shadows of Random Sets*, Beijing Normal University Press, Beijing.
12. Wang, P.Z., Zhang, H.M., Ma, X.W. and Xu, W. (1991). Fuzzy set-operations represented by falling shadow theory, In *Fuzzy Engineering toward Human Friendly Systems, Proc. Intern. Fuzzy Engineering Symposium'91*, Yokohama, Vol.1, 82-90.

13. Wang, P.Z., Huang, M. and Zhang, D.Z. (1992). Reexamining fuzziness and randomness using falling shadow theory, *Proc. 10th International Conference on Multiple Criteria Decision Making*, Taipei, 101-110.

The structure of fuzzy measure families induced by upper and lower probabilities

Andrew G. Bronevich[1] and Alexander N. Karkishchenko[1]

Taganrog State University of Radio-Engineering Nekrasovskij bystreet, 44
Taganrog, RUSSIA

Abstract. This paper contains researches in the fuzzy measure theory. Convex families of measures are considered, among them upper and lower probabilities, superadditive measures, and it is found that these measures can be represented by sums of primitive measures. It gives a possibility us to get important results due to algebraic structure of fuzzy measures on the finite algebra and to generalize the well-known theorems of the probability theory.

The main aim of this paper is to construct and to investigate a new approach in the fuzzy measure theory (Sugeno, 1972) that consists in the following conception. The random set theory enables (Dempster, 1967, Shafer, 1976, Dubois and Prade, 1992) us to understand a value of fuzzy measure as a lower or upper estimate of probability. Taking this into account the family of all fuzzy measures can be divided into three classes:

1) lower probabilities;

2) upper probabilities;

3) contradictory measures that can not be interpreted as lower or upper probabilities.

This paper introduces some results concerning the algebraic structure of the set of fuzzy measures and its subfamilies pointed above. We also make an assertion that our investigation is a development of the random set theory but in the wider class of fuzzy measures.

1 The main notions and theoretical constructions

The set function g on the finite algebra $\Im$ of the measurable space X is called a *fuzzy measure* (Sugeno, 1972) if the following conditions hold:

1) $g(\emptyset) = 0$, $g(X) = 1$ (norming);

2) $g(A) \geq 0$ for all $A \in \Im$ (non-negativeness);

3) $g(A) \leq g(B)$, if $A \subseteq B$ (monotonicity).

Lemma 1. *Let g be a fuzzy measure upon algebra $\Im$ of the space X then the set function $q(A) = 1 - g(\bar{A})$ is a fuzzy measure upon $\Im$ too.*

The fuzzy measure q from Lemma 1 is called usually as a *dual* one to the generating measure g. It is clear that the duality relation among fuzzy measures is symmetric, i.e. fuzzy measure g from Lemma 1 is dual to q.

The fuzzy measure Π on the algebra $\Im$ is a *possibility measure* (Dubois and Prade, 1992) if

1) $\Pi(\emptyset) = 0$, $\Pi(X) = 1$;
2) $\Pi(A) \geq 0$ for an arbitrary $A \in \Im$;
3) $\Pi(A \cup B) = \max(\Pi(A), \Pi(B))$.

When the space X is finite, i.e. $X = \{x_1, x_2, \ldots, x_N\}$, the function $\pi(x_i) = \Pi(\{x_i\})$ is called possibility distribution. It is easy to show

$$\Pi(A) = \max_{x \in A} \pi(x) \text{ for any } A \neq \emptyset. \tag{1}$$

It means the possibility distribution determines the possibility measure Π uniquely. In the opposite direction: if we have a function $\pi(x)$ on X and the following conditions are provided

1) $\pi(x) \geq 0$ for all $x \in X$;
2) $\max_{x \in X} \pi(x) = 1$;

then we determine a possibility measure by means of the formula (1). The fuzzy measure N being a dual one to the possibility measure is called a *necessity measure*

Consider now *belief* and *plausibility* measures. To do this we suggest that a non-negative function m is determined on $\Im$ and the following condition holds:

$$\sum_{A \in \Im} m(A) = 1.$$

Then the set function

$$Cr(B) = \sum_{A \subseteq B} m(A) \tag{2}$$

is called a belief measure and the corresponding dual measure is a plausibility measure:

$$Pl(B) = \sum_{A \cap B \neq \emptyset} m(A). \tag{3}$$

The following result (Shafer, 1976) shows the connection between the possibility and plausibility measures.

Theorem 1. *Let $\{A_i | m(A_i) \neq 0\} \subseteq \Im$ be a set of focal elements of $\Im$. Then plausibility measure Pl (belief measure Cr) will be a possibility measure (necessity measure) if and only if the set of focal elements is coherent, i.e. linearly ordered by means of the set-theoretical inclusion relation.*

2 Representation of fuzzy measure by the convex sum of primitive fuzzy measures

We establish some simple facts about the structure of the set of fuzzy measures on a finite algebra $\Im$ of a type 2^X (the space X has a finite number of elements). Let g_1 and g_2 be two fuzzy measures upon $\Im$. Then their convex combination $g = \alpha_1 g_1 + \alpha_2 g_2$, $\alpha_1 + \alpha_2 = 1$, $\alpha_1, \alpha_2 \geq 0$, is also a fuzzy measure. Indeed, $g(\emptyset) = 0$, $g(X) = 1$ and if $A \subseteq B$, then $g(A) = \alpha_1 g_1(A) + \alpha_2 g_2(A) \leq \alpha_1 g_1(B) + \alpha_2 g_2(B) = g(B)$. Thus, the set of all fuzzy measures is convex.

Introduce the following definition. The fuzzy measure is called a *primitive measure* if its values belong to the set $\{0, 1\}$ for an arbitrary event.

For more detailed description of the set of fuzzy measures we need some definitions of the theory of partially ordered sets (or posets). We shall consider the algebra $\Im$ as the poset with respect to the ordinary set-theoretical inclusion. Filter is such a subset $\mathbf{f}$ of $\Im$ that if $A \in \mathbf{f}$ and $A \subseteq B$ then $B \in \mathbf{f}$. We take by definition that neither filter contains $\emptyset$. However, each filter contains a set of minimal elements $\{A_1, A_2, \ldots, A_k\}$, i.e. such mutually incomparable elements that $\mathbf{f} = \{A \in \Im | \exists A_i \subseteq A\}$. Clearly minimal elements generate a filter $\mathbf{f}$: this fact is noted by $\mathbf{f} = \langle A_1, A_2, \ldots, A_k \rangle$. The filter $\mathbf{f}$ is principal one if it is generated by one minimal element, i.e. $\mathbf{f} = \langle A \rangle$.

Lemma 2. *Let* $\mathbf{f}$ *be an arbitrary filter in* $\Im$ *and* $\eta_{\mathbf{f}}$ *be a characteristic function of the filter* $\mathbf{f}$, *i.e.*

$$\eta_{\mathbf{f}}(A) = \begin{cases} 1, & A \in \mathbf{f}, \\ 0, & A \notin \mathbf{f}, \end{cases}$$

then $\eta_{\mathbf{f}}$ *is a primitive measure. Conversely each primitive measure is associated with a certain filter* $\mathbf{f}$.

Notice that among primitive measures we can extract primitive necessity measures being associated with principal filters $\mathbf{f} = \langle A \rangle$ in $\Im$. Sometimes these measures are called Dirac measures. In particular, when $\langle A \rangle = \langle \{x_i\} \rangle$ $x_i \in X$, Dirac measure will be a probability measure concentrated in the point x_i.

The following theorems describe important characteristics of the convex set of fuzzy measures.

Theorem 2. *Any fuzzy measure* g *can be represented as a convex combination of primitive measures.*

Proof. Let g_0 be an arbitrary fuzzy measure. Construct the filter $\mathbf{f}_0 = \{A \in \Im | g_0(A) > 0\}$ and denote $t_0 = \min_{A \in \mathbf{f}_0} g_0(A_i)$. If $t_0 = 1$, then the representation required has been found because $g_0 = \eta_{\mathbf{f}_0}$. In other case, if $t_0 < 1$, it is easy to show that the set function $g_1(A) = \dfrac{g_0(A) - t_0 \eta_{\mathbf{f}_0}}{1 - t_0}$ will be a fuzzy measure. By analogy for the fuzzy measure g_1 we can get the filter $\mathbf{f}_1$, the value t_1, the function g_2 and so on. As a result of this procedure we have the finite consequence of embedded filters $\mathbf{f}_0 \supset \mathbf{f}_1 \supset \mathbf{f}_2 \supset \ldots \supset \mathbf{f}_K$ and the consequence of fuzzy measures that satisfy the following conditions:

$$\begin{cases} g_1(A) = \dfrac{g_0(A) - t_0\eta_{\mathfrak{f}_0}}{1 - t_0}, \\ \vdots \\ g_{i+1}(A) = \dfrac{g_i(A) - t_i\eta_{\mathfrak{f}_i}}{1 - t_i}, \ldots \\ \vdots \\ g_{K+1}(A) = \eta_{\mathfrak{f}_K} \end{cases} \tag{4}$$

Solving the equation system (4) we shall find the representation of the fuzzy measure g_0 by the convex sum of primitive measures. ■

Theorem 3. *Neither primitive measure can be represented by a convex sum of other primitive measures.*

Proof. Suppose the contradictory statement that for a certain primitive measure $\eta_{\mathfrak{f}}$ the following representation exists $\eta_{\mathfrak{f}} = \alpha_1 g_1 + \alpha_2 g_2$, $\alpha_1 + \alpha_2 = 1$, $0 < \alpha_2 < 1$, $g_1 \neq \eta_{\mathfrak{f}}$. Then we can find event A, $A \in \Im$ with the following property $g_1(A) \neq \eta_{\mathfrak{f}}(A)$. From this we should consider only two possible cases:

1) $0 = \eta_{\mathfrak{f}}(A) < g_1(A)$. In this case we infer the contradiction: $\eta_{\mathfrak{f}}(A) = \alpha_1 g_1(A) + \alpha_2 g_2(A) > 0$.
2) $1 = \eta_{\mathfrak{f}}(A) > g_1(A)$, then $\eta_{\mathfrak{f}}(A) = \alpha_1 g_1(A) + \alpha_2 g_2(A) < \alpha_1 g_1(A) + \alpha_2 < 1$.

Thus, the supposition has been made is fault and the theorem has been proved. ■

One can easily prove that representation of fuzzy measure by the convex combination of primitive measures is not unique. This can be shown by examples (appendix).

We also point out that plausibility (belief) measures can be represented in a form of a convex linear combination of primitive possibility (necessity) measures. Actually, consider the primitive necessity measure

$$\eta_{\langle A \rangle}(B) = \begin{cases} 1, \ A \subseteq B, \\ 0, \ A \not\subseteq B, \end{cases} \quad A, B \in \Im.$$

Then, using expression (2), we obtain equality

$$Cr(B) = \sum_{A \in \Im} m(A)\eta_{\langle A \rangle}(B). \tag{5}$$

By analogy the formula for the plausibility measure can be derived. It is based on the convex sum of primitive possibility measures. One can prove (Shafer, 1976) that the representation (5) is determined uniquely.

3 The probability interpretation of fuzzy measures

The papers by Dempster (1967) and Shafer (1976) give us possibility to interpret fuzzy measures as upper or lower probabilities. Taking this in our account we introduce the following definitions.

The fuzzy measure g is called a *lower probability* if there is a probability measure P upon $\Im$ with the following property: for all $A \in \Im$ $g(A) \leq P(A)$.

By analogy, the fuzzy measure q is an *upper probability* if there is probability measure P upon $\Im$ that satisfies the following condition: for all $A \in \Im$ $q(A) \geq P(A)$.

It is easy to notice that upper and lower probabilities are connected with the dual relation: if g is a lower probability then the fuzzy measure q being dual to g is an upper probability. This lets us carry out further investigation only for lower probabilities. Properties of upper probabilities can be easily inferred by means of the dual relation.

Theorem 4. *Fuzzy measure g is a lower probability if it satisfies a property of a weak superadditivity, i.e.*

$$g(A \cup B) \geq g(A) + g(B) - g(A \cap B) \text{ for any } A, B \in \Im. \tag{6}$$

Proof. Let the space X contains a finite number of elements, i.e. $X = \{x_1, x_2, \ldots, x_n\}$, and the fuzzy measure g is determined on the algebra $\Im = 2^X$. We will prove this theorem by induction. If $n = 1, 2$, the theorem is obviously valid. For example, if $n = 2$, we set the probabilities in the following way: $P\{x_1\} = g\{x_1\}$, $P\{x_2\} = 1 - g\{x_1\}$, then using inequality $g\{x_1\} + g\{x_2\} \leq g\{X\} = 1$, we get $g\{x_2\} \leq P\{x_2\}$.

Further we assume that theorem is valid for all n, $n \leq m - 1$ and using it will prove this theorem for $n = m$. We assign the probability $P\{x_m\} = 1 - g(X \setminus \{x_m\})$, then under the weak superadditivity property we get $g\{x_m\} \leq P\{x_m\}$.

If $P\{x_m\} = 1$, then probability measure required has been found: $P(A) = \begin{cases} 1, x_m \in A \\ 0, x_m \notin A \end{cases}$ because under the monotonicity condition $g(A) = 0$ if $A \subseteq X \setminus x_m$, i.e. $g \leq P$.

If $P\{x_m\} < 1$ consider a fuzzy measure $g_1(A) = \dfrac{g(A)}{1 - P\{x_m\}}$ on the algebra $\Im_1 = 2^{X_1}$ of the space $X_1 = X \setminus \{x_m\}$. It is easy to show that this measure satisfies the weak superadditivity property and, hence, by the inductive supposition is a lower probability, i.e. there is a such probability measure P_1 that $g_1(A) \leq P_1(A)$ for all $A \in \Im_1$. Using the measure P_1 we construct the probability measure P on $\Im$ as follows:

$$P(A) = P(A \cap \{x_m\}) + P_1(A \setminus \{x_m\})(1 - P\{x_m\}).$$

It is easy to notice that this measure satisfies the all properties that are required:

1) if $x_m \notin A$ then $P(A) \geq g_1(A)(1 - P\{x_m\}) = g(A)$;

2) if $A = \{x_m\} \cup B$, $\{x_m\} \cap B = \emptyset$ then $P(A) \geq 1 - g(X \setminus \{x_m\}) + g_1(B)(1 - P\{x_m\}) = 1 - g(X \setminus \{x_m\}) + g(B) \geq g(A)$.

The last inequality has been derived using the weak superadditivity property. ■

Remark 1. The belief measure is a lower probability because it satisfies the superadditivity property (Shafer, 1976):

$$g(A_1 \cup A_2 \cup \ldots \cup A_n) \geq \sum_{i=1}^{n} g(A_i) - \sum_{i<j} g(A_i \cap A_j) +$$

$$+ \sum_{i<j<k} g(A_i \cap A_j \cap A_k) - \ldots + \ldots (-1)^{n-1} g(A_1 \cap A_2 \cap \ldots \cap A_n).$$

In particular, if $n = 2$, we obtain the property of the weak superadditivity:

$$g(A_1 \cup A_2) \geq g(A_1) + g(A_2) - g(A_1 \cap A_2);$$

for $n = 3$ we have inequality

$$g(A_1 \cup A_2 \cup A_3) \geq g(A_1) + g(A_2) + g(A_3) - g(A_1 \cap A_2) -$$
$$- g(A_1 \cap A_3) - g(A_2 \cap A_3) + g(A_1 \cap A_2 \cap A_3).$$

that is more strong then weak superadditivity.

It is natural to put a question - what kind of primitive measures are lower probabilities and what fuzzy measures can be represented by the convex sum of such measures. The following results give us the answer.

Lemma 3. *Let $\eta_{\mathbf{f}}$ be a primitive measure generated by a certain filter $\mathbf{f} = \langle A_1, A_2, \ldots, A_k \rangle$. Then $\eta_{\mathbf{f}}$ will be a lower probability if and only if $\bigcap_{i=1}^{k} A_i \neq \emptyset$.*

Proof. Let $\eta_{\mathbf{f}}$ be a lower probability then there is such a probability measure P that $\eta_{\mathbf{f}}(A) \leq P(A)$ for all $A \in \Im$. In particular one can find that $P(A_1) = P(A_2) = \ldots = P(A_k) = 1$, i.e. all minimal elements of the filter $\mathbf{f}$ are certain events. Hence the event $A_1 \cap A_2 \cap \ldots \cap A_k$ will be a certain event too, i.e. $P(A_1 \cap A_2 \cap \ldots \cap A_k) = 1$. Last equation lets us assert that $A_1 \cap A_2 \cap \ldots \cap A_k \neq \emptyset$.

Conversely, if $A_1 \cap A_2 \cap \ldots \cap A_k \neq \emptyset$, we choose probability measure P concentrated in the point $x \in A_1 \cap A_2 \cap \ldots \cap A_k$, i.e.

$$P(A) = \begin{cases} 1, x \in A, \\ 0, x \notin A. \end{cases}$$

It is clear that $\eta_{\mathbf{f}}(A) \leq P(A)$ for all $A \in \Im$. ■

Up to the end primitive measures that belong to the lower probability family we call coherent primitive measures. Also for the sake of brevity we denote $g \leq q$ if $g(A) \leq q(A)$ for all $A \in \Im$.

Lemma 4. *Let g_1 and g_2 be lower probabilities then their convex combination $g = \alpha_1 g_1 + \alpha_2 g_2$, $\alpha_1 + \alpha_2 = 1$, $\alpha_1, \alpha_2 \geq 0$ will be also the lower probability.*

Proof. Suppose that $g_1 \leq P_1$, $g_2 \leq P_2$, where P_i, $i = 1, 2$, are probability measures, then $g = \alpha_1 g_1 + \alpha_2 g_2 \leq \alpha_1 P_1 + \alpha_2 P_2 = P$, i.e. measure g is also a lower probability. ∎

Let $\chi(x) = \begin{cases} x, x \geq 0, \\ 0, x < 0. \end{cases}$. The fuzzy measure g is called monotonic in a probability measure P, if for any α, $\alpha \in [0,1]$ the set function $q(A) = \chi[g(A) - \alpha P(A)]$ has a monotonic property.

Lemma 5. *Any monotonic fuzzy measure g in a probability measure P is a lower probability.*

Proof. Let g be monotonic fuzzy measure in a probability measure P. Consider the set function $q(A) = \chi[g(A) - P(A)]$. It is easy to notice that $q(X) = 0$, and, because q is monotonic function, it is identical to zero, hence $P \geq g$. ∎

Lemma 6. *The fuzzy measure g is monotonic in a probability measure P if and only if for arbitrary events $A, B \in \Im$ such that $g(A) > 0$, $A \subseteq B$, the following inequality is provided*

$$g(A) - \alpha P(A) \leq g(B) - \alpha P(B) \text{ for any } \alpha \in \left[0, \frac{g(A)}{P(A)}\right].$$

Proof. Lemma 6 is valid for the pointed inequality should be only fulfilled by definition for non-negative values of difference $g(S) - \alpha P(A)$. ∎

Corollary 1. *The fuzzy measure g is monotonic in a probability measure P if and only if for arbitrary such events $A, B \in \Im$, that $A \subseteq B$, the following inequality holds: $P(A)g(B) - P(B)g(A) \geq 0$.*

By definition the fuzzy measure g is *weakly additive* if the there are probability measure P and the filter $\mathbf{f}$ that describe the measure g in the following way:

$$g(A) = \begin{cases} P(A), \ A \in \mathbf{f}, \\ \quad 0, \quad F \notin \mathbf{f}. \end{cases} \tag{7}$$

Denote the weakly additive measure g from the expression (7) by $P_{\mathbf{f}}$. The following theorem indicates the important properties of weakly additive measures.

Theorem 5. *Let P be a probability measure. Then fuzzy measure family $\mathcal{M}_P = \{\mu_i\}$, $\mu_i \leq P$ that contains monotonic fuzzy measures in a probability measure P is convex.*

1) Any measure $\mu \in \mathcal{M}_P$ can be represented by the convex combination of weakly additive measures.

2) Neither measure $P_{\mathfrak{f}}$ can be represented by the convex combination of other weakly additive measures.

Proof. It is easy to show that the set $\mathcal{M}_P$ is convex. At first we shall prove the statement 1). Suppose $g_0 \in \mathcal{M}_P$ and consider the filter $\mathbf{f}_0 = \{A \in \Im | g_0(A) > 0\} = \langle A_1, A_2, \dots, A_k \rangle$, where $\{A_1, A_2, \dots, A_k\}$ is a set of minimal elements, and a value $t_0 = \min_{i=1,2,\dots,k} \frac{g_0(A_i)}{P(A_i)}$. Then by the supposition the set function $q_0(A) = g_0(A) - t_0 P_{\mathfrak{f}_0}(A)$ is monotonic.

If $t_0 = 1$ the representation required has been found for under the monotonicity condition the function q_0 should be identical to zero, i.e. $g_0(A) = P_{\mathfrak{f}_0}(A)$ for all $A \in \Im$.

If $t_0 < 1$ then the set function $g_1(A) = \frac{q_0(A)}{1-t_0} = \frac{g_0(A) - t_0 P_{\mathfrak{f}_0}(A)}{1-t_0}$ is a fuzzy measure. We shall prove that this measure should be monotonic in a probability measure. Actually,

$$q(A) = \chi\left[g_1(A) - \alpha P(A)\right] = \chi\left[\frac{g_0(A) - t_0 P_{\mathfrak{f}_0}(A)}{1-t_0} - \alpha P(A)\right] =$$

$$= \frac{\chi\left[g_0(A) - (t_0 + (1-t_0)\alpha)P(A)\right]}{1-t_0.}.$$

By analogy for the fuzzy measure g_1 we can get the filter $\mathbf{f}_1$, the value t_1, the function g_2 and so on. As a result of this procedure we have the finite consequence of embedded filters $\mathbf{f}_0 \supset \mathbf{f}_1 \supset \mathbf{f}_2 \supset \dots \subset \mathbf{f}_K$ and the consequence of fuzzy measures that satisfy the following conditions:

$$\begin{cases} g_1(A) = \dfrac{g_0(A) - t_0 P_{\mathfrak{f}_0}}{1-t_0}, \\ \vdots \\ g_{i+1}(A) = \dfrac{g_i(A) - t_i P_{\mathfrak{f}_i}}{1-t_i}, \\ \vdots \\ g_{K+1}(A) = P_{\mathfrak{f}_K}. \end{cases} \tag{8}$$

Solving the equation system (8) we shall find the representation of the fuzzy measure by the convex sum of weakly additive measures.

The statement 2) of the theorem is obvious for any convex combination $g = \alpha_1 g_1 + \alpha_2 g_2$, $\alpha_1 + \alpha_2 = 1$, $\alpha_1, \alpha_2 > 0$, $g_1 \neq g_2$, of fuzzy measures $g_1, g_2 \in \mathcal{M}_P$ can not be a weakly additive measure. ∎

Lemma 7. *Any weakly additive measure can be represented by the convex sum of coherent primitive measures.*

Proof. Let $P_{\mathbf{f}}$ be a weakly additive measure, generated by the probability measure P and the filter $\mathbf{f}$. Denote $\mathcal{N}_{\mathbf{f}} = \{\mathbf{f}_1, \mathbf{f}_2, \ldots, \mathbf{f}_n\}$ the set of coherent filters $\mathbf{f}_i = \{A \in \mathbf{f} | A \cap \{x_i\} \neq 0\}$. Show that

$$P_{\mathbf{f}} = \sum_{i=1}^{K} P\{x_i\}\eta_{\mathbf{f}_i}. \tag{9}$$

Assume $A \in \mathbf{f}$. Then $A \cap \{x_i\} = \begin{cases} \emptyset, & A \notin \mathbf{f}_i, \\ \{x_i\}, & A \in \mathbf{f}_i. \end{cases}$ It means that according to the formula (9) we get $P_{\mathbf{f}}(A) = \sum\limits_{x_i \in A} P\{x_i\}$ if $A \in \mathbf{f}$ and $P_{\mathbf{f}}(A) = 0$ if $A \notin \mathbf{f}$. Thus the formula (9) is valid and the lemma has been proved. ∎

The direct consequence of Lemma 5 and Theorem 4 can be found in the following theorem.

Theorem 6. *Any lower probability that is monotonic in a probability measure P can be represented by the convex sum of coherent primitive measures.*

To derive more accurate inferences on the set of fuzzy measures we should use fuzzy measures that give us more precise estimates of probabilities. From this point of view it is useful to apply fuzzy measures that we call *accurate* upper (lower) probabilities.

Let g be a lower probability. Then this measure will be an accurate lower probability if for any $A \in \Im$ there is such a probability measure P, $g \leq P$, that $P(A) = g(A)$. By definition the fuzzy measure P being dual to the accurate lower probability will be an *accurate upper probability.* It is not difficult to notice that for any lower probability g we can construct the corresponding accurate lower probability g_{ac}. Actually, for the pointed measure g consider a probability measure family $\Xi = \{P | g \leq P\}$ then we derive this fuzzy measure in the following way:

$$g_{ac}(A) = \inf_{P \in \Xi} P(A), A \in \Im.$$

The following lemmas describe some properties of accurate lower probabilities.

Lemma 8. *The set $\mathcal{M}_{ac}$ of lower probabilities is convex. Let g be an accurate lower probability. Then*

1) $g(A \cup B) \geq g(A) + g(B)$, $A \cap B = \emptyset$, $A, B \in \Im$;
2) the set functions
$g_1(A) = \dfrac{g(A \cup B) - g(B)}{1 - g(B)}$, $g(B) < 1$, $B \in \Im$,
$g_2(A) = \dfrac{g(A \cap B)}{g(B)}$, $g(B) > 0$, $B \in \Im$,
are lower probabilities.

Proof. It is easy to show that the set $\mathcal{M}_{ac}$ is convex. Actually, let g_1, g_2 be accurate lower probabilities then for any $B \in \Im$ there are such probability measures P_1, P_2 that $P_1 \geq g_1$, $P_2 \geq g_2$ and $P_1(B) = g_1(B)$, $P_2(B) = g_2(B)$. The fuzzy measure $g = \alpha_1 g_1 + \alpha_2 g_2$, $\alpha_1 + \alpha_2 = 1$, $\alpha_1, \alpha_2 \geq 0$, will be an accurate lower probability because $g \leq \alpha_1 P_1 + \alpha_2 P_2 = P$ and $P(B) = g(B)$.

At first we will prove the statement 1). Assume $A \cap B \neq \emptyset$. By the supposition g is an accurate lower probability it lets us choose the probability measure P with the following property $P \geq g$ and $P(A \cup B) = g(A \cup B)$, then $g(A \cup B) = P(A) + P(B) \geq g(A) + g(B)$.

Consider the proof of the statement 2). It is easy to show that the set functions g_1 and g_2 are fuzzy measures. Further we choose the probability measure P that the following conditions: $P \geq g$ and $P(B) = g(B)$ hold. Then considering probability measures $P_1(A) = \dfrac{P(A \cup B) - P(B)}{1 - P(B)}$ and $P_2(A) = \dfrac{P(A \cap B)}{P(B)}$ one can notice that $P_1 \geq g_1$ and $P_2 \geq g_2$. The lemma has been completely proved. ■

Lemma 9. *The family of weakly superadditive fuzzy measures is convex. Any weakly superadditive measure is an accurate lower probability.*

Proof. It is easy to show that the family of weakly superadditive measures is convex. Further we will prove that an arbitrary weakly superadditive measure is an accurate lower probability. To do this one can notice that fuzzy measures g_1 and g_2 from the Lemma 8 in given case are weakly superadditive measures. Actually,

$$g_1(A \cup C) = \frac{g(A \cup C \cup B) - g(B)}{1 - g(B)} \geq$$

$$\geq \frac{g(A \cup B) + g(C \cup B) - g((A \cap C) \cup B) - g(B)}{1 - g(B)} =$$

$$= g_1(A) + g_1(C) - g_1(A \cap C),$$

$$g_2(A \cup C) = \frac{g((A \cup C) \cap B)}{g(B)} \geq \frac{g(A \cap B) + g(C \cap B) - g(A \cap C \cap B)}{g(B)} =$$

$$= g_2(A) + g_2(C) - g_2(A \cap C).$$

Thus, g_1, g_2 are weakly superadditive measures and according to Theorem 4 they are also lower probabilities. If g is an accurate lower probability then for any $B \in \Im$ one can find a probability measure P that $P \geq g$ and $P(B) = g(B)$. It is sufficient to consider three cases:

1) if $g(B) = 1$, then we choose measure P, $P \geq g$, that satisfies the all pointed properties;

2) if $g(B) = 0$, then we take measure P, $P \geq g_1$ and it is clear that $g \leq g_1$ and $P(B) = 0$;

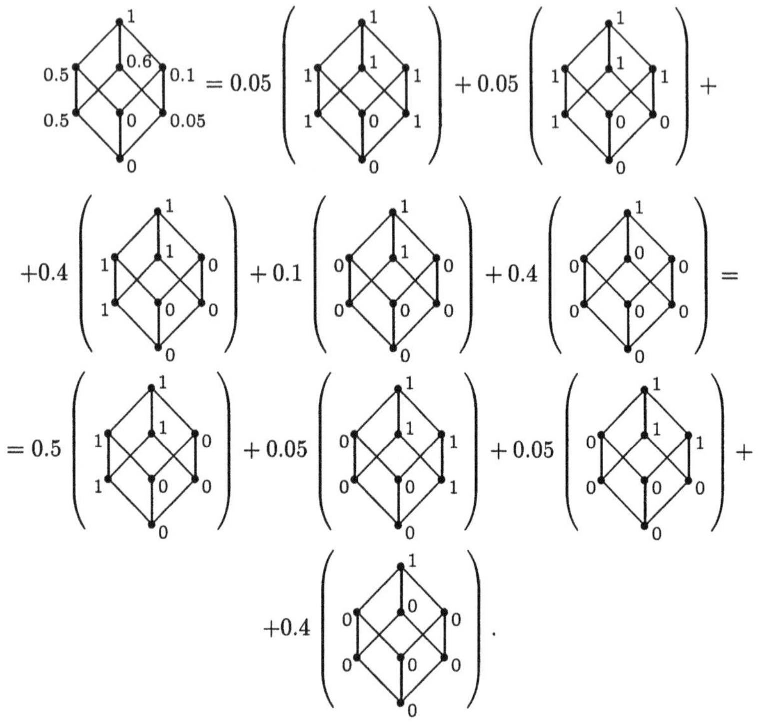

Fig. 1.

3) if $0 < g(B) < 1$, then we choose measures P_1 and P_2 with the following properties: $P_1 \geq g_1$, $P_2 \geq g_2$. With the help of these measures determine the probability measure $P = (1 - g(B))P_1 + g(B)P_2$. In this case $P(B) = g(B)$, because $P_1(B) = 0$, $P_2(B) = 1$. We will prove that $P \geq g$:

$$P(A) = [1 - g(B)]\, P_1(A) + g(B)P_2(A) \geq g(A \cup B) - g(B) + g(A \cap B) \geq g(A).$$

The last inequality has been obtained by means of the weak superadditivity property. ■

It is easy to show that an arbitrary accurate lower probability does not satisfy weak superadditivity property in a common case.

4 Appendix

A) The example of different representations of a fuzzy measure by the convex sum of primitive measures (see Figure 1).

B) The example of a accurate lower probability that is not weakly superadditive. Generate the lower probability g_{ac} using probability measures

P_1, P_2, P_3, P_4 that is determined on the algebra $\Im = 2^X$ of the space $X = \{x_1, x_2, x_3, x_4\}$ as follows:

$$g_{ac}(A) = \min_{i=1,2,3,4} P_i(A).$$

The values of the probability measures and the fuzzy measure g_{ac} are shown in Table 1.

x_1	x_2	x_3	x_4	P_1	P_2	P_3	P_4	g_{ac}
0	0	0	0	0	0	0	0	0
1	0	0	0	1/3	0	0	2/3	0
0	1	0	0	1/3	0	2/3	0	0
1	1	0	0	2/3	0	2/3	0	0
0	0	1	0	1/3	2/3	0	0	0
1	0	1	0	2/3	2/3	0	2/3	0
0	1	1	0	2/3	2/3	2/3	0	0
1	1	1	0	1	2/3	2/3	2/3	2/3
0	0	0	1	0	1/3	1/3	1/3	0
1	0	0	1	1/3	1/3	1/3	1	1/3
0	1	0	1	1/3	1/3	1	1/3	1/3
1	1	0	1	2/3	1/3	1	1	1/3
0	0	1	1	1/3	1	1/3	1/3	1/3
1	0	1	1	2/3	1	1/3	1	1/3
0	1	1	1	2/3	1	1	1/3	1/3
1	1	1	1	1	1	1	1	1

Table 1. The values of fuzzy measures

One can notice that the fuzzy measure derived does not satisfy the weak superadditive property because

$$g\{x_1, x_4\} + g\{x_2, x_4\} - g\{x_4\} > g\{x_1, x_2, x_4\}.$$

References

1. Sugeno, M. (1972). Fuzzy measure and fuzzy integral, *Trans. SICE* **8**, 95–102.
2. Dempster, A.P. (1967). Upper and lower probabilities induced by multivalued mapping, *Ann. Math. Statist.* **38**, 325–339.

3. Shafer, G. (1976). *A mathematical theory of evidence*, Princeton University Press, Princeton.
4. Dubois, D., and Prade, H. (1992). When upper probabilities are possibility measures, *Fuzzy Sets and Systems* **24**, 279–300.

Statistical classes and fuzzy set theoretical classification of probability distributions

Andrew G. Bronevich[1] and Alexander N. Karkishchenko[1]

Taganrog State University of Radio-Engineering Nekrasovskij bystreet, 44
Taganrog, RUSSIA

Abstract. This paper presents a method of classifying statistical classes that can be considered as empirical images of probability spaces. For statistical classes we introduce level sets with the help of which set-theoretic operations over statistical classes are defined. Particularly, we define measures of mutual inclusion and equality of arbitrary classes that provide the classification problem solution. We also show that these measures are well justified and consider connection between introduced structure and traditional axiomatic theory of fuzzy sets. Further we algebraically generalize the concept of statistical classes and investigate the connection of generalized classes with the possibility theory.

Keywords: Statistical class; Probability theory and statistics; Possibilistic inclusion; Inclusion measure; Lower and upper probabilities and membership functions

1 Introduction

The very important problems of identification and classification usually arise when analysing a large-scale statistical information. It is shown by the experience of intelligent, control and decision support systems working out that the classical methods of a multidimensional statistical analysis are often useless because of their computational complexity and restrictions of the classical probability scheme (Fine, 1973). It is also accounted by the fact that when analysing inaccurate and conflicting information a developer is often under necessity to model other kinds of uncertainty that are not probabilistic ones. In these cases it is expedient to use non-traditional methods of data processing that are in most cases based on the theory of fuzzy sets, particularly, on fuzzy measures of Sugeno (1972) and the possibility theory of Zadeh (1978).

However, there is a deep connection between the probability theory and the theory of fuzzy sets. Thus, any fuzzy set can be identified with a random set (Nguyen, 1978), a membership function of the fuzzy set coinciding with a cover function of a random set. The probabilistic interpretation can also be given to upper and lower belief measures in a set-possibilistic model. Also, we point out the works by Dubois and Prade (1983, 1986ab) where they, developing the Dempster-Shafer theory of evidence, offer methods for constructing fuzzy sets upon statistical data. A set-theoretic method of statistical data investigation proposed in this paper requires partial ordering of the set of probabilistic distributions. This algebraic structure is formalized with the help of an idea of a statistical class, introduced below, which can be considered as an empirical or a posteriori image of a certain probability space. On a set of the statistical classes inclusion and equality relations as well as inclusion and equality measures are defined. It enables us to solve the main problem of the given paper - to classify the statistical classes.

The basic way for describing the statistical classes is using level sets, which numerous properties are investigated in the first part of the article. Furthermore, it will be shown that any statistical class can be adequately defined by the aggregate of such sets. Owing to this the defined operations on the statistical classes are natural expansions of the corresponding operations on the level sets.

In the second part of the article we will investigate the connection between the described structure of the statistical classes and the traditional structure of fuzzy sets. Also, we consider an alternative of possibilistic defining the inclusion relation and inclusion measure, that enables us to interpret the notion of the statistical class in the framework of conceptual constructions of the possibility theory.

2 Basic definitions and problem statement

Let X be a measurable space of elementary events together with a σ -algebra of events $\mathfrak{A}$. We will assume that there is a volume measure V upon $\mathfrak{A}$ satisfying the usual requirements of non-negativity and additivity. For any $A \in \mathfrak{A}$ the meaning of $V(A)$ can be interpreted as a power of the set A or as a closure of the event A to an elementary event. For instance, if X consists of a finite number of elements then for any $x \in X$ we may put $V(x) = 1$. If $X = R^n$, then in the capacity V of we may use the Lebesgue measure for all measurable sets in this space.

We are given a probability measure P. The triple $F = (X, \mathfrak{A}, P)$ will be called as a statistical class on X. Thus, any statistical class on X is completely defined by assigning the probability measure P. We subsequently denote by $\mathfrak{F} = \{F_i \mid i = 1, 2, ...\}$ a finite or infinite aggregate of the statistical classes. Let us assume that there is a certain subset $S = \{S_1, S_2, \ldots, S_r\}$ of classes in $\mathfrak{A}$, which are said to be *standard.* Then the problem of classification of the statistical classes on the standard ones consists in constructing the inclusion measure $\psi(F_1 \subseteq F_2)$ upon $\mathfrak{F}$ with the help of which for any $F \in \mathfrak{F}$ there can be found a classifying vector

$$(\psi(F \subseteq S_1), \psi(F \subseteq S_2), \ldots, \psi(F \subseteq S_r)).$$

Consider a set $\mathcal{A}(p) = \{A \in \mathfrak{A} | P(A) = p\}$ of p-probable events for a certain class in the space X. An event $E \in \mathcal{A}(p)$ is called the *minimal event* (m.e.) for a class F if the following condition holds $V(E) = \inf_{A \in \mathcal{A}(p)} V(A)$.

The set of all m.e. for a class F will be denoted as $\mathfrak{M}$. It should be emphasized that minimal events give us in a certain sense the most "accurate" description of the statistical class.

We will subsequently assume that any probability measure P is absolutely continuous regarding a volume measure V. It means that for any class $F \in \mathfrak{F}$ it is possible to construct the probability density function $h(x)$, $x \in X$ connecting the volume and probability measures, i.e. for any $A \in \mathfrak{A}$

$$P(A) = \int_A h(x) dV(x).$$

Also, we will assume that the volume measure V is continuous in regard to its values, that is if $A \in \mathfrak{A}$ and $V(A) = a$ then for arbitrary $b \in [0, a]$ there exists $B \in \mathfrak{A}$ such as $B \subseteq A$ and $V(B) = b$. The continuous property of a probability measure regarding a volume measure gives us a possibility to introduce definitions of an inclusion and an equality of events in a volume measure. Namely, $A \subseteq B$ in measure V for any $A, B \in \mathfrak{A}$ if, and only if, $V(A \backslash B) = 0$. Then $A = B$ in measure V when both inclusions $A \subseteq B$ and $B \subseteq A$ are valid in measure V. Note that if inclusion or equality property holds in volume measure then it takes place in any probability measure for a class F because of continuity of the probability measure regarding the volume one. On the basis of the defined properties of the volume measure V it can be shown that for any $F \in \mathfrak{F}$ the probability measure P also possesses the continuous property in regard to its values.

3 Description of the minimal events set

We first establish a lemma that gives us an important characteristic property of m.e.

Lemma 1. *Let $E \in \mathfrak{M}$, $V(E) = a$ and $\mathcal{B}(a)$ is the set of events of the volume a , i.e. $\mathcal{B}(a) = \{A \in \mathfrak{A} | V(A) = a\}$. Then, $P(E) = \sup_{A \in \mathcal{B}(A)} P(A)$.*

Proof. Suppose that there is an event A such as $V(A) = V(E)$ and $P(E) < P(A)$. Since the probability measure P is continuous in regard to its values then there is such an event $C \subset A$ for which the equality $P(C) = P(E)$ holds. Because of the volume measure additivity we write $V(A) = V(C) + V(A \backslash C)$. Since $P(A \backslash C) > 0$, we have $V(A \backslash C) > 0$. So we deduce from this that $V(E) > V(C)$, that is the event E is not minimal one by the definition. From this contradiction we conclude that lemma is valid. ■

Thus, the result of lemma shows that m.e. are the most probable among the events of the same volume. The following lemma establishes an important class of m.e.

Lemma 2. *For any $\alpha > 0$ the event $E = \{x \in X | h(x) \geq \alpha\}$ is minimal.*

Proof. Let us assume that, on the contrary, there is $B \in \mathfrak{A}$ such as $P(B) = P(E)$ and $V(B) < V(E)$. Then by definition it means that E is not m.e. To prove lemma we note that from one hand $V(B) - V(E) = V(B \backslash E) - V(E \backslash B) < 0$. From another hand because of supposition we have $P(B) - P(E) = P(B \backslash E) - P(E \backslash B) = 0$. Taking into account the formula, connecting the volume and probability measures, we can get $P(E \backslash B) \geq \alpha V(E \backslash B)$ and $P(B \backslash E) \leq \alpha V(B \backslash E)$. Hence, $0 = P(B \backslash E) - P(E \backslash B) \leq \alpha V(B \backslash E) - \alpha V(E \backslash B)$ and $V(B \backslash E) - V(E \backslash B) = V(B) - V(E) \geq 0$. This inequality contradicts the supposition. ■

We now introduce an important definition. The *fundamental set* of m.e. is the set of events $E(\alpha) = \{x \in X | h(x) \geq \alpha\}$, $\alpha > 0$. It is not hard to see that the events $E(\alpha)$ are linearly ordered regarding to the inclusion operation, i.e. for any $\alpha_1, \alpha_2 > 0$ the inclusion $E(\alpha_1) \subseteq E(\alpha_2)$ follows from the inequality $\alpha_1 \geq \alpha_2$.

The concept of the fundamental set of m.e. causes a natural question: does the fundamental set of m.e. coincide with the set of all m.e. or not? That is whether it is possible to represent an arbitrary m.e. in the form $\{x \in X | h(x) \geq \alpha\}$ with a certain α. The following results answer this question.

Consider an event $C \in \mathfrak{A}$. The least upper bound $Lub(C)$ is defined as $Lub(C) = \sup\{\alpha | C \subseteq E(\alpha)\}$. Note that in this definition $C \subseteq E(\alpha)$ means the inclusion into in measure V, i.e. $V(C \backslash E(\alpha)) = 0$. We also introduce the following notations: $intE(\alpha) = \{x \in X | h(x) > \alpha\}$; $bdE(\alpha) = \{x \in X | h(x) = \alpha\}$. Thus, the equality $E(\alpha) = intE(\alpha) \cup bdE(\alpha)$ holds.

Lemma 3. *Let for m.e. C $Lub(C) = \alpha$. Then $intE(\alpha) \subseteq C$ in measure V.*

Proof. To prove the lemma it is sufficiently to show that for any $\epsilon > 0$ the inclusion $E(\alpha + \epsilon) \subseteq C$ in measure V holds, i.e. $V(E(\alpha + \epsilon) \backslash C) = 0$.

Assume, on the contrary, $V(E(\alpha + \epsilon) \backslash C) > 0$, then $P(E(\alpha + \epsilon) \backslash C) > 0$. On the other hand, as $\alpha = Lub(C)$, then $V(C \backslash E(\alpha + \epsilon)) > 0$. Besides $P(C \backslash E(\alpha + \epsilon)) > 0$, because otherwise the event C would not be minimal.

Consider events $E(\alpha + \epsilon) \backslash C$ and $C \backslash E(\alpha + \epsilon)$. As the probability measure is continuous regarding its values we choose events A_1, A_2, A_3 so that the following conditions hold:

1) $A_1 \subseteq E(\alpha + \epsilon) \backslash C$;

2) $A_2, A_3 \subseteq C \backslash E(\alpha + \epsilon)$, and $A_2 \cup A_3 = C \backslash E(\alpha + \epsilon)$, $A_2 \cap A_3 = \emptyset$;

3) $P(A_1) = P(A_2) = \min\{P(E(\alpha + \epsilon) \backslash C), P(C \backslash E(\alpha + \epsilon))\}$.

It follows from the condition 2) that $C = A_2 \cup A_3 \cup C \cap E(\alpha+\epsilon)$. Along with this we consider event $\tilde{C} = A_1 \cup A_3 \cup C \cap E(\alpha+\epsilon)$. It is easy to see that because of the condition 3) the events $\tilde{C}$ and C are equiprobable. Thus,

$$P(\tilde{C}) - P(C) = P(A_1) - P(A_2) = 0.$$

Together with it, using the relation between the probability and the volume measures, we can get $P(A_1) \geq (\alpha+\epsilon)V(A_1)$, $P(A_2) < (\alpha+\epsilon)V(A_2)$, hence,

$$P(A_1) - P(A_2) > (\alpha+\epsilon)V(A_1) - (\alpha+\epsilon)V(A_2).$$

Consequently, $V(A_1) - V(A_2) < 0$. On other hand, $V(A_1) - V(A_2) = V(\tilde{C}) - V(C)$, therefore $V(C) > V(\tilde{C})$. It means that by the definition the event C is not minimal. ∎

The following theorem gives a description of the structure of an arbitrary m.e.

Theorem 1. *Let $C \in \mathfrak{M}$ and $Lub(C) = \alpha$. Then*

1) if $\alpha > 0$ then $C = intE(\alpha) \cup A$, where $A \subseteq bdE(\alpha)$; conversely, for any $\alpha > 0$ the event $C = intE(\alpha) \cup A$ is m.e.;
2) if $\alpha = 0$ then $C = intE(0)$; conversely, $intE(0)$ is m.e..

Proof. Let us first prove the direct statements for cases 1) and 2). Since by condition $C \in \mathfrak{M}$ and $Lub(C) = \alpha$, then in view of Lemma 3 and by definition of the least upper bound we have $intE(\alpha) \subseteq C \subseteq E(\alpha)$. Thus, the direct statement for the case $\alpha > 0$ is proved. If $\alpha = 0$, then there is no A such as $V(A) > 0$ and the event $C = intE(\alpha) \cup A$ is minimal. Indeed, in this case for any $x \in A$ $h(x) = 0$ and $P(A) = 0$, hence $P(intE(\alpha) \cup A) = P(intE(\alpha))$ and $V(intE(\alpha)) < V(intE(\alpha) \cup A)$. If $V(A) = 0$ then $C = intE(\alpha)$ in measure V.

Now we prove the validity of the inverse statements for the cases 1) and 2). Let us assume that C is not m.e., i.e. there is $B \in \mathfrak{M}$ for which $P(B) = P(C)$ and $V(B) < V(C)$. Let $Lub(B) = \beta$; consider possible cases:

1) $\alpha > \beta$, then $C \subseteq B$ and C is a minimal event;

2) $\alpha < \beta$, then $B \subseteq C$ and from the equiprobability of the events B and C the equality $P(C \setminus B) = 0$ follows. Hence, $C = B$ in measure V;

3) $\alpha = \beta$, there are in its turn two cases: $\alpha = 0$ and $\alpha > 0$. When $\alpha = 0$, then the stated result immediately follows. If $\alpha > 0$, then because of the first part of the given theorem we have $B = intE(\alpha) \cup \tilde{A}$, where $\tilde{A} \subseteq bdE(\alpha)$. However, $P(C) - P(B) = P(intE(\alpha)) + P(A) - P(intE(\alpha)) - P(\tilde{A}) = \alpha V(A) - \alpha V(\tilde{A}) = 0$. So we get $V(A) = V(\tilde{A})$. Consequently, $V(C) - V(B) = V(intE(\alpha)) + V(A) - V(intE(\alpha)) - V(\tilde{A}) = 0$. This contradicts the supposition $V(B) < V(C)$. ∎

Corollary 1. (Corollary of Lemma 3) *If $C_1, C_2 \in \mathfrak{M}$ and $Lub(C_1) > Lub(C_2)$, then $C_1 \subseteq C_2$ in measure V.*

Now we establish a feature of equiprobable m.e. that we state as a lemma.

Lemma 4. *Let $C_1, C_2 \in \mathfrak{M}$ and $P(C_1) = P(C_2)$, then there is the equality $Lub(C_1) = Lub(C_2)$.*

Proof. Let for instance $Lub(C_1) > Lub(C_2)$, then $C_1 \subseteq C_2$. If with it $V(C_2 \setminus C_1) > 0$, then C_2 is not m.e., therefore $C_1 = C_2$ and $Lub(C_1) = Lub(C_2)$. ■

Let $p \in [0,1]$, then as the probability measure is continuous regarding its values the set $\mathcal{A}(p)$ of the equiprobable events is not empty. Consequently, the set of the equiprobable m.e. is also not empty. The following theorem establishes conditions for the p-probable m.e. to be unique in measure V for any $p \in [0,1]$.

Theorem 2. *Any m.e. is determined by its probability uniquely if, and only if, for any α the equality $P(bdE(\alpha)) = 0$ holds.*

Proof. We first prove the sufficient condition. Let $C_1, C_2 \in \mathfrak{M}$, $P(C_1) = P(C_2)$, but $C_1 \neq C_2$ in the measure V. For, in view of Lemma 4, $Lub(C_1) = Lub(C_2) = \alpha$, then by Theorem 2 $C_1 \setminus C_2 \cup C_2 \setminus C_1 \subseteq bdE(\alpha)$. In view of supposition we have $P(bdE(\alpha)) = 0$, hence $C_1 = C_2$ in measure V, that is we get a contradiction.

To prove the necessity we show that if for any concrete α $P(bdE(\alpha)) = a > 0$ then there are events $C_1, C_2 \in \mathfrak{M}$ for which $P(C_1) = P(C_2)$ and with it $C_1 \neq C_2$ in measure V. Because of the continuous property of the probability measure it is possible to choose an event A so that $A \subseteq bdE(\alpha)$ and $P(A) = a/2$. Introducing into consideration an event $intE(\alpha)$ we can get due to Theorem 1 that events $\tilde{C}_1 = intE(\alpha) \cup A$ and $\tilde{C}_2 = intE(\alpha) \cup (bdE(\alpha) \setminus A)$ would be m.e. Besides we have $P(\tilde{C}_1) = P(\tilde{C}_2)$ and $P(\tilde{C}_1 \setminus \tilde{C}_2 \cup \tilde{C}_2 \setminus \tilde{C}_1) = a > 0$. From this we obtain $\tilde{C}_1 \neq \tilde{C}_2$ in measure V. ■

It is obvious, that when satisfying the conditions of the theorem the set of all m.e. coincides with the fundamental set. We introduce the following definition. A statistical class $F \in \mathfrak{F}$ is called *regular* if each m.e. in it is defined by its probability uniquely. Note that the set of m.e. $\mathfrak{M}$ of the regular class F coincides with the fundamental set.

A special significance of the fundamental set follows from Theorem 3.

Theorem 3. *The fundamental set of m.e. determines each statistical class uniquely.*

Proof. We will show that it is possible to uniquely reconstruct the probability measure upon σ-algebra $\mathfrak{A}$ of the space X by the fundamental set of m.e. Note that a function $h(x)$ of the class F is integrable, hence it is a measurable function. Therefore, we may use functions $\tilde{h}_n(x) = \frac{k}{n}$, when $\frac{k-1}{n} \leq h(x) < \frac{k}{n}$, as simple integrable functions uniformly converging to $h(x)$, k and n belonging to the set of integer numbers. It is also easy to see that the sequence of functions $\underset{\sim n}{h}(x) = \frac{k-1}{n}$, when $\frac{k-1}{n} \leq h(x) < \frac{k}{n}$, will uniformly converge to $h(x)$ as well

$$E\left(\frac{k-1}{n}\right) \setminus E\left(\frac{k}{n}\right) = \left\{x \in X \mid \frac{k-1}{n} \leq h(x) < \frac{k}{n}\right\}.$$

Consider function

$$h_n(x) = \frac{P\left\{E\left(\frac{k-1}{n}\right) \setminus E\left(\frac{k}{n}\right)\right\}}{V\left\{E\left(\frac{k-1}{n}\right) \setminus E\left(\frac{k}{n}\right)\right\}}, \textit{ when } \frac{k-1}{n} \leq h(x) < \frac{k}{n}.$$

It is obvious that $\underset{\sim n}{h}(x) \leq h_n(x) \leq \tilde{h}_n(x)$, therefore the sequence of the simple integrable functions $h_n(x)$ uniformly converges to $h(x)$. However, values of these functions can be straightforward calculated by the fundamental set of m.e. Reconstructability of the function $h(x)$ means in its turn the reconstructability of the probability measure P upon $\mathfrak{A}$. ∎

4 Set-theoretic operations on statistical classes. Inclusion and equality measures

We shall first consider regular statistical classes. Let F_1 and F_2 be regular statistical classes from $\mathfrak{F}$; $A_1(p)$ and $A_2(p)$ - equiprobable with the probability p m.e. that correspond to the classes F_1 and F_2, that is $P_1[A_1(p)] = p$, $P_2[A_2(p)] = p$. Note here that informally the probability p gives us representativity estimation of the m.e. $A_1(p)$ and $A_2(p)$ when describing the classes F_1 and F_2. These events possess an important extremal property since they are the least events with a given probability. That is why when defining set-theoretic operations and relations on statistical classes it is natural to take as a basis the corresponding operations on m.e.

Put in by definition that $F_1 \subseteq F_2$ if, and only if, for any $p \in [0,1]$ there is the inclusion $A_1(p) \subseteq A_2(p)$ in measure V; $F_1 = F_2$ if $A_1(p) = A_2(p)$ in measure V for any $p \in [0,1]$. Now we define the basic operations on classes.

The *union* of classes F_1 and F_2 is a class $F_3 = F_1 \cup F_2$ such that for any $p \in [0,1]$ there is $A_3(p) = A_1(p) \cup A_2(p)$. The *intersection* of classes F_1 and F_2 is a class $F_3 = F_1 \cap F_2$ such that the equality $A_3(p) = A_1(p) \cap A_2(p)$ holds for any $p \in [0,1]$.

We now begin to solve the previously stated problem of statistical classes classification. To start we construct an appropriate inclusion measure ψ for statistical classes. Let $F_1, F_2 \in \mathfrak{F}$ be regular statistical classes. Given $p \in [0,1]$ we define *p-local inclusion measure* of F_1 into F_2 as $\psi_p(F_1 \subseteq F_2) = P_1[A_2(p)|A_1(p)]$, that is the conditional probability of the event $A_2(p)$ occurrence provided that the event $A_1(p)$ has taken place in measure P_1. Note that the p-local measure possesses all the required properties of a measure, namely $\psi_p(F_1 \subseteq F_2) = 1$ if $A_1(p) \subseteq A_2(p)$ in measure V; $\psi_p(F_1 \subseteq F_2) = 0$ if $V(A_1(p) \cap A_2(p)) = 0$.

Mean while, it is more useful to introduce an *integral inclusion measure* (for the sake of brevity we will further call it simply as an inclusion measure) $\psi(F_1 \subseteq F_2)$, that can be defined on the basis of the p-local measure for various p:

$$\psi(F_1 \subseteq F_2) = \int_0^1 \psi_p(F_1 \subseteq F_2)(2p)\,dp = 2\int_0^1 P_1[A_2(p)|A_1(p)]\,pdp,$$

where $2p$ is a normalization factor. The *equality measure* we define such that

$$\psi(F_1 = F_2) = \min\{\psi(F_1 \subseteq F_2),\ \psi(F_2 \subseteq F_1)\}.$$

The following theorem shows that the introduced measures are reasonable.

Theorem 4. *For arbitrary regular statistical classes* $F_1, F_2 \in \mathfrak{F}$ *we have:*
1) $\psi(F_1 \subseteq F_2) = 1$ *if, and only if,* $F_1 \subseteq F_2$;
2) $\psi(F_1 = F_2) = 1$ *if, and only if,* $F_1 = F_2$.

Proof. Let us prove the first statement. The sufficiency is obvious. To prove the necessity we assume that the contrary condition is valid, that is $F_1 \not\subseteq F_2$, i.e. for a certain p_0 the value $V[A_1(p_0) \setminus A_2(p_0)] > 0$. From this we get $P_1[A_1(p_0) \setminus A_2(p_0)] = \varepsilon > 0$, since otherwise the event $A_1(p_0)$ would not be minimal.

Let us transform the expression of the inclusion measure

$$\psi(F_1 \subseteq F_2) = 2\int_0^1 pP_1[A_2(p)|A_1(p)]\,dp = 2\int_0^1 P_1[A_1(p) \cap A_2(p)]\,dp =$$

$$= 2\int_0^1 P_1[A_1(p)]\,dp - 2\int_0^1 P_1[A_1(p) \setminus A_2(p)]\,dp = 1 - 2\int_0^1 P_1[A_1(p) \setminus A_2(p)]\,dp.$$

Bounding the value

$$P_1[A_1(p_0 - \Delta p) \setminus A_2(p_0 - \Delta p)] \geq P_1[A_1(p_0 - \Delta p) \setminus A_2(p_0)] \geq \varepsilon - \Delta p,$$

we can get

$$\int_0^\varepsilon P_1[A_1(p_0 - \Delta p) \setminus A_2(p_0 - \Delta p)]\,d\Delta p \geq \int_0^\varepsilon (\varepsilon - \Delta p)\,d\Delta p = \frac{\varepsilon^2}{2}.$$

Therefore, $\psi(F_1 \subseteq F_2) \leq 1 - \varepsilon^2 < 1$, thus the first part of the theorem is proved. The second statement follows from the first one and the expression for the equality measure of regular statistical classes F_1 and F_2. ■

Let us transform the formula of the inclusion measure to the form that is more convenient for practical calculations. To do this, we shall first introduce a function

$$\chi(z) = \begin{cases} 1\,, & z \geq 0\,, \\ 0\,, & z < 0\,. \end{cases}$$

Then by definition of the fundamental set of m.e. we can write

$$\chi\left[h_1(y) - h_1(x)\right] = \begin{cases} 1\,, & y \in E_1\left[h_1(x)\right]\,, \\ 0\,, & y \notin E_1\left[h_1(x)\right]\,. \end{cases}$$

Taking into account this representation we introduce a value

$$\mathcal{P}_1(x) = P_1\left[E_1\left(h_1(x)\right)\right] = \int\limits_X h_1(y)\chi\left[h_1(y) - h_1(x)\right] dV(y)$$

that has a sense of the probability of m.e. from $\mathfrak{M}$, whose density is not less than $h_1(x)$. By analogy define $\mathcal{P}_2(x) = P_2\left[E_2\left(h_2(x)\right)\right]$ for the class F_2.

As for regular statistical classes the set of m.e. coincides with the fundamental set, then

$$\chi\left[p - \mathcal{P}_1(y)\right] = \begin{cases} 1, \, if \;\; y \in A_1(p), \\ 0, \, if \;\; y \notin A_1(p). \end{cases}$$

Therefore, the expression for the inclusion measure can be rewritten as follows

$$\psi\left(F_1 \subseteq F_2\right) = 2\int\limits_0^1 P_1\left[A_1(p) \cap A_2(p)\right] dp =$$

$$= 2\int\limits_0^1\int\limits_X h_1(x)\chi\left[p - \mathcal{P}_1(x)\right]\chi\left[p - \mathcal{P}_2(x)\right] dV(x)dp =$$

$$= 2\int\limits_X h_1(x)\int\limits_0^1 \chi\left[p - \mathcal{P}_1(x)\right]\chi\left[p - \mathcal{P}_2(x)\right] dpdV(x)$$

Note that

$$\chi\left[p - \mathcal{P}_1(x)\right]\chi\left[p - \mathcal{P}_2(x)\right] = \begin{cases} 1\,, & p \geq \max\left\{\mathcal{P}_1(x),\ \mathcal{P}_2(x)\right\}\,, \\ 0\,, & p < \max\left\{\mathcal{P}_1(x),\ P_2(x)\right\}\,. \end{cases}$$

Thus, it follows from this

$$\int\limits_0^1 \chi\left[p - \mathcal{P}_1(x)\right]\chi\left[p - \mathcal{P}_2(x)\right] dp = 1 - \max\left\{\mathcal{P}_1(x),\ \mathcal{P}_2(x)\right\} =$$

$$= \min\left\{1 - \mathcal{P}_1(x),\ 1 - \mathcal{P}_2(x)\right\}.$$

Taking account of this result we can write the final expression for the inclusion measure

$$\psi(F_1 \subseteq F_2) = 2 \int_X h_1(x) \min\{1 - \mathcal{P}_1(x),\ 1 - \mathcal{P}_2(x)\}\, dV(x)$$

or, with notations $\mu_1(x) = 1 - \mathcal{P}_1(x)$, $\mu_2(x) = 1 - \mathcal{P}_2(x)$,

$$\psi(F_1 \subseteq F_2) = 2 \int_X h_1(x) \min\{\mu_1(x),\ \mu_2(x)\}\, dV(x).$$

The function $\mu(x)$ has a profound sense when representing statistical class F by a fuzzy set. This will be discussed below.

We now show a statistical method of calculating $\psi(F_1 \subseteq F_2)$ under assumption that the class F_1 is given by a learning sample $\{x_i \mid i = 1, 2, ..., N\}$, whose elements are subjected to a probability distribution on X with probability measure P_1. As $\psi(F_1 \subseteq F_2)$ can be considered as a mathematical expectation of the function $2\min\{\mu_1(x),\ \mu_2(x)\}$ then the value of the inclusion measure can be estimated statistically as follows

$$\psi(F_1 \subseteq F_2) = \frac{2}{N} \sum_{i=1}^{N} \min\{\mu_1(x_i),\ \mu_2(x_i)\}.$$

It is known that this estimation converges in probability to the correct value $\psi(F_1 \subseteq F_2)$. Thus, this formula provides us with a convenient method of calculating the inclusion measure.

5 Fuzzy representation of regular statistical classes

Consider the function $\mu(x) = 1 - \mathcal{P}(x)$ of a regular statistical class $F \in \mathfrak{F}$. The following lemma shows that values of this function are closely related with the set of m.e.

Lemma 5. *For any $p > 0$ $A(1-p) = \{x \in X \mid \mu(x) \geq p\}$.*

Proof. We would remind that by definition $\mathcal{P}(x) = P\{E[h(x)]\}$ It means $A[\mathcal{P}(x)] = E[h(x)]$. It is easy that $E[h(x)] = \{y \in X \mid \mathcal{P}(y) \leq \mathcal{P}(x)\}$ in measure V, then $A[\mathcal{P}(x)] = \{y \in X \mid \mathcal{P}(y) \leq \mathcal{P}(x)\}$. We now put $\mathcal{P}(x) = 1-p$ Thus $A(1-p) = \{y \in X \mid \mathcal{P}(y) \leq 1-p\}$ or $A(1-p) = \{y \in X \mid \mu(y) \geq p\}$ and we obtain the required result. ∎

Corollary 2. *The function $\mu(x)$ determines each statistical class uniquely.*

We now show that each statistical class $F \in \mathfrak{F}$ can be considered as a fuzzy subset of the space with a membership function $\mu(x)$. Let $F_1, F_2 \in \mathfrak{F}$.

Then, if $F_3 = F_1 \cup F_2$ then $\mu_3(x) = \max\{\mu_1(x), \mu_2(x)\}$, and if $F_3 = F_1 \cap F_2$ then $\mu_3(x) = \min\{\mu_1(x), \mu_2(x)\}$. Thus, stated above set-theoretic operations on statistical classes coincide with the traditional operations in the theory of fuzzy sets (Kaufmann, 1975).

Due to Sugeno (1972) we can consider probabilities of fuzzy events in the measurable space X with the probability measure P. For a fuzzy event with a membership function $\mu(x) : X \to [0,1]$ the probability $P(F)$ is defined by the formula

$$P(F) = \int_X \mu(x) dP(x).$$

Let $F_1, F_2 \in \mathfrak{F}$ be regular statistical classes. Then it is not hard to see that $P_1(F_1) = 0.5$ and the inclusion measure $\psi(F_1 \subseteq F_2)$ can be written in a form

$$\psi(F_1 \subseteq F_2) = 2\int_X \min\{\mu_1(x), \mu_2(x)\} dP_1(x) = 2P_1(F_1 \cap F_2) =$$

$$= \frac{P_1(F_1 \cap F_2)}{P_1(F_1)} = P_1(F_2 | F_1).$$

Thus, the inclusion measure has a sense of the conditional probability of the fuzzy event F_2 occurrence provided that the fuzzy event F_1 has taken place in measure P_1.

6 Inclusion relation and inclusion measure for irregular statistical classes

In the section 4 we defined set-theoretic operations and relations for regular statistical classes, i.e. such classes for which each m.e. is uniquely determined by its probability. However, as it is often the case, when solving rather complicated problems of data processing and control, real probability distributions provide statistical classes that do not possess the mentioned property. Such classes we will call *irregular* classes. As examples of probability distributions that provide irregular statistical classes we point to discrete, discrete-continuous distributions and even purely continuous distributions of random values, whose probability density function has constancy domains of a non- zero measure. The irregularity of a statistical class becomes apparent when its m.e. are determined not uniquely, or do not exist for certain probabilities[1]. Further we shall consider a general method of defining set-theoretic operations and relations on the whole set $\mathfrak{F}$ of statistical classes (including irregular classes).

[1] Here and below it is not assumed that the volume measure is obligaory continuous in regard to its values (*authors' remark*).

Let us extend the set of m.e. $\mathfrak{M}$ with fuzzy events, i.e. with such events E with a membership function $\mu_E(x): X \to [0,1]$, whose probability $P(E)$ and volume $V(E)$ are defined by the following expressions,

$$P(E) = \int_X \mu_E(x) dP(x), \qquad V(E) = \int_X \mu_E(x) dV(x),$$

and together with it fuzzy an event E is called minimal one for a class $F = (X, \mathfrak{A}, P)$, $P(E) = p$, if it has a minimal volume among all fuzzy equiprobable events, i.e.

$$V(E) = \inf_{\substack{over\ all\ p-probable \\ fuzzy\ events\ A}} V(A).$$

The following lemma describes membership functions of fuzzy m.e. from $\mathfrak{M}$.

Lemma 6. *The event $E(\alpha, q)$ with the membership function*

$$\mu_E(x) = \begin{cases} 1, & x \in intE(\alpha), \\ q, & x \in \mathrm{bd}E(\alpha), \ \alpha > 0, q \in [0,1], \\ 0, & x \notin E(\alpha), \end{cases}$$

is the fuzzy minimal event for a certain q.

Proof. For the sake of simplicity, we denote $E = E(\alpha, q)$. Let us assume that there is $B \in \mathfrak{A}$ such as $P(B) = P(E)$ and $V(B) < V(E)$. Then by definition it will mean that E is not m.e. We will divide the space X into three domains:

$$\Omega_1 = \{x \in X \mid \mu_E(x) = \mu_B(x)\},$$

$$\Omega_2 = \{x \in X \mid \mu_E(x) > \mu_B(x)\},$$

$$\Omega_3 = \{x \in X \mid \mu_E(x) < \mu_B(x)\}.$$

Since by supposition $P(E) = P(B)$, then

$$P(E) - P(B) = \int_X \mu_E(x) h(x) dV(x) - \int_X \mu_B(x) h(x) dV(x) =$$

$$= \int_{\Omega_2} [\mu_E(x) - \mu_B(x)]\, h(x) dV(x) - \int_{\Omega_3} [\mu_B(x) - \mu_E(x)]\, h(x) dV(x) = 0$$

Note, that $h(x) \geq \alpha$ if $x \in \Omega_2$ and $h(x) \leq \alpha$ if $x \in \Omega_3$, therefore

$$\alpha \int_{\Omega_2} [\mu_E(x) - \mu_B(x)]\, dV(x) - \alpha \int_{\Omega_3} [\mu_B(x) - \mu_E(x)]\, dV(x) \leq 0.$$

As $\int\limits_{\Omega_1} \mu_E(x)dV(x) = \int\limits_{\Omega_1} \mu_B(x)dV(x)$, then dividing the last inequality by $\alpha > 0$, we obtain

$$\int\limits_{\Omega_1} \mu_E(x)dV(x) + \int\limits_{\Omega_2} \mu_E(x)dV(x) + \int\limits_{\Omega_3} \mu_E(x)dV(x) -$$

$$- \int\limits_{\Omega_1} \mu_B(x)dV(x) - \int\limits_{\Omega_2} \mu_B(x)dV(x) - \int\limits_{\Omega_3} \mu_B(x)dV(x) \leq 0$$

that is $V(E) \leq V(B)$. This inequality contradicts the supposition that was made above. ■

It is not hard to see that if $\mathrm{bd}E(\alpha)$ equals to zero in measure V, then the proved lemma gives us a description of ordinary (not fuzzy) m.e. from the fundamental set of m.e. of the class F.

We will by analogy consider the minimal events $E(\alpha, q)$ having been described in Lemma 6 to belong to the fundamental set of fuzzy m.e. The following lemma shows that these events can be taken as a basis to define set- theoretic operations on arbitrary statistical class.

Lemma 7. *The fuzzy m.e. with probability* $p \in [0,1]$ *from the fundamental set of fuzzy m.e. is determined uniquely.*

Proof. We will show a constructive method of choosing such an event. Let us choose α such that $P\{intE(\alpha)\} \leq p \leq P\{E(\alpha)\}$. It is obviously, that this choice can be done uniquely. If $P\{\mathrm{bd}E(\alpha)\} = 0$ then the event satisfies all the required conditions, otherwise $q = (p - P\{intE(\alpha)\})/P\{\mathrm{bd}E(\alpha)\}$ and m.e. E from Lemma 6 has the probability p. ■

Because of proved uniqueness of constructing fuzzy m.e. from the fundamental set each such an event is completely determined by its probability p Therefore, it can be denoted as $A(p)$. In its turn this allows us to introduce an inclusion relation and inclusion measure of statistical classes in just the same way as in the section 4, but however taking into account that m.e. might be fuzzy. For definiteness we will use the classical **min**- and **max**-operations on fuzzy sets.

Let $\Phi_1 = \{A_1(p)\}$ and $\Phi_2 = \{A_2(p)\}$ be the fundamental sets of fuzzy m.e. corresponding to statistical F_1 and F_2 respectively. We shall take by definition that $F_1 \subseteq F_2$ if, and only if, for any $p \in [0,1]$ the inclusion $A_1(p) \subseteq A_2(p)$ takes place in measure V. The inclusion measure of statistical classes is defined by the formula

$$\psi(F_1 \subseteq F_2) = 2\int\limits_0^1 P_1[A_2(p) \mid A_1(p)]\,pdp = 2\int\limits_0^1 P_1[A_1(p) \cap A_2(p)]\,dp.$$

Let us now find an expression being more convenient for practical calculating the measure $\psi(F_1 \subseteq F_2)$. Taking into account that the integrand in the last expression is the probability of fuzzy sets intersection we write by definition (Zadeh, 1968)

$$P_1\{A_1(p) \cap A_2(p)\} = \int_X \mu_{A_1(p)\cap A_2(p)}(x) h_1(x) dV(x) =$$

$$= \int_X \min\{\mu_{A_1(p)}(x), \mu_{A_2(p)}(x)\} h_1(x) dV(x).$$

Putting this expression into the formula of the inclusion measure, we obtain

$$\psi(F_1 \subseteq F_2) = 2\int_0^1 \int_X \min\{\mu_{A_1(p)}(x), \mu_{A_2(p)}(x)\} h_1(x) dV(x) dp =$$

$$= \int_X h_1(x)\left(2\int_0^1 \min\{\mu_{A_1(p)}(x), \mu_{A_2(p)}(x)\} dp\right) dV(x). \tag{1}$$

Denote the expression in the round brackets by $I(x)$. Thus, the inclusion measure $\psi(F_1 \subseteq F_2)$ represents the mathematical expectation of the function $I(x)$ and can be statistically estimated as a sample mean of a learning sample $(x_1, x_2, ..., x_N)$ of the statistical class F_1

$$\psi(F_1 \subseteq F_2) \approx \frac{1}{N}\sum_{i=1}^{N} I(x_i). \tag{2}$$

To obtain a convenient calculation formula we will analyse $I(x)$ in more details. To do this we shall need the following lemma that gives us an expression of the membership function in the form of the parameter p dependence.

Lemma 8. *Let $F = (X, \mathfrak{A}, P)$ be a statistical class. Denote by $\mathcal{P}(x,q)$ the probability of a fuzzy set $E(h(x), q)$, $q \in [0,1]$. Then the membership function of a p-probable ($p < 1$) fuzzy m.e. $A(p)$ has a form*

$$\mu_{A(p)}(x) = \begin{cases} 1, & \mathcal{P}(x,1) \leq p, \\ \dfrac{p - \mathcal{P}(x,0)}{\mathcal{P}(x,1) - \mathcal{P}(x,0)}, & \mathcal{P}(x,0) \leq p < \mathcal{P}(x,1), \\ 0, & p < \mathcal{P}(x,0). \end{cases} \tag{3}$$

Proof. We at first establish two additional propositions that will be further required.

Proposition 1. *Let* $\alpha = \inf\{\beta | P\{intE(\beta)\} \leq p\}$*, then* $P\{intE(\alpha)\} \leq p$.

Proof. It is sufficient to show that the function $P\{intE(\alpha)\}$ is continuous on the right, i.e. $P\{intE(\alpha)\} = P\{intE(\alpha+0)\}$. Since the events $A_n = \{x \in X \mid \alpha < h(x) \leq \alpha + \frac{1}{n}\} \to \emptyset$ with $n \to \infty$, then because of continuity of the probability measure we obtain the following

$$\lim_{n\to\infty} P(A_n) = \lim_{n\to\infty}\left[P\{intE(\alpha)\} - P\left\{intE\left(\alpha + \frac{1}{n}\right)\right\}\right] = 0,$$

and so the result is proved. ∎

Proposition 2. *The function* $f(\alpha) = P\{E(\alpha)\}$ *is continuous on the left.*

Proof. Since the events $B_n = \{x \in X \mid \alpha - \frac{1}{n} \leq h(x) < \alpha\} \to \emptyset$ with $n \to \infty$, then again taking into account continuity of the probability measure we obtain $\lim_{n\to\infty} P(B_n) = \lim_{n\to\infty}\left[P\{intE(\alpha - \frac{1}{n})\} - P\{intE(\alpha)\}\right] = 0$, i.e. $P\{E(\alpha - 0\} = P\{E(\alpha)\}$. ∎

To prove the lemma it is sufficient to show that $A(p) = E(\alpha, q)$ for certain values α and q, i.e.

$$\mu_{A(p)}(x) = \begin{cases} 1, & x \in intE(\alpha)\ , \\ q, & x \in \mathrm{bd}E(\alpha)\ , \quad q \in [0,1], \\ 0, & x \notin E(\alpha)\ \ , \end{cases} \tag{4}$$

and $P\{A(p)\} = p$. Show that α and q can be calculated by formulas

$$\alpha = \inf\{\beta \mid P\{intE(\beta)\} \leq p\}, \tag{5}$$

$$q = \begin{cases} 1, & P\{\mathrm{bd}E(\alpha)\} = 0 \\ \dfrac{p - P\{intE(\alpha)\}}{P\{\mathrm{bd}E(\alpha)\}}, & P\{\mathrm{bd}E(\alpha)\} > 0. \end{cases} \tag{6}$$

To prove the equivalence of the representations (3) and (4) we consider in consecutive order each condition defining the membership function.

1. Prove that if $x \in intE(\alpha)$ then $\mathcal{P}(x,1) \leq p$. If $x \in intE(\alpha)$ then $h(x) > \alpha$ and $E(h(x)) \subseteq intE(\alpha)$. At the same time because of Proposition 1 the formula (5) shows that $P\{intE(\alpha)\} \leq p$. Therefore, $\mathcal{P}(x,1) = P\{E(h(x))\} \leq P\{intE(\alpha)\} \leq p$.

2. Prove that if $x \notin E(\alpha)$ then $\mathcal{P}(x,0) > p$. Let us assume that the contrary inequality holds, i.e. $\mathcal{P}(x,0) \leq p$, however, $x \notin E(\alpha)$. Let $\mathcal{P}(x,0) = P\{intE(h(x))\} \leq p$, then we have from (5) that $\alpha \leq h(x)$, i.e. $x \in E(\alpha)$. This contradicts the stated assumption, hence $\mathcal{P}(x,0) > p$.

3. To finish proving the lemma we show that the condition $x \in \mathrm{bd}E(\alpha)$ implies the validity of the inequalities $P\{intE(\alpha)\} \leq p \leq P\{E(\alpha)\}$, and with it q in (4) is defined by the expression (6). It is not hard to see that

these inequalities guarantee the validity of the required conditions $q \in [0,1]$ and $P\{A(p)\} = p$.

Note that the inequality $P\{intE(\alpha)\} \leq p$ is implied by (5) because of Proposition 1. To prove the validity of the inequality $p \leq P\{E(\alpha)\}$ we assume the contrary, that is $P\{E(\alpha)\} < p$. Then according to Proposition 2 the function $f(\alpha) = P\{E(\alpha)\}$ is continuous on the left, i.e. there is such an $\varepsilon > 0$ that $P\{E(\alpha - \varepsilon)\} < p$. However, this contradicts the condition (5) of choosing the parameter α. ∎

We can now replace the membership functions $\mu_{A_1(p)}(x)$ and $\mu_{A_2(p)}(x)$ in (1) by expressions that are given by Lemma 8. Then it is straightforward to show that

$$I(x) = 2 - \max\left[\mathcal{P}_1(x,0), \mathcal{P}_2(x,0)\right] - \max\left[\mathcal{P}_1(x,1), \mathcal{P}_2(x,1)\right] + \Delta(x),$$

where $\Delta(x) = \max\left\{0, \dfrac{\left[\mathcal{P}_1(x,0) - \mathcal{P}_2(x,0)\right]\left[\mathcal{P}_2(x,1) - \mathcal{P}_1(x,1)\right]}{|\mathcal{P}_1(x,0) - \mathcal{P}_2(x,0)| + |\mathcal{P}_2(x,1) - \mathcal{P}_1(x,1)|}\right\}$.

Denote $\underline{\mu}_i(x) = 1 - \mathcal{P}_i(x,1)$ and $\bar{\mu}_i(x) = 1 - \mathcal{P}_i(x,0)$, then

$$I(x) = \min\left[\underline{\mu}_1(x), \underline{\mu}_2(x)\right] + \min\left[\bar{\mu}_1(x), \bar{\mu}_2(x)\right] + \Delta(x), \tag{7}$$

where $\Delta(x) = \max\left\{0, \dfrac{\left[\underline{\mu}_1(x) - \underline{\mu}_2(x)\right]\left[\bar{\mu}_2(x) - \bar{\mu}_1(x)\right]}{\left|\underline{\mu}_1(x) - \underline{\mu}_2(x)\right| + |\bar{\mu}_2(x) - \bar{\mu}_1(x)|}\right\}$.

By analogy with the function $\mu(x)$ we will call the functions $\underline{\mu}(x)$ and $\bar{\mu}(x)$ as lower and upper membership functions of the statistical class F. If F is a regular statistical class then, obviously, $\underline{\mu}(x) = \bar{\mu}(x)$. When $\underline{\mu}(x) \neq \bar{\mu}(x)$ we can consider that the proper function $\mu(x)$ is located within the segment $\left[\underline{\mu}(x), \bar{\mu}(x)\right]$. In particular, if $\underline{\mu}_1(x) < \underline{\mu}_2(x)$, $\bar{\mu}_1(x) < \bar{\mu}_2(x)$, i.e. the segments $\left[\underline{\mu}_1(x), \bar{\mu}_1(x)\right]$, $\left[\underline{\mu}_2(x), \bar{\mu}_2(x)\right]$ being not included one into another, we have $\Delta(x) = 0$ and $I(x) = \min\left[\underline{\mu}_1(x) + \bar{\mu}_1(x), \underline{\mu}_2(x) + \bar{\mu}_2(x)\right]$. In this case $\mu(x)$ is taken as the arithmetic mean of the lower and upper bounds, i.e. $\mu_i(x) = \dfrac{\underline{\mu}_i(x) + \bar{\mu}_i(x)}{2}$.

With the help of the functions $\underline{\mu}(x)$ and $\bar{\mu}(x)$ one could express the relations and the algebraic operations on statistical classes. In particular, $F_1 \subseteq F_2$ if $\underline{\mu}_1(x) \leq \underline{\mu}_2(x)$, $\bar{\mu}_1(x) \leq \bar{\mu}_2(x)$ for any $x \in X$. For arbitrary statistical classes there are theorems analogous to Theorem 3 and Lemma 4 showing that the definitions of the inclusion relation and the inclusion measure are correct.

Let us now consider an example of finding the lower and the upper membership functions and calculating the inclusion measures for discrete probability distributions when X consists of a finite number of elements, i.e. $X = \{x_1, x_2, ..., x_k\}$. For the sake of simplicity we put $V(x_i) = 1$ and denote

$P(x_i) = p_i$, $i = 1, 2, ..., k$, for a class F. In the given case the probabilities p_i have a sense of densities, i.e. $h(x_i) = p_i$.

The expressions of the membership functions $\underline{\mu}(x)$ and $\bar{\mu}(x)$ and the inclusion measure $\psi(F_1 \subseteq F_2)$ can be in this case written as

$$\underline{\mu}(x_i) = \sum_{j \mid p_j < p_i} p_j, \quad \bar{\mu}(x_i) = \sum_{j \mid p_j \le p_i} p_j, \tag{8}$$

$$\psi(F_1 \subseteq F_2) = \sum_{i=1}^{k} I(x_i) P_1(x_i). \tag{9}$$

The values of the function $I(x_i)$ used in the last formula can be calculated by (7).

	$P_k(x_1)$	$P_k(x_2)$	$P_k(x_3)$
F_1	0.3	0.4	0.3
F_2	0	0	0
F_3	1/3	1/3	1/3
F_4	0.4	0.3	0.3

Table 1.

	$\underline{\mu}_k(x_1)$	$\bar{\mu}_k(x_1)$	$\underline{\mu}_k(x_2)$	$\bar{\mu}_k(x_2)$	$\underline{\mu}_k(x_3)$	$\bar{\mu}_k(x_3)$
F_1	0	0.6	0.6	1	0	0.6
F_2	0	0	0	1	0	0
F_3	0	1	0	1	0	1
F_4	0.6	1	0	0.6	0	0.6

Table 2.

Let us now assume that we are given in the space $X = \{x_1, x_2, x_3\}$ four statistical classes F_1, F_2, F_3, F_4 with their probabilities $P_k(x_i)$, $i = 1, 2, 3$, $k = 1, 2, 3, 4$, whose values are put down in the Table 1. By means of the formulas (8) and (9) one could calculate the lower and the upper membership functions and the mutual inclusion measures. All the results of calculations are shown in Tables 2, 3.

$\subseteq$	F_1	F_2	F_3	F_4
F_1	1	0.4	0.76	0.6
F_2	1	1	1	0.6
F_3	0.733	1/3	1	0.733
F_4	0.6	0.18	0.76	1

Table 3.

7 Generalization of the statistical class concept

We defined above set theoretic operations on statistical classes: the union and the intersection. It should be now especially emphasized that these operations might lead out of the set of statistical classes $\mathfrak{F}$. For example, for arbitrary classes $F_1, F_2 \in \mathfrak{F}$ the existence of a statistical class $F_3 = F_1 \cap F_2$ is not guaranteed, as the probability measure inducing this class can be constructed not in every case. In other words, the set of statistical classes $\mathfrak{F}$ is not closed regarding the union and the intersection of classes. Thus, these operations lead to coming into being new objects which we will also call statistical classes.

Formal introducing such classes has rather profound pragmatical sense when analysing and processing statistical data. For reiterative observations of the same random process, which generating mechanism being not sufficiently understood, we obtain samples that lead in general to distinct statistical classes. It is partly accounted by the fact that factors affecting the proceeding of the random process have different influence at different times. That is why every sample reflects only some part of properties of a real process. In this sense the union and the intersection of statistical classes lead to objects that possess all or at least one of the properties of these classes respectively. Analysis of the set F enables us to establish the structure of these objects (which we have arranged to call also statistical classes). In order to take into account the property inheritance of statistical classes we introduce the concept of the generalized statistical class that is defined as

$$\mathfrak{f}_F = \{F_i \in \mathfrak{F} \mid F_i \subseteq F\},$$

i.e. $\mathfrak{f}_F$ is a set of statistical classes F_i that include into the class F. In this case, obviously, the generalized statistical class is an ordinary subset of $\mathfrak{F}$. Further generalization of the concept is closely associated with using the previously introduced inclusion measure of classes. Namely, the *generalized fuzzy statistical class* is the fuzzy subset

$$\tilde{\mathfrak{f}}_F = \{F_i \mid \psi(F_i \subseteq F)\}.$$

Thus $\tilde{\mathfrak{f}}_F$ is a fuzzy subset in $\mathfrak{F}$ with the membership function $\mu_F(F_i) = \psi(F_i \subseteq F)$. It is not hard to see that 1-level (Kaufmann, 1975) of the set

$\tilde{got f}_F$ is just $\mathfrak{f}_F$. Note that $\tilde{\mathfrak{f}}_F$ is uniquely determined by the membership function $\mu(x)$ of the statistical class F.

Taking into consideration what was said above we will further consider generalized fuzzy statistical classes $\tilde{\mathfrak{f}}_F$ given by *an arbitrary* measurable membership function $\mu(x)$ (not obligatory induced by any probability measure) of the class F. With this there remains valid the formula of the inclusion measure defining the membership function of the given generalized class

$$\mu_F(F_i) = \psi(F_i \subseteq F) = 2 \int_X \min\left[\mu_i(x), \mu(x)\right] dP_i(x).$$

Introducing the generalized classes allows us to extend the definition of the intersection and the union on statistical classes given by arbitrary measurable membership functions

$$F_3 = F_1 \cap F_2 \quad \Leftrightarrow \quad \forall x \in X \quad \mu_3(x) = \min\left[\mu_1(x), \mu_2(x)\right],$$

$$F_3 = F_1 \cup F_2 \quad \Leftrightarrow \quad \forall x \in X \quad \mu_3(x) = \max\left[\mu_1(x), \mu_2(x)\right].$$

The set $\mathfrak{F}$ is obviously a distributive lattice regarding the defined operations.

8 Relation and measure of a possibilistic inclusion

The previous results allow to elucidate the connection between the concept of the generalized statistical class and the theory of possibility (Zadeh, 1978). Note that every generalized statistical class determines a family of probability measures $\mathcal{P}(\mathfrak{f}) = \{P_i \mid F_i = (X, \mathfrak{A}, P_i) \in \mathfrak{f}\}$. Analogous families appear in the possibility theory when probabilistic interpreting fuzzy necessity and possibility measures (Dempster, 1967, Shafer, 1976). It is known that such measures can be fixed with the help of a normal fuzzy set (i.e. such a set A that $\sup_{x \in X} \mu_A(x) = 1$) given on an arbitrary measurable space X. As the generalized statistical class is also fixed with a membership function $\mu(x)$ then a problem of analysing these families and extracting their common properties (if they exist) immediately arises.

Let a fuzzy set F be normal. Then in the possibilistic model the function $\mu(x)$ is interpreted as a function of a possibility distribution. With the help of this function for any $A \in \mathfrak{A}$ we can express the necessity measure $NESS(A) = \inf_{x \notin A} (1 - \mu(x))$ and the possibility measure $POSS(A) = \sup_{x \in A} \mu(x)$.

Taking into consideration that the necessity and the possibility measures have a sense of the lower and upper probabilities (Dempster, 1967, Shafer,

1976) we draw the conclusion that the function $\mu(x)$ determines a family $\mathcal{P} = \{P_i\}$ of probability measures P_i such that

$$NESS\,(A) \leq P_i(A) \leq POSS\,(A) \;\; for\ any\ A \in \mathfrak{A}. \tag{10}$$

In connection with this conclusion we introduce the concept of the possibilistic inclusion "$\overset{pos}{\subseteq}$" of an arbitrary statistical class F_i into statistical class F. We define $F_i \overset{pos}{\subseteq} F$ if the condition (10) holds.

The condition (10) is obviously not convenient for practical application. The following theorem enables us to express this condition by m.e. $\{A(p)\}$, $A(p) = \{x \in X \mid 1 - \mu(x) < p\}$, of the statistical class F.

Theorem 5. *The inclusion $F_i \overset{pos}{\subseteq} F$ is valid if, and only if, for any $p \in [0,1]$ the inequality $P_i\,[A(p)] \geq p$ holds.*

Proof. We at first prove the necessity. One can easily see that $\mu(x) \leq 1 - p$ for any $x \notin A(p)$, therefore $NESS\,[A(p)] = \inf_{x \notin A(p)} [1 - \mu(x)] \geq p$. Thus, if the condition (10) holds then $P_i\,[A(p)] \geq p$.

To prove the sufficiency we assume that the condition of the theorem holds, i.e. $P_i\,[A(p)] \geq p$ for any $p \in [0,1]$, then we show that the inequalities (10) take place. Let us consider an arbitrary event $B \in \mathfrak{A}$ and denote

$$p_1 = \inf_{x \notin B} [1 - \mu(x)] = NESS(B).$$

Prove that $P_i(B) \geq p_1$. It follows from the last expression that for any $x \notin B$ $1-\mu(x) \geq p_1$, therefore $X \backslash B \subseteq \{x \in X \mid 1 - \mu(x) \geq p_1\}$, and $A(p_1) = \{x \in X \mid 1 - \mu(x) < p_1\} \subseteq B$. Hence, $p_1 \leq P_i\,[A(p_1)] \leq P_i(B)$. That is the left inequality in (10) has proved.

We now show that the right inequality in (10) also holds. To do this we consider the value

$$p_2 = \sup_{x \in B} \mu(x) = POSS(B).$$

Let us prove that $P_1\,(B) \leq p_2$. It follows from the last expression that for any $x \in B$ $\mu(x) \leq p_2$, thus, $B \subseteq \{x \in X \mid \mu(x) \leq p_2\}$, $\{x \in X \mid \mu(x) > p_2\} = A\,(1 - p_2) \subseteq X \backslash B$. From this we obtain $P_i\,[A(1 - p_2)] \leq 1 - P_i(B)$ and by the sufficiency condition of the theorem we can write $1 - p_2 \leq 1 - P_i(B)$ and $P_i(B) \leq p_2$. This proves the right inequality in (10) and the theorem on the whole. ■

Corollary 3. *The relation of the set-theoretic inclusion "$\subseteq$" implicates the relation of possibilistic inclusion "$\overset{pos}{\subseteq}$", i.e. $F_i \overset{pos}{\subseteq} F$ follows $F_i \subseteq F$, but the contrary statement is false in general. In other words the inclusion "$\subseteq$" is stronger than "$\overset{pos}{\subseteq}$".*

Proof. Let $F_i \subseteq F$, then by definition $A_i(p) \subseteq A(p)$ for any $p \in [0,1]$ and consequently $P_i[A(p)] \geq P_i[A_i(p)] = p$. ■

Corollary 4. *The relations "$\subseteq$" and "$\overset{pos}{\subseteq}$" are equivalent if m.e. of statistical classes $F = (X, \mathfrak{A}, P)$ and $F_i = (X, \mathfrak{A}, P_i)$ coincide, i.e. $F_i \subseteq F \Leftrightarrow F_i \overset{pos}{\subseteq} F$, if $\Phi = \Phi_i$.*

Proof. Let us suppose that the conditions of this corollary hold, i.e. for statistical classes $F_i, F \in \mathfrak{F}$ the equality $\{A_i(p_1)\} = \{A(p_2)\}$, $p_1, p_2 \in [0,1]$, and the inclusion $F_i \overset{pos}{\subseteq} F$ take place. By virtue of Lemma 5 the possibilistic inclusion $F_i \overset{pos}{\subseteq} F$ means that for any $p_1 \in [0,1]$ the inequality $P[A_i(p_1)] = p_2 \leq p_1$ holds. From this and the condition $\Phi = \Phi_i$ it follows that $A_i(p_1) = A(p_2)$. Furthermore, as $p_2 \leq p_1$ then $A(p_2) \subseteq A(p_1)$. These conditions imply the inclusion $A_i(p_1) \subseteq A(p_1)$ for any $p_1 \in [0,1]$, that is by definition $F_i \subseteq F$. ■

Corollary 4 enables us to project an approach to constructing the possibilistic inclusion measure $\psi(F_i \overset{pos}{\subseteq} F)$. Let the fundamental sets of m.e. of statistical classes F_i and F coincide. Then because of Corollary 4 it is naturally to require the values of set-theoretic and possibilistic $\psi(F_i \overset{pos}{\subseteq} F) = \psi(F_i \subseteq F)$. As the following lemma shows in this case we can get a simpler expression for $\psi(F_i \subseteq F)$.

Lemma 9. *Let the fundamental sets $\Phi_i = \{A_i(p)\}$ and $\Phi = \{A(p)\}$ of m.e. of statistical classes F_i and F coincide, i.e. for any $p \in [0,1]$ there is such a $p_i \in [0,1]$ that $A(p) = A_i(p_i)$. Then*

$$\psi(F_i \subseteq F) = 2 \int_0^1 \min\left[p, P_i\left[A(p)\right]\right] dp. \tag{11}$$

Proof. Let us write again the expression of the set-theoretic inclusion $\psi(F_i \subseteq F) = 2 \int_0^1 P_i[A_i(p) \cap A(p)]\, dp$. Show that the integrands of the last expressions are equal, i.e. $P_i[A(p) \cap A_i(p)] = \min[p, P_i[A(p)]]$. Taking into account that the fundamental sets of m.e. coincide by constructing, i.e. $\Phi = \Phi_i$, and represent a family of included sets, it is sufficient to analyse only two possible cases. In the first case $A_i(p) \subseteq A(p)$, therefore $P_i[A(p) \cap A_i(p)] = P_i[A_i(p)] = p$. In the second one $A(p) \subseteq A_i(p)$ thus $P_i[A(p) \cap A_i(p)] = P_i[A(p)]$. ■

The lemma having been proved enables us to define a measure of possibilistic inclusion on the whole set of statistical classes. Consider the statistical class F, fixed by the membership function $\mu(x)$, and the class $F_i = \{X, \mathfrak{A}, P_i\}$, fixed by a probability measure P_i. One can construct the fundamental set of m.e. $\Phi = \{A(p)\}$,$A(p) = \{x \in X \mid 1 - \mu(x) < p\}$, of the statistical class F by the function $\mu(x)$.

We choose conditionally the set of m.e. Φ_i for the statistical class F_i in such a way that $\Phi = \Phi_i$. In this case to define the inclusion measure $\psi(F_i \subseteq F)$ we could use the representation (11). We will consider that this formula is also valid for possibilistic inclusion measure (even in a case of sets whose m.e. of statistical classes F and F_i are not coincide), i.e.

$$\psi(F_i \overset{pos}{\subseteq} F) = 2 \int_0^1 \min \left[p, P_i \left[A(p)\right]\right] dp. \tag{12}$$

The following theorem shows that this measure is reasonable.

Theorem 6. *$\psi(F_i \overset{pos}{\subseteq} F) = 1$ if, and only if, $F_i \overset{pos}{\subseteq} F$.*

This theorem is proved by arguments analogous to those used in the proof of Lemma 4. One can show that it is possible to use the formulas (2) and (7) for the set-theoretic inclusion to obtain a statistical estimation of the possibilistic inclusion measure $\psi(F_i \overset{pos}{\subseteq} F)$. With it the upper and the lower membership functions of the statistical class F_i are found from the following expressions,

$$\bar{\mu}_i(x) = 1 - P_i \left\{y \in X \mid \mu(y) > \mu(x)\right\},$$

$$\underline{\mu}_i(x) = 1 - P_i \left\{y \in X \mid \mu(y) \geq \mu(x)\right\}.$$

9 Conclusion

The set-theoretic representation of probability distributions with the purpose of classifying them, proposed in the article, is the idea that has some merits. It enables us to establish in a rather natural way the relation of a partial order on the set of probability distributions and to define algebraic operations that can be interpreted as finding the least upper (the greatest lower) bounds in a partially ordered set. With this it should be mentioned that the method of probability distributions classification essentially differs from the analogous methods in the statistical theory of pattern recognition. The main restriction of these methods is that, as a rule, a rigorous classification is considered. That is observations of a certain random process or emergence are required to be strictly associated with one of the some classes. Such an approach does not take into account really existing "intersection" of the classes that could considerably distort the picture of being observed emergency. In this sense the idea of statistical classes gives a possibility to avoid the obstacles mentioned above.

In contrast to the classical Bayesian methods the classes in the proposed classification scheme might have fuzzy boundaries, the requirement of their disjointness and completeness does not appear to be obligatory. Formally to realise the proposed classification scheme we at first have to construct fuzzy sets describing the standard classes. Then when classifying unknown observations we should use the measures of set-theoretic or possibilistic inclusion. Reasonable constructing of such fuzzy sets by probabilistic distributions in abstract sample spaces with dominating statistical structure is just the problem under consideration in the first sections of this article. It must be especially emphasized that similar mathematical constructions are come out from quite another consideration in Dubois and Prade (1983) when approximating the probability distribution by the possibility one. The analysis executed in the present paper showed that it could be found the formal ground of applying the theory of fuzzy sets and possibility theory for classification and identification of statistical data.

Acknowledgements

This work is supported by Russian Fund of fundamental investigations (RFFI), grant N 98-01-00013.

References

1. Dempster, A.P. (1967). Upper and lower probabilities induced by multivalued mapping, *Ann. Math. Statist.* **38**, 325–339.
2. Dubois, D., and Prade, H. (1983). Unfair coins and necessity measures: Towards a possibilistic interpretation of histograms, *Fuzzy Sets and Systems* **10**, 15–20.
3. Dubois, D., and Prade, H. (1986a). Fuzzy sets and statistical data, *Europ. J. Oper. Res.* **25**, 345–356.
4. Dubois, D., and Prade, H. (1986b). A set-theoretic view of belief functions. Logical operations and approximations by fuzzy sets, *Intern. J. Gen. Syst.*, **12**, 193–226.
5. Dubois, D., and Prade, H. (1988). *Theorie des possibilités. Applications a la representation des connaissances en informatique*, Masson, Paris.
6. Fine, T.L. (1973). *Theories of Probability: An Examination of Foundations*, Academic Press, New York.
7. Kaufmann, A. (1975). *Introduction to the theory of fuzzy subsets 1*, Academic Press, New York.
8. Nguyen, H.T. (1978). On random sets and belief functions, *J. Math. Anal. Appl.* **65**, 531–542.
9. Shafer, G. (1976). *A Mathematical Theory of Evidence*, Princeton University Press, Princeton.
10. Sugeno, M. (1972). Fuzzy measure and fuzzy integral, *Trans. SICE* **8**, 95–102.
11. Zadeh, L.A. (1968). Probability measures of fuzzy events, *J. Math. Anal. Appl.* **23**, 421–427.
12. Zadeh, L.A. (1978). Fuzzy sets as a basis for a theory of possibility, *Fuzzy Sets and Systems* **1** 3–28.

Part 4

STATISTICS AND FUZZY DATA ANALYSIS

Statistics with one-dimensional fuzzy data

Reinhard Viertl[1]

Institut für Statistik, Technische Uniersität Wien, Austria

Abstract. In this paper a theory to model main elements in statistical problems concerning fuzzy data is presented.

1 Fuzzy data as fuzzy numbers

Real observations of one-dimensional continuous quantities are not precise real numbers but more or less *non-precise.* This kind of imprecision is different from measurement errors. In practical situations different kinds of uncertainty are present but usually only error models are considered. In applications three kinds of uncertainty – fuzziness, errors, and statistical variation – are present. In this chapter errors are not considered but statistical inference procedures for the case of non-precise data.

In order to generalize statistical analyses to the situation of non-precise data the description of non-precise observations with special fuzzy subsets of the real numbers $\mathbb{R}$ is the best up to date mathematical model. The corresponding fuzzy numbers $x^\star$ are characterized by a special form of membership functions, so called *characterizing functions* denoted by $\xi(\cdot)$.

Definition 1. *Characterizing functions* are real functions $\xi : \mathbb{R} \longrightarrow [0,1]$ with the following properties:

(1) $\exists\, x_0 \in \mathbb{R}\colon \xi(x_0) = 1$
(2) $\forall\, \alpha \in (0,1]$ the so-called α*-cut* $C_\alpha(\xi(\cdot)) = \{x \in \mathbb{R}\colon \xi(x) \geq \alpha\} = [a_\alpha, b_\alpha]$ is a closed finite interval

The set of all *fuzzy numbers* is denoted by $\mathcal{F}_c(\mathbb{R})$.

Remark 1. The definition above contains also the concept of fuzzy intervals from the literature. This is necessary to include interval data.

An important problem is how to obtain the characterizing function for a fuzzy observation. There are no general rules how to obtain $\xi(\cdot)$ of a fuzzy observation $x^\star$ but some methodology exists which is best seen from real examples.

Example 1. Color intensity pictures are often results of measurement procedures. For example X-ray pictures which are gray level pictures. If the data "point" is given by the boundary between two regions with different light intensity as depicted in Figure 1, the characterizing function of this fuzzy data point is obtained by differentiating the light intensity function $h(\cdot)$ and normalizing the result.

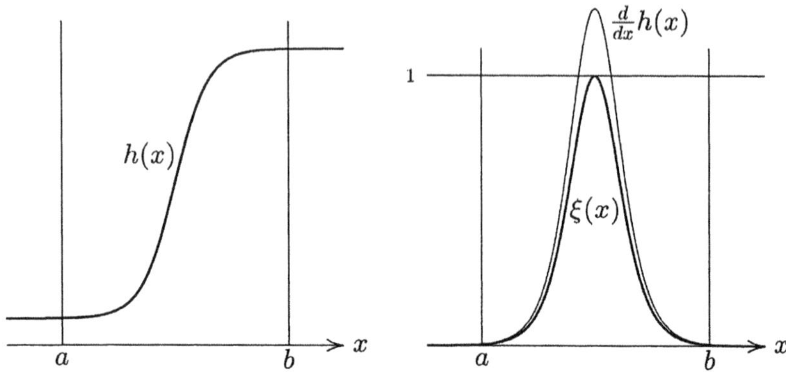

Fig. 1. Light intensity and corresponding characterizing function

Remark 2. In case the function $\frac{d}{dx}h(x)$ doesn't fulfill condition (2) of characterizing functions the so-called convex hull of $\frac{d}{dx}h(x)$ can be taken. The α-cuts of the convex hull of a function $\eta(\cdot)$ with α-cuts in form of unions of intervals $B_{\alpha,i}$, i. e. $B_\alpha = \bigcup_{i=1}^{k_\alpha} B_{\alpha,i} = \bigcup_{i=1}^{k_\alpha} [a_{\alpha,i}, b_{\alpha,i}]$ the α-cuts of the convex hull $\xi(\cdot)$ of $\eta(\cdot)$ are given by

$$C_\alpha := \left[\min_{i=1(1)k_\alpha} a_{\alpha,i} \, , \max_{i=1(1)k_\alpha} b_{\alpha,i} \right] \qquad \forall\, \alpha \in (0,1].$$

By the representation lemma of fuzzy numbers the characterizing function $\xi(\cdot)$ is given by

$$\xi(x) = \max_{\alpha \in (0,1]} \alpha \cdot I_{C_\alpha}(x) \qquad \forall\, x \in \mathbb{R},$$

where $I_{C_\alpha}(\cdot)$ denotes the *indicator function* of the subset C_α of $\mathbb{R}$.

Example 2. In environmental statistics life times of different units are important. These life times are often non-precise, for example the life time of a tree. Such life times can be described by fuzzy numbers with corresponding characterizing functions.

Other examples are given in the monograph by Viertl (1996).

2 Fuzzy samples and combined fuzzy sample

In considering samples of *stochastic quantities* X in case of continuous quantities such samples are n fuzzy numbers $x_1^\star, \cdots, x_n^\star$ with corresponding characterizing functions $\xi_1(\cdot), \cdots, \xi_n(\cdot)$. In classical statistical inference with precise data $x_1, \cdots, x_n$ where each x_i is an element of the *observation space* M_X of the stochastic quantity X, the combined sample $(x_1, \cdots, x_n)$ is an element of the *sample space* M_X^n. For one-dimensional quantities the observation space is a subset of $\mathbb{R}$, i. e., $M_X \subseteq \mathbb{R}$, and the sample space based on n observations is a subset of $\mathbb{R}^n$, i. e., $M_X^n \subseteq \mathbb{R}^n$.

The basis for statistical inference is in classical statistics the combined sample $\underline{x} = (x_1, \cdots, x_n) \in \mathbb{R}^n$.

In case of fuzzy samples $x_1^\star, \cdots, x_n^\star$ the fuzzy sample has to be combined to a fuzzy element $\underline{x}^\star$ of the sample space where $\underline{x}^\star$ is a *fuzzy vector* which is determined by a so-called *vector characterizing function* $\zeta(\cdot, \cdots, \cdot)$ which is a real function of n variables $\zeta\colon \mathbb{R}^n \longrightarrow [0,1]$ with the following properties:

(1) $\exists \underline{x}_0 \in \mathbb{R}^n\colon \zeta(\underline{x}_0) = 1$,
(2) $\forall\, \alpha \in (0,1]$ the so-called *α-cut* $C_\alpha(\zeta(\cdot, \cdots, \cdot)) := \{\underline{x} \in \mathbb{R}^n\colon \zeta(\underline{x}) \geq \alpha\}$ is a star shaped compact subset of $\mathbb{R}^n$.

Remark 3. An n-dimensional fuzzy vector $\underline{x}^\star$ is different from a vector $(x_1^\star, \cdots, x_n^\star)$ of n fuzzy numbers $x_1^\star, \cdots, x_n^\star$

In order to combine n fuzzy observations with corresponding characterizing functions $\xi_1(\cdot), \cdots, \xi_n(\cdot)$ so called *combination rules* are used. Such combination rules $K_n(\cdot, \cdots, \cdot)$ have to fulfill the following:

(1) $K_1(\xi(\cdot))\ \ \forall\, \xi(\cdot) \in \mathcal{F}_c(\mathbb{R})$,
(2) $K_n\Big(I_{[a_1,b_1]}(\cdot), \cdots, I_{[a_n,b_n]}(\cdot)\Big) = I_{[a_1,b_1]\times\cdots\times[a_n,b_n]}(\cdot, \cdots, \cdot)$,
(3) $K_n(\xi_1(\cdot), \cdots, \xi_n(\cdot))$ has to be a vector characterizing function $\forall n \in \mathbb{N}$.

There are two combination rules which are frequently used:

2.1 Minimum combination rule

This rule is defined by obtainig the vector characterizing function

$$\zeta(\cdot, \cdots, \cdot) = K_n\Big(\xi_1(\cdot), \cdots, \xi_n(\cdot)\Big)$$

of the combined fuzzy sample in the following way. The values $\zeta(x_1, \cdots, x_n)$ of this vector characterizing function are given by

$$\zeta(x_1, \cdots, x_n) := \min\{\xi_1(x_1), \cdots, \xi_n(x_n))\} \qquad \forall\, (x_1, \cdots, x_n) \in \mathbb{R}^n.$$

For the *minimum combination rule* the following is valid for the α-cuts $C_\alpha\Big(\zeta(\cdot, \cdots, \cdot)\Big)$ of the combined fuzzy sample:

$$C_\alpha\Big(\zeta(\cdot, \cdots, \cdot)\Big) = \mathop{\times}_{i=1}^{n} C_\alpha\Big(\xi_i(\cdot)\Big) \qquad \forall\, \alpha \in (0,1].$$

For this combination rule the α-cuts of the combined fuzzy sample are simple structured sets which are useful in statistical inference for fuzzy data.

The fuzzy vector $\underline{x}^\star$ whose vector characterizing function $\zeta(\cdot, \cdots, \cdot)$ is obtained by a combination of the characterizing functions $\xi_1(\cdot), \cdots, \xi_n(\cdot)$ of the fuzzy sample $x_1^\star, \cdots, x_n^\star$ is the basis for the generalization of statistical inference procedures to the situation of fuzzy data.

2.2 Product combination rule

Another possibiliy to form a combined fuzzy sample is the so-called product rule which is given via the values $\zeta(x_1, \cdots, x_n)$ of the vector characterizing function of $\underline{x}^\star$ by

$$\zeta(x_1, \cdots, x_n) = \prod_{i=1}^{n} \xi_i(x_i) \qquad \forall\, (x_1, \cdots, x_n) \in \mathbb{R}^n.$$

It can be proved that the α-cuts of $\zeta(\cdot, \cdots, \cdot)$ are star shaped compact subsets of $\mathbb{R}^n$ for all $\alpha \in (0, 1]$.

3 Point estimators based on fuzzy data

Point estimators for *parameters* θ in *stochastic models* $X \sim f(\cdot|\theta), \theta \in \Theta$ can be generalized to the situation of fuzzy data $x_1^\star, \cdots, x_n^\star$ in the following way:

Let $\vartheta(X_1, \cdots, X_n)$ be a statistically reasonable estimator for θ. For given sample $x_1, \cdots, x_n$ an estimation $\hat{\theta}$ for the true parameter is obtained by

$$\hat{\theta} = \vartheta(x_1, \cdots, x_n\ ) \in \Theta.$$

Let M_X be the observation space of X, then $\vartheta\colon M_X^n \longrightarrow \Theta$ is a function on the sample space M_X^n of X. Therefore for combined fuzzy sample $\underline{x}^\star$ the *extension principle* can be used to construct a generalized *fuzzy estimate* $\hat{\theta}^\star$ for the true parameter value.

Definition 2. Let $\zeta(\cdot, \cdots, \cdot)$ be the vector characterizing function of $\underline{x}^\star$ and using the denotation $\underline{x} = (x_1, \cdots, x_n)$, the characterizing function $\varphi(\cdot)$ of the *fuzzy estimate* $\hat{\theta}^\star$ is given by its values

$$\varphi(\theta) := \left\{ \begin{array}{ll} \sup\left\{\zeta(\underline{x})\colon \vartheta(\underline{x}) = \theta\right\} & \text{if } \exists\ \underline{x}\colon \vartheta(\underline{x}) = \theta \\ 0 & \text{otherwise} \end{array} \right\} \quad \forall\, \theta \in \Theta.$$

Remark 4. This is a reasonable generalization which takes care of the fuzziness of data in form of a fuzzy estimate for the true parameter.

Remark 5. The calculation of the values $\varphi(\theta)$ can be complicated. In applications a finite number of α-cuts are calculated for the characterizing function $\varphi(\cdot)$. This is possible relatively easy if the funtion $\vartheta(\cdot, \cdots, \cdot)$ is continuous. In this case the following proposition gives the basis for the calculations.

Proposition 1. *Let $\underline{x}^\star \in \mathcal{F}_c(\mathbb{R}^n)$ be a fuzzy combined sample with vector characterizing function $\zeta(\cdot, \cdots, \cdot)$ with supp$\Big(\zeta(\cdot, \cdots, \cdot)\Big) \subseteq M$ and*

$$g\colon M \longrightarrow \mathbb{R}$$

a continuous function, then the function $\varphi(\cdot)$ from Definition 2 is a characterizing function in the sense of the definition from Section 1. The α-cuts $B_\alpha(y^\star)$ of the fuzzy value $y^\star = g(\underline{x}^\star)$ are the following intervals

$$B_\alpha(y^\star) = \left[\min_{\underline{x}\in B_\alpha(\underline{x}^\star)} g(\underline{x}), \max_{\underline{x}\in B_\alpha(\underline{x}^\star)} g(\underline{x})\right]$$

where $B_\alpha(\underline{x}^\star]$ are the α-cuts of $\underline{x}^\star$.

The proof can be found in Viertl (1996).

An example of a fuzzy estimate $\hat{\mu}^\star$ of the expectation μ of a one-dimensional stochastic quantity is given in Figure 2.

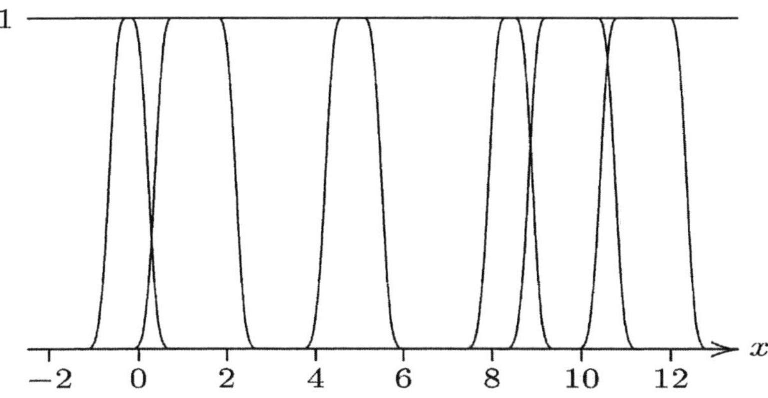

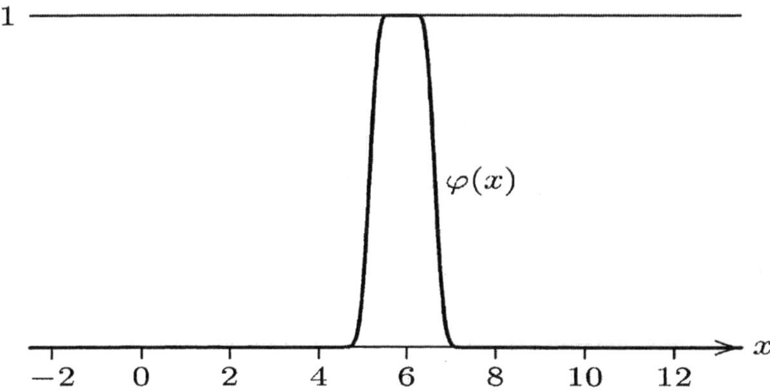

Fig. 2. Fuzzy data and corresponding fuzzy estimate of the expectation

4 Fuzzy confidence regions

In case of precise data $x_1, \cdots, x_n$ confidence sets for the parameter θ in a stochastic model $X \sim f(\cdot|\theta), \theta \in \Theta$ are crisp subsets $\Theta_0 \subset \Theta$. Such a Θ_0 is also called *confidence region.* Such confidence sets are obtained by so-called *confidence functions* $\kappa(X_1, \cdots, X_n)$ which are functions from the sample space M_X^n to the power set $\mathcal{P}(\Theta)$ such that

$$Pr\Big\{\theta \in \kappa(X_1, \cdots, X_n)\Big\} = 1 - \delta \qquad \forall\, \theta \in \Theta, \quad \text{with } 0 < \delta \ll 1.$$

For observed sample $X_1 = x_1, \cdots, X_n = x_n$ a classical subset

$$\Theta_0 = \kappa(x_1, \cdots, x_n) \subset \Theta$$

is obtained called *confidence set* for the true parameter θ_0 with *confidence level* $1 - \delta$.

In case of fuzzy data $x_1^\star, \cdots, x_n^\star$ of X a corresponding confidence set is a fuzzy subset $\Theta^\star$ of Θ whose characterizing function $\varphi(\cdot)$ is given in the following way:

Let $\zeta(\cdot, \cdots, \cdot)$ denote the vector characterizing function of the fuzzy combined sample $\underline{x}^\star$ then the values $\varphi(\theta)$ are given by

$$\varphi(\theta) = \left\{ \begin{array}{ll} \sup\{\zeta(\underline{x}) : \theta \in \kappa(\underline{x})\} & \text{if } \exists\, \underline{x} : \theta \in \kappa(\underline{x}) \\ 0 & \text{if } \nexists\, \underline{x} : \theta \in \kappa(\underline{x}) \end{array} \right\} \quad \forall\, \theta \in \Theta.$$

Remark 6. For the above given construction the following is valid:

$$\varphi(\theta) = 1 \qquad \text{for all } \theta \in \bigcup_{\underline{x} : \zeta(\underline{x}) = 1} \kappa(\underline{x})$$

where $\kappa(\cdot, \cdots, \cdot)$ is a classical confidence function.

Example 3. In case of normally distributed stochastic quantities $X \sim N(\mu, \sigma^2)$ a generalized confidence set based on a fuzzy sample consisting of fuzzy observations in form of triangular characterizing functions can be computed by an algorithm described in Dutter an Viertl (1998).

The resulting membership function $\varphi(\mu, \sigma^2)$ of the generalized confidence set for the vector parameter $\theta = (\mu, \sigma^2)$ is presented in Figure 3.

5 Statistical tests based on fuzzy data

Testing statistical hypotheses based on fuzzy data can become a problem because the values of test statistics $T = t(X_1, \cdots, X_n)$ become fuzzy. Therefore it is frequently not possible to decide if the value is in the acceptance region or in the rejection region for given error probability.

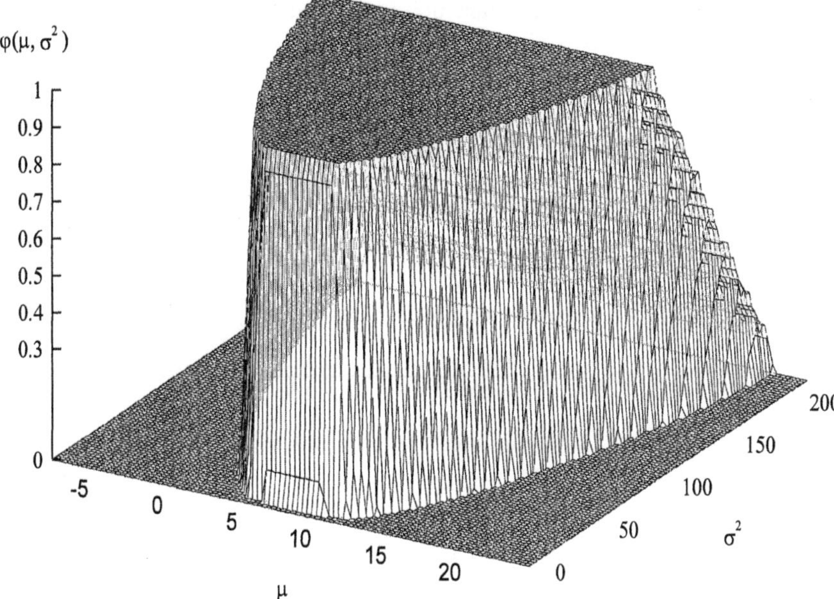

Fig. 3. Fuzzy confidence region for the parameter of a normal distribution

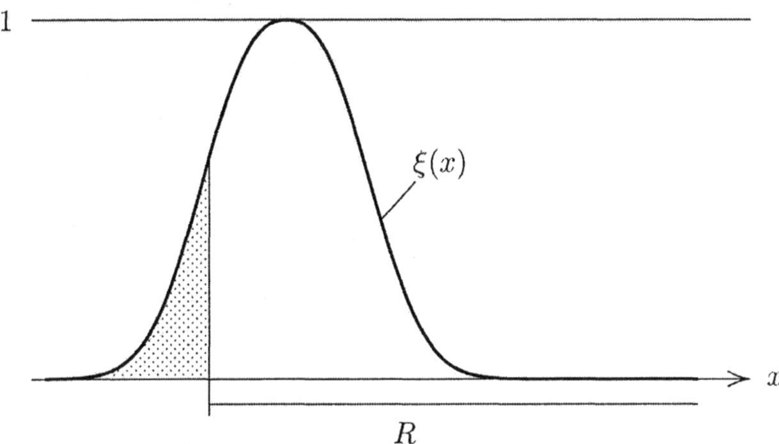

Fig. 4. Fuzzy value of a test statistic

This problem can be overcome using the concept of *p-values* (compare Lehmann, 1986).

Looking at the characterizing function $\xi(\cdot)$ of the fuzzy value $t^\star$ of a test statistic in order to make a decision the support of $\xi(\cdot)$ is considered. In case of a one-sided test the lower boundary point of the support of $\xi(\cdot)$ corresponds to a p-value. An example is given in Figure 4. This p-value can be taken as the basis for the test.

6 Bayesian inference with fuzzy data

6.1 Bayes' Theorem for non-precise data

For continuous stochastic model $X \sim f(\cdot \,|\, \theta), \quad \theta \in \Theta$ with continuous parameter space Θ and *a priori* density $\pi(\cdot)$ of the parameter and observation space M_X of X, Bayes' theorem for precise data $x_1, \ldots, x_n$ is

$$\pi(\theta \,|\, x_1, \ldots, x_n) = \frac{\pi(\theta) l(\theta; x_1, \ldots, x_n)}{\int_\Theta \pi(\theta) l(\theta; x_1, \ldots, x_n) d\theta} \qquad \forall\ \theta \in \Theta,$$

where $l(.; x_1, \ldots, x_n)$ is the likelihood function. In the most simple situation of complete data the likelihood function is given by

$$l(\theta; x_1, \ldots, x_n) = \prod_{i=1}^{n} f(x_i \,|\, \theta) \qquad \forall\ \theta \in \Theta.$$

Remark 7. Using the abbreviation $\underline{x} = (x_1, \ldots, x_n)$ Bayes' theorem can be stated in the form

$$\pi(\theta \,|\, \underline{x}) \ \propto\ \pi(\theta) \cdot l(\theta; \underline{x}) \qquad \forall\ \theta \in \Theta,$$

where $\propto$ stands for "proportional" since the right hand of the formula is a non-normalized function which is – after normalization – a density on the parameter space Θ.

For non-precise data $D^\star = (x_1^\star, \ldots, x_n^\star)$ with corresponding characterizing functions $\xi_1(\cdot), \ldots, \xi_n(\cdot)$ the non-precise combined sample $\underline{x}^\star$ with characterizing function $\zeta(\cdot, \ldots, \cdot)$,

$$\zeta : M_X^n \to [0, 1],$$

is the basis for a generalization of Bayes' theorem to the situation of non-precise data in the following way.

For all $\underline{x} \in supp(\zeta(\cdot, \ldots, \cdot))$ the value $\pi(\theta|\underline{x})$ of the *a posteriori* density $\pi(\cdot \,|\, \underline{x})$ is calculated using Bayes' theorem for precise data.

To every θ by variation of $\underline{x}$ in the support of the non-precise combined sample $\underline{x}^\star$ a family

$$\left(\pi(\theta|\underline{x}); \ \ \underline{x} \in supp(\underline{x}^\star)\right)$$

of values is obtained and the characterizing function $\psi_\theta(\cdot)$ of this fuzzy value is obtained via the characterizing function $\zeta(\cdot, \ldots, \cdot)$ of the non-precise combined sample element by its values

$$\psi_\theta(y) := \left\{ \begin{array}{ll} \sup \left\{\zeta(\underline{x}) \colon \ \pi(\theta \,|\, \underline{x}) = y\right\} & \text{for } \exists\ \underline{x} \colon \pi(\theta \,|\, \underline{x}) = y \\ 0 & \text{otherwise} \end{array} \right\} \quad \forall\, y \geq 0,$$

where the supremum has to be taken over the sample space M_X^n.

Definition 3. The family $\big(\psi_\theta(\cdot),\ \theta \in \Theta\big)$ of non-precise values of the *a posteriori density* is containing the imprecision of the observations $x_i^\star$, $i = 1, \ldots, n$ and is called *fuzzy a posteriori density* $\pi^\star(\cdot \mid D^\star)$, i.e.,

$$\pi^\star(\cdot \mid D^\star) := \big(\psi_\theta(\cdot);\ \theta \in \Theta\big).$$

A graphical presentation of the fuzzy *a posteriori* density is the drawing of so-called *α-level curves.* These α-level curves are the curves which connect the ends of the α-cuts of $\psi_\theta(\cdot)$ as functions of θ. An example of the presentation of a fuzzy *a posteriori* density is given in Figure 5.

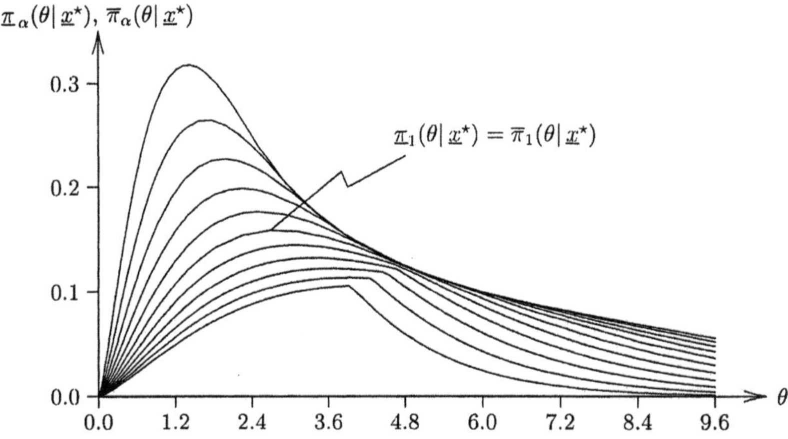

Fig. 5. Fuzzy a posteriori density

Example 4. Let X be a stochastic quantity having exponential distribution with density

$$f(x \mid \theta) = \frac{1}{\theta}\, e^{-x/\theta} I_{(0,\infty)}(x) \quad \text{with } \theta \in (0, \infty) = \Theta,$$

and the *a priori* density $\pi(\cdot)$ a gamma-density

$$\pi(\theta) = \frac{\theta^{\alpha-1} e^{-\theta/\beta}}{\Gamma(\alpha)\beta^\alpha} I_{(0,\infty)}(\theta) \quad \text{with } \alpha > 0,\ \beta > 0,$$

$\Gamma(\cdot)$ denoting the gamma-function.

Assume $n = 10$ non-precise observations $x_1^\star, \ldots, x_{10}^\star$ in order to obtain the fuzzy *a posteriori* density $\pi^\star(\cdot | x_1^\star, \ldots, x_n^\star)$ the minimum combination-rule is used to obtain the characterizing function $\zeta(\cdot, \ldots, \cdot)$ of the non-precise combined sample element $\underline{x}^\star$, i.e.,

$$\zeta(x_1, \ldots, x_n) = \min_{i=1(1)n} \xi_i(x_i) \qquad \forall\ (x_1, \ldots, x_n) \in \mathbb{R}^n.$$

Using this, the characterizing functions $\psi_\theta(\cdot)$ of the fuzzy values of the *a posteriori* density are obtained as explained before Definition 3 for every $\theta \in \Theta = (0, \infty)$.

In Figure 5 some α-level curves of the fuzzy *a posteriori* density $\pi^\star(\cdot \,|\, x_1^\star, \ldots, x_{10}^\star)$ based on the fuzzy data from Figure 6 are depicted.

Remark 8. The *fuzzy a posteriori density* can be used for estimations, predictions, and decisions.

6.2 Generalized HPD-regions

The Bayesian analog to confidence regions are highest *a posteriori density* regions, abbreviated by HPD-regions, for the parameter θ of a stochastic model $X \sim f(\cdot \,|\, \theta)$, $\theta \in \Theta$. For continuous parameter θ and precise data $D = \underline{x} = (x_1, \ldots, x_n)$ HPD-regions are defined using the *a posteriori* density $\pi(\cdot \,|\, \underline{x})$.

The *HPD-region* for θ with confidence level $1 - \delta$ based on *a posteriori* density $\pi(\cdot|\underline{x})$ is the subset Θ_0 of Θ, for which holds

(1) $$\int_{\Theta_0} \pi(\theta \,|\, \underline{x})\, d\theta \;=\; 1 - \delta$$

and

(2) $\pi(\theta \,|\, \underline{x}) \geq C$ for all $\theta \in \Theta_0$,
where C is the largest possible constant for which (1) and (2) hold.

Remark 9. HPD-regions are Bayesian confidence regions which are as small as possible and mostly unique.

The generalization of HPD-regions to the situation of fuzzy *a posteriori* densities is possible in the following way.

For non-precise data $D^\star = (x_1^\star, \ldots, x_n^\star)$ let $\zeta(\cdot, \ldots, \cdot)$ be the characterizing function of the non-precise combined sample $\underline{x}^\star$. Then for every precise vector $\underline{x}$ in $supp(\zeta(\cdot, \ldots, \cdot))$ the *a posteriori* density $\pi(\cdot \,|\, \underline{x})$ can be calculated.

Definition 4. In order to construct a generalized HPD-region which is a fuzzy subset of the parameter space Θ, for all $\theta \in \Theta$ and $\underline{x} \in supp(\zeta(\cdot))$, a classical subset $B(\theta, \underline{x})$ of Θ is defined by

$$B(\theta, \underline{x}) := \{\theta' \in \Theta : \ \pi(\theta' \,|\, \underline{x}) \geq \pi(\theta \,|\, \underline{x})\}.$$

The generalized *fuzzy HPD-region* for θ with confidence level $1-\delta$ is the *fuzzy subset* of Θ defined by its characterizing function $\varphi(\cdot)$ given by its values

$$\varphi(\theta) = \begin{cases} \sup\{\zeta(\underline{x}) : \int_{B(\theta,\underline{x})} \pi(\theta'|\underline{x})\, d\theta' \leq 1-\delta\} & \text{if } \exists\, \underline{x}: \int_{B(\theta,\underline{x})} \pi(\theta'|\underline{x}) \leq 1-\delta \\ 0 & \text{otherwise,} \end{cases}$$

where the supremum has to be taken over the set $supp(\zeta(\cdot))$.

Remark 10. The construction of generalized HPD-regions applied to precise data, describing the precise data by its one-point indicator functions, yields as result the indicator function of the classical HPD-region for precise data.

Example 5. For stochastic quantity X with exponential distribution having density

$$f(x|\theta) = \frac{1}{\theta}\, e^{-x/\theta} I_{(0,\infty)}(x)$$

a gamma density is used as *a priori* density. For ten non-precise observations $x_1^\star, \ldots, x_{10}^\star$ whose characterizing functions are presented in Figure 2, the minimum combination-rule is used to obtain the characterizing function of the non-precise combined sample.

Application of the definition of the characterizing function $\varphi(\cdot)$ from Definition 4 yields the fuzzy HPD-region, here a fuzzy interval, whose characterizing function is shown in Figure 6.

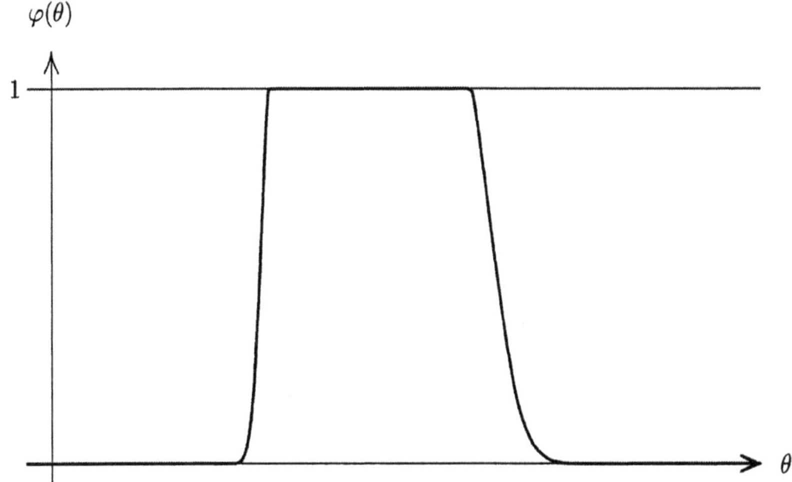

Fig. 6. Fuzzy HPD-interval for θ

6.3 Fuzzy predictive distributions

Information on future values of stochastic quantities X with observation space M_X and parametric stochastic model $f(\cdot\,|\,\theta)$, $\theta \in \Theta$ is provided by the *predictive density.*

In case of precise data $\underline{x} = (x_1, \ldots, x_n)$ and corresponding *a posteriori* density $\pi(\cdot|\underline{x})$ for the parameter $\tilde{\theta}$ the predictive density $g(\cdot|\underline{x})$ for X based on data $\underline{x}$ is the conditional density of X, i.e.

$$g(x\,|\,\underline{x}) = \int_\Theta f(x\,|\,\theta)\,\pi(\theta\,|\,\underline{x})\,d\theta \quad \text{for all} \quad x \in M_X.$$

For fuzzy data $D^\star = (x_1^\star, \ldots, x_n^\star)$ the non-precise combined sample $\underline{x}^\star$ with characterizing function $\zeta(\cdot, \ldots, \cdot)$ is used for the generalization of the concept of predictive densities. This generalization is defined by a family of fuzzy values for the predictive density.

Definition 5. For fixed $x \in M_X$ the vector $\underline{x}$ is varying in $supp(\zeta(\cdot, \ldots, \cdot))$. The characterizing function $\psi_x(\cdot)$ of the *fuzzy value of the predictive density* is given by its values

$$\psi_x(y) = \left\{ \begin{array}{ll} \sup\{\zeta(\underline{x})\colon \underline{x} \in M_X^n,\ g(x|\underline{x}) = y\} & \text{if } \exists\ \underline{x}\colon g(x|\underline{x}) = y \\ 0 & \text{otherwise} \end{array} \right\} \forall\ x \in M_X,$$

where $g(x|\underline{x})$ is the value of the classical predictive density based on precise data $\underline{x}$ and the supremum is to be taken over the set $supp(\zeta(\cdot, \ldots, \cdot))$. The family $(\psi_x(\cdot),\ x \in M_X)$ of non-precise values of the predictive density is called *fuzzy predictive density*

$$g^\star(\cdot\,|\,D^\star) = \Big(\psi_x(\cdot);\ \ x \in M_X\Big).$$

Remark 11. A graphical representation of fuzzy predictive densities can be given using α-level curves which are described in Subsection 6.1.

In Figure 7 an example of a fuzzy predictive density of an exponential distribution is depicted.

For precise data $\underline{x} = (x_1, \ldots, x_n)$ with $\xi_i(\cdot) = I_{\{x_i\}}(\cdot)$ the resulting characterizing functions are

$$\psi_x(\cdot) = I_{\{g(x\,|\,\underline{x})\}}(\cdot),$$

where $g(x|\underline{x})$ is the predictive density based on precise data $\underline{x}$.

Therefore the concept is a reasonable generalization of the classical predictive density.

Remark 12. There is also another approach to fuzzy predictive distributions using the integration of a fuzzy valued function. For details compare Viertl (1999).

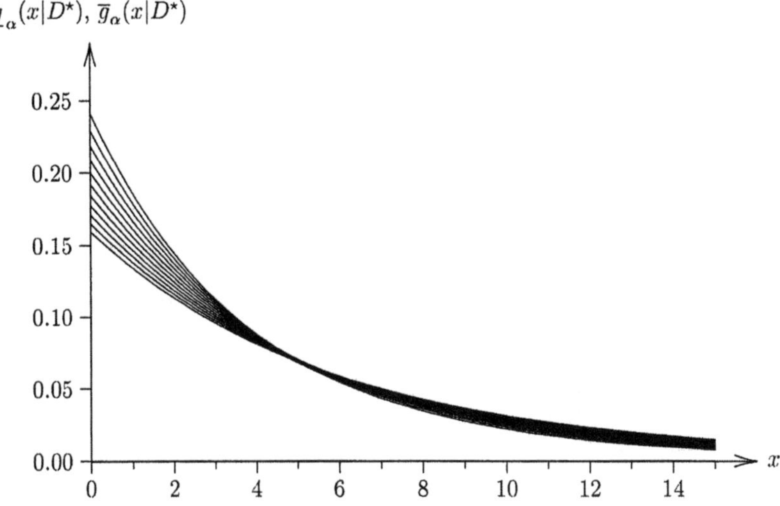

Fig. 7. Fuzzy predictive density of an exponential distribution

6.4 Fuzzy a priori distributions

Using precise a priori distributions $\pi(\cdot)$ for parameters θ in stochastic models

$$X \sim f(\cdot|\theta), \quad \theta \in \Theta$$

is a topic of critical discussions. Allowing a more general formulation of *a priori knowledge*, general agreement could be possible on reasonable use of a priori information on parameters.

Looking at the resulting fuzzy a posteriori density in Subsection 6.1 in natural way fuzzy a priori distributions in form of fuzzy densities $\pi^\star(\cdot)$ can be used. Such fuzzy densities are given by the family $\pi^\star(\theta)$, $\theta \in \Theta$ of fuzzy values of the density with characterizing functions $\varphi_\theta(\cdot)$, i.e.,

$$\pi^\star(\cdot) = \Big(\pi^\star(\theta);\ \ \theta \in \Theta\Big) \ \hat{=}\ \Big(\varphi_\theta(\cdot);\ \ \theta \in \Theta\Big).$$

This model is also necessary to describe the sequential information process which is obtained by additional data. Therefore, the approach from Subsection 6.1 has to be generalized.

Let $x_1^\star, \ldots, x_n^\star$ be n fuzzy observations with corresponding characterizing functions $\xi_1(\cdot), \ldots, \xi_n(\cdot)$ and $\pi^\star(\cdot)$ a fuzzy a priori density with fuzzy values $\pi^\star(\theta)$ and corresponding characterizing functions $\varphi_\theta(\cdot)$. Then the imprecision of the a priori distribution has to be combined with the imprecision of the fuzzy combined sample $\underline{x}^\star$.

This combination and the generalization yields a fuzzy a posteriori distribution for $\tilde{\theta}$.

In case of fuzzy a priori distributions formed by fuzzy hyperparameters of the a priori distribution and for conjugate families the analysis is relatively simple. This is explained in Frühwirth-Schnatter (1993) and the monograph by Viertl (1996).

References

1. Bandemer, H. (Ed.) (1993). *Modelling Uncertain Data.* Akademie Verlag, Berlin.
2. Dubois, D. and Prade, H. (1986). Fuzzy sets and statistical data. *European Journal of Operational Research* **25**, 345-356.
3. Dutter, R. and Viertl, R. (1998). Computation of Confidence Regions Based on Non-Precise Data. *COMPSTAT Proc. in Comput. Statist., Short Communications*, IACR-Rothamsted.
4. Frühwirth-Schnatter, S. (1993). On fuzzy Bayesian inference. *Fuzzy Sets and Systems* **60**, 41-58.
5. Gil, M.A., Corral, N. and Gil, P. (1988). The minimum inaccuracy estimates in χ^2 tests for goodness of fit with fuzzy observations. *Journal of Statistical Planning and Inference* **19**, 95-115.
6. Kacprzyk, J. and Fedrizzi, M. (Eds.) (1988). *Combining Fuzzy Imprecision with Probabilistic Uncertainty in Decision Making.* Lecture Notes in Economics and Mathematical Systems, Vol. 310, Springer-Verlag, Berlin.
7. Kalman, R.E. (1982). Identification from real data. In *Current Developments in the Interface: Economics, Econometrics, Mathematics*, (M. Hazewinkel and A.H.G. Rinnoy Kan, Eds.). D. Reidel Publ., Dordrecht, 161-196.
8. Kruse, R. and Meyer, K.D. (1987). *Statistics with Vague Data*, D. Reidel Publ., Dordrecht.
9. Lehmann, E.L. (1986). *Testing Statistical Hypotheses*, 2nd ed., John Wiley & Sons, New York.
10. Niculescu, S.P. and Viertl, R. (1992). A comparison between two fuzzy estimators for the mean. *Fuzzy Sets and Systems* **48**, 341-350.
11. Römer, C. and Kandel, A. (1995). Statistical tests for fuzzy data. *Fuzzy Sets and Systems* **72**, 1-26.
12. R. Viertl (Ed.) (1987). *Probability and Bayesian Statistics.* Plenum Press, New York.
13. Viertl, R. (1996). *Statistical Methods for Non-Precise Data.* CRC Press, Boca Raton, Florida.
14. Viertl, R. (1999). On Fuzzy Predictive Densities, *Tatra Mt. Math. Publ.* (Fuzzy Sets, Part II) **16**, 379-382.

Testing fuzzy hypotheses with vague data

Przemysław Grzegorzewski[1,2]

[1] Systems Research Institute, Polish Academy of Sciences,
Newelska 6, 01-447 Warsaw, Poland
[2] Faculty of Math. and Inf. Sci., Warsaw University of Technology,
Plac Politechniki 1, 00-661 Warsaw, Poland

Abstract. The problem of testing fuzzy hypotheses in the presence of vague data is considered. A new method based on the necessity index of strict dominance (NSD) is suggesteded. An example how to apply the proposed test in statistical quality control is shown.

1 Introduction

Testing hypotheses is one of the primary purposes of statistical inference. In classical statistics all model parameters, i.e. data, hypotheses and test requirements should be precise. However in real life we meet very often vague data, like "about ten", "more or less between five and seven", "rather greater than 100", etc. Moreover, sometimes we are quite satisfied in verifying fuzzy hypothesis like "the mean μ is about 40" instead of the crisp one $\mu = 40$. We may also test a hypothesis on a significance level "not greater than α" instead of the crisp value of α. Such statistical test under fuzzy constraints were considered by Arnold (1995). The problem of testing hypotheses with vague data was considered by Casals *et al.* (1986a, 1986b) and Grzegorzewski (2000, 2001). Testing fuzzy hypotheses was discussed by Delgado *et al.* (1985), Saade and Schwarzlander (1990), Saade (1994), Watanabe and Imaizumi (1993) and Arnold (1996).

The present paper is devoted to testing fuzzy hypotheses in presence of vague data. This problem was slightly touched in the excellent book by Kruse and Meyer (1987). Unfortunately, their solution reveals many disadvantages and gaps (see Grzegorzewski and Hryniewicz, 1997). Below, we propose other approach utilizing the Dubois-Prade necessity index of strict dominance, so popular in the possibility theory. An example of possible aplication in statistical quality control is also shown.

2 Basic concepts and notation

Assume that the investigated phenomenon is described by a probability distribution P_θ which belongs to a family of distributions $\mathcal{P} = \{P_\theta : \theta \in \Theta\}$. We consider the null hypothesis $H : \theta \in \Theta_H$ concerning the parameter θ, with the alternative hypothesis $K : \theta \in \Theta_K$, where Θ_H and Θ_K are subsets of Θ

such that $\Theta_H \cap \Theta_K = \emptyset$. We assume that if θ were known one would also know whether or not the hypothesis is true.

In the hypothesis testing problem we observe a random sample $V_1, \ldots, V_n$, and this observation can lead to one of two possible decisions: either to reject H (and to accept K), or to not reject H (usually identified with accepting H). Traditionally, rejection of H is denoted by zero and acceptance of H by one. Hence, a decision rule, called a statistical test, can be defined as a function $\varphi : \mathbb{R}^n \to \{0,1\}$. Each statistical test divides the sample space $\mathbb{R}^n$ into two exclusive subsets: $\{(v_1, \ldots, v_n) \in \mathbb{R}^n : \varphi(v_1, \ldots, v_n) = 0\}$ - the set of the acceptance of H, and $\mathcal{K} = \{(v_1, \ldots, v_n) \in \mathbb{R}^n : \varphi(v_1, \ldots, v_n) = 1\}$ - the set of the rejection of H which is also called a *critical region*. In practice, we compute a certain test statistic $T(V_1, \ldots, V_n)$ (i.e. a function of the observations), then we find a critical region $\mathcal{K}$, and finally we reject the considered hypothesis if $T(V_1, \ldots, V_n) \in \mathcal{K}$ or accept it otherwise. A typical statistical test has a following form:

$$\varphi(V_1, \ldots, V_n) = \begin{cases} 1 & \text{if } \; T(V_1, \ldots, V_n) \in \mathcal{K}, \\ 0 & \text{otherwise.} \end{cases} \tag{1}$$

The critical region depends on a preselected upper bound of the probability of the type I error (i.e. rejecting H when it is actually true), called a *significance level* δ . Thus we have

$$P\{\varphi(V_1, \ldots, V_n) = 1 \mid H\} \leq \delta. \tag{2}$$

For more details concerning the traditional theory of testing statistical hypotheses we refer the reader to Lehmann (1986).

Now let $X_1, \ldots, X_n$ denote a fuzzy sample which is a fuzzy perception of the usual random sample $V_1, \ldots, V_n$, from the population with the distribution P_θ. Assume that each X_i is a fuzzy number, i.e. X_i is a normal, fuzzy convex and bounded fuzzy subset of the real line $\mathbb{R}$ with an upper semicontinuous membership function $\mu_{X_i} : \mathbb{R} \to [0,1]$ (see, e.g., Dubois and Prade, 1980). A space of all fuzzy numbers will be denoted by $\mathcal{F}_c(\mathbb{R})$. Of course, $\mathcal{F}_c(\mathbb{R}) \subset \widetilde{\mathcal{F}}(\mathbb{R})$, where $\widetilde{\mathcal{F}}(\mathbb{R})$ denotes the space of all fuzzy sets on the real line. A useful tool for dealing with fuzzy numbers are their α-cuts. The α-cut, $\alpha \in (0,1]$, of a fuzzy number X with its membership function μ_X is a crisp set defined as

$$X_\alpha = \{t \in \mathbb{R} : \mu_X(t) \geq \alpha\}. \tag{3}$$

Every α-cut of a fuzzy number is a closed interval. A following notation will be useful below: $X_\alpha = [X_\alpha^L, X_\alpha^U]$, where

$$X_\alpha^L = \inf\{t \in \mathbb{R} : \mu_X(t) \geq \alpha\}, \tag{4}$$

$$X_\alpha^U = \sup\{t \in \mathbb{R} : \mu_X(t) \geq \alpha\}. \tag{5}$$

The precise definitions of a fuzzy random variable and fuzzy sample can be found in Kwakernaak (1978, 1979), Kruse (1982) and Puri and Ralescu

(1986). In this paper we adopt the approach similar to those of Kwakernaak and Kruse.

Suppose a random experiment is described as usual by a probability space $(\Omega, \mathcal{A}, P)$, where Ω is a set of all possible outcomes of the experiment, $\mathcal{A}$ is a σ-algebra of subsets of Ω (the set of all possible events) and P is a probability measure

Definition 1. A mapping $X : \Omega \rightarrow \mathcal{F}_c(\mathbb{R})$ is called a *fuzzy random variable* if it satisfies the following properties:

(a) $\{X_\alpha(\omega) : \alpha \in [0,1]\}$ is a set representation of $X(\omega)$ for all $\omega \in \Omega$,

(b) for each $\alpha \in [0,1]$ both $X_\alpha^L = X_\alpha^L(\omega) = \inf X_\alpha(\omega)$ and $X_\alpha^U = X_\alpha^U(\omega) = \sup X_\alpha(\omega)$, are usual real-valued random variables on $(\Omega, \mathcal{A}, P)$.

Thus a fuzzy random variable X is considered as a perception of an unknown usual random variable $V : \Omega \rightarrow \mathbb{R}$, called an *original* of X. Let χ denote a set of all possible originals of X. If only vague data are available, it is of course impossible to show which of the possible originals is the true one. Therefore we can define a fuzzy set on χ, with a membership function $\nu : \chi \rightarrow [0,1]$ given as follows:

$$\nu(V) = \inf\{\mu_{X(\omega)}(V(\omega)) : \omega \in \Omega\}, \tag{6}$$

which corresponds to the grade of acceptability that a fixed random variable V is the original of the fuzzy random variable in question (see Kruse and Meyer, 1987).

In the presence of fuzzy data we cannot observe the parameter θ directly but only its vague image. Moreover, we cannot estimate θ better than to its fuzzy perception $\Lambda(\theta)$ defined as

$$\mu_{\Lambda(\theta)}(t) = \sup\left\{\nu(V_1, \ldots, V_n) : (V_1, \ldots, V_n) \in \chi^n, \theta(V_1) = t\right\}, \quad t \in \mathbb{R}, \tag{7}$$

where χ^n is a set of all possible originals of that fuzzy random sample with membership function

$$\nu(V_1, \ldots, V_n) = \min_{i=1,\ldots,n} \inf\{\mu_{X_i(\omega)}(V_i(\omega)) : \omega \in \Omega\}. \tag{8}$$

One can easily obtain α-cuts of $\Lambda(\theta)$:

$$\begin{aligned}(\Lambda(\theta))_\alpha = \{t \in \mathbb{R} : \exists (V_1, \ldots, V_n) \in \chi^n, \theta(V_1) = t, \quad &\text{such that} \\ V_i(\omega) \in (X_i(\omega))_\alpha \text{ for } \omega \in \Omega \text{ and for } i = 1, \ldots, n\}&.\end{aligned} \tag{9}$$

For more information we refer the reader to Kruse and Meyer (1987).

Kruse and Meyer concluded that since in estimation with vague data the best they could do was to find fuzzy perception $\Lambda(\theta)$ of the parameter θ under study, they also restricted hypotheses testing to hypotheses about $\Lambda(\theta)$. Thus actually they considered the problem of testing fuzzy hypotheses with fuzzy data. They suggested a following definition of a test:

Definition 2. Let $X_1, \ldots, X_n$ denote a fuzzy sample, then a function $\phi : [\mathcal{F}_c(\mathrm{I\!R})]^n \to \{0,1\}$ such that

$$P\{\omega \in \Omega : \phi(X_1(\omega), \ldots, X_n(\omega)) = 1 \mid \Lambda(\theta) = \Lambda_0\} \leq \delta \tag{10}$$

is called a *test* for testing hypothesis $H : \Lambda(\theta) = \Lambda_0$, where $\Lambda_0 \in \mathcal{F}_c(\mathrm{I\!R})$, on the significance level $\delta \in (0,1)$.

It is easily seen that the definition given above is a natural generalization of definition of the classical test and (10) reduces to (2) if all the data are crisp, i.e. $X_i = V_i$ and we consider crisp hypothesis $\theta = \theta_0$.

Kruse and Meyer also proposed how to construct such a test for verifying hypothesis $H : \Lambda(\theta) = \Lambda_0$ against two-sided and one-sided alternatives. Unfortunately their method has many drawbacks. For example, the statement "$\Lambda(\theta)$ is greater than Λ_0" they use for one-sided alternative hypothesis, has no sense in the case of fuzzy numbers where there is no unique linear order. Any time we say that one fuzzy number is "greater" than the second one we have to explain what does it mean, i.e. we have to mention how we order fuzzy numbers. This and other problems connected with Kruse and Meyer's approach to hypotheses testing were described in Grzegorzewski and Hryniewicz (1997).

3 Hypotheses using NSD

In this section we propose a new method of testing fuzzy hypotheses with vague data which can be used both for one-sided and two-sided alternatives and which satisfies the Kruse-Meyer definition (10). To express this alternatives we use the necessity index of strict dominance (NSD) due to Dubois and Prade (1983). Let us recall that for any fuzzy numbers A and B with membership functions μ_A and μ_B, respectively, we can evaluate the degree of necessity to which the relation $A > B$ is fulfilled

$$Ness(A > B) = 1 - \sup_{x,y:x \leq y} \min\{\mu_A(x), \mu_B(y)\}. \tag{11}$$

Dubois and Prade proposed also the possibility of strict dominance index and other indices. However, we decided to use NSD index because of its natural interpretation and effectiveness in solving real-life problems (see, e.g. Hryniewicz, 1992, 1994).

Let us begin with the problem of testing the null hypothesis $H : \Lambda(\theta) = \Lambda_0$ against one-sided alternative $K : Ness(\Lambda(\theta) > \Lambda_0) \geq \xi$, where ξ is a fixed number from the interval $[0,1]$. Although we have both fuzzy data and fuzzy hypotheses the order based on NSD leads to a very simple statistical test. Before we show the construction of that test a following lemma should be stated.

Lemma 1. *Let $X, Y \in \mathcal{F}_c(\mathbb{R})$. The following conditions are equivalent:*

(i) $Ness(X > Y) \geq \xi$,

(ii) $X^L_{1-\xi} \geq Y^U_{1-\xi}$,

(iii) $X^L_\alpha \geq Y^U_\alpha \quad \forall \alpha \in [0,1],\ \alpha \geq 1-\xi$.

Proof: *(i)* $\Longrightarrow$ *(ii)*

Let μ_X and μ_Y denote membership functions of fuzzy numbers X and Y, respectively. By definition (11) we get

$$\begin{aligned} &Ness(X > Y) \geq \xi \Leftrightarrow \\ &\Leftrightarrow 1 - \sup_{u,v:u \leq v} \min\{\mu_X(u), \mu_Y(v)\} \geq \xi \Leftrightarrow \\ &\Leftrightarrow \sup_{u,v:u \leq v} \min\{\mu_X(u), \mu_Y(v)\} \leq 1 - \xi \Leftrightarrow \\ &\Leftrightarrow \forall (u,v : u \leq v) \quad \min\{\mu_X(u), \mu_Y(v)\} \leq 1 - \xi. \end{aligned} \tag{12}$$

Since X is a fuzzy number then it is normal, i.e. there exist such u_0 that $\mu_X(u_0) = 1$. Hence $X_{1-\xi} \neq \emptyset$, because- in particular - $u_0 \in X_{1-\xi}$. Let $X^L_{1-\xi} = \inf\{u : u \in X_{1-\xi}\} = u_1$. Since $Ness(X > Y) \geq \xi$ then by (12) $\mu_Y(v) \leq 1 - \xi\ \forall v \geq u_1$. Thus $v \notin Y_{1-\xi}\ \forall v \geq u_1$ which means that $Y^U_{1-\xi} \leq u_1$. And therefore $Y^U_{1-\xi} \leq X^L_{1-\xi}$.

(ii) $\Longrightarrow$ *(iii)*

Since $X, Y \in \mathcal{F}_c(\mathbb{R})$ then X and Y are fuzzy convex. Hence $\forall \alpha \in [0,1], \alpha \geq 1 - \xi$ we have $X_\alpha \subseteq X_{1-\xi}$ and $Y_\alpha \subseteq Y_{1-\xi}$. And because $X^L_{1-\xi} \geq Y^U_{1-\xi}$ it follows that $X^L_\alpha \geq Y^U_\alpha\ \forall \alpha \geq 1 - \xi$.

(iii) $\Longrightarrow$ *(i)*

Since $X^L_\alpha \geq Y^U_\alpha\ \forall \alpha \geq 1 - \xi$ then, in particular, $X^L_{1-\xi} \geq Y^U_{1-\xi}$. Let us take any $u \in \mathbb{R}$. If $u \in X_{1-\xi}$ then $v \notin Y_{1-\xi}\ \forall v \geq u$, because $X^L_{1-\xi} \geq Y^U_{1-\xi}$, and consequently $\mu_Y(v) \leq 1 - \xi\ \forall v \geq u$. On the other hand, if $u \notin X_{1-\xi}$ then $\mu_X(u) \leq 1 - \xi$. Thus for any $u \in \mathbb{R}$ and for any $v \geq u$ we get $\min\{\mu_X(u), \mu_Y(v)\} \leq 1 - \xi$. By (12) it means that $Ness(X > Y) \geq \xi$, which completes the proof. ■

Therefore, according to the lemma, it is enough to consider only one α-level in order to check whether the relation $Ness(\Lambda(\theta) > \Lambda_0) \geq \xi$ holds. This conclusion makes the starting point for our test construction. A following proposition holds

Proposition 1. *Let $X_1, \ldots, X_n$, denote a fuzzy random sample, where $X_i \in \mathcal{F}_c(\mathbb{R})$ for $i = 1, \ldots, n$, from the distribution with unknown real parameter θ and let $\xi \in [0,1]$. Let $\Lambda(\theta) \in \mathcal{F}_c(\mathbb{R})$ denote a fuzzy perception of θ and let $[\pi_1, +\infty)$ be the upper one-sided confidence interval for the parameter θ on the confidence level $1 - \delta$. Then a function $\phi : (\mathcal{F}_c(\mathbb{R}))^n \to \{0,1\}$ such that*

$$\phi(X_1, \ldots, X_n) = \begin{cases} 1 & \text{if } (\Lambda_0)^U_{1-\xi} < \Pi^L_{1-\xi} \\ 0 & \text{otherwise,} \end{cases} \tag{13}$$

where

$$\Pi_{1-\xi}^{L} = \Pi_{1-\xi}^{L}(X_1,\ldots,X_n;\delta) = \inf\{t \in \mathbb{R} : \forall i \in \{1,\ldots,n\} \\ \exists x_i \in (X_i)_{1-\xi} \text{ such that } \pi_1(x_1,\ldots,x_n) \le t\}, \tag{14}$$

is a test for hypothesis $H : \Lambda(\theta) = \Lambda_0$ against one-sided alternative $K : Ness(\Lambda(\theta) > \Lambda_0) \ge \xi$, on the significance level δ.

Proof: We take advantage from the well known fact that there is an equivalence between the totality of parameters for which the null hypothesis is accepted and the structure of confidence intervals. More precisely, there is one-to-one correspondence between the acceptance region of the test for the hypothesis $H : \theta = \theta_0$ against $K : \theta > \theta_0$ on the significance level δ and one-sided confidence interval $[\pi_1, +\infty)$ for the parameter θ on the confidence level $1-\delta$, where $\pi_1 = \pi_1(V_1,\ldots,V_n;\delta)$.

Kruse and Meyer (1987) introduced the notion of fuzzy confidence interval for the unknown parameter θ. They also shown how to construct such fuzzy confidence interval in the presence of vague data. Namely, a fuzzy number $\Pi(\omega)$ with α-cuts $\Pi_\alpha(\omega) = [\Pi_\alpha^L(\omega), +\infty)$, $\alpha \in (0,1]$, where

$$\Pi_\alpha^L(\omega) = \Pi_\alpha^L(X_1(\omega),\ldots,X_n(\omega);\delta) = \inf\{t \in \mathbb{R} : \forall i \in \{1,\ldots,n\} \\ \exists x_i \in (X_i)_\alpha \text{ such that } \pi_1(x_1,\ldots,x_n) \le t\}, \tag{15}$$

is the upper fuzzy confidence interval for θ on the confidence level $1-\delta$, i.e.

$$P\{\omega \in \Omega : (\Lambda(\theta))_\alpha \subseteq \Pi_\alpha(\omega)\} \ge 1-\delta, \quad \forall \alpha \in (0,1], \tag{16}$$

where $\Lambda(\theta)$ is a fuzzy perception of θ given by (...).

By (14), (15) and (16) $P\{\omega \in \Omega : (\Lambda(\theta))_{1-\xi} \subseteq \Pi_{1-\xi}(\omega)\} \ge 1-\delta$ for $\xi \in [0,1)$ and, in concequence, $P\{\omega \in \Omega : (\Lambda(\theta))_{1-\xi} \nsubseteq \Pi_{1-\xi}(\omega)\} < \delta$. From (13) we conclude that

$$P\{\omega \in \Omega : \phi(X_1(\omega),\ldots,X_n(\omega)) = 1 \mid \Lambda(\theta) = \Lambda_0\} = \\ = P\{\omega \in \Omega : (\Lambda_0)_{1-\xi}^U < \Pi_{1-\xi}^L(\omega) \mid \Lambda(\theta) = \Lambda_0\}. \tag{17}$$

Since $\{\omega \in \Omega : (\Lambda_0)_{1-\xi}^U < \Pi_{1-\xi}^L(\omega)\} \subseteq \{\omega \in \Omega : (\Lambda_0)_{1-\xi} \nsubseteq \Pi_{1-\xi}(\omega)\}$ then finaly

$$P\{\omega \in \Omega : \phi(X_1(\omega),\ldots,X_n(\omega)) = 1 \mid \Lambda(\theta) = \Lambda_0\} < \\ < P\{\omega \in \Omega : (\Lambda_0)_{1-\xi} \nsubseteq \Pi_{1-\xi}(\omega)\} < \delta. \tag{18}$$

From (10) it follows that the test given by (13) and (14) is on the significance level δ and the proof is complete. ∎

Similarly, using one-to-one correspondence between the acceptance region of the test for the hypothesis $H : \theta = \theta_0$ against $K : \theta < \theta_0$ on the significance level δ and one-sided confidence interval $(-\infty, \pi_2]$ for the parameter θ on the confidence level $1-\delta$, where $\pi_2 = \pi_2(V_1,\ldots,V_n;\delta)$, we get a test for the opposite one-sided alternative fuzzy hypothesis.

Proposition 2. *Let $X_1, \ldots, X_n$, denote a fuzzy random sample, where $X_i \in \mathcal{F}_c(\mathbb{R})$ for $i = 1, \ldots, n$, from the distribution with unknown real parameter θ and let $\xi \in [0,1]$. Let $\Lambda(\theta) \in \mathcal{F}_c(\mathbb{R})$ denote a fuzzy perception of θ and let $(-\infty, \pi_2]$ be lower one-sided confidence interval for the parameter θ on the confidence level $1-\delta$. Then a function $\phi : (\mathcal{F}_c(\mathbb{R}))^n \to \{0,1\}$ such that*

$$\phi(X_1, \ldots, X_n) = \begin{cases} 1 & \text{if } \ (\Lambda_0)^L_{1-\xi} > \Pi^U_{1-\xi} \\ 0 & \text{otherwise,} \end{cases} \tag{19}$$

where

$$\begin{aligned} \Pi^U_{1-\xi} = \Pi^U_{1-\xi}(X_1, \ldots, X_n; \delta) = \sup \{ t \in \mathbb{R} : \forall i \in \{1, \ldots, n\} \\ \exists x_i \in (X_i)_{1-\xi} \ \text{ such that } \ \pi_2(x_1, \ldots, x_n) \geq t \}, \end{aligned} \tag{20}$$

is a test for hypothesis $H : \Lambda(\theta) = \Lambda_0$ against one-sided alternative $K : Ness(\Lambda_0 > \Lambda(\theta)) \geq \xi$, on the significance level δ.

The proof runs as before.

We can also use NSD index for testing our null hypothesis against two-sided alternative. Firstly we will define a following relation

Definition 3. Let $X, Y \in \mathcal{F}_c(\mathbb{R})$ and let $\xi \in [0,1]$. Then

$$Ness(X \neq Y) \geq \xi \Leftrightarrow (Ness(X > Y) \geq \xi \quad \text{or} \quad Ness(Y > X) \geq \xi). \tag{21}$$

Now, keeping in mind that there is one-to-one correspondence between the acceptance region of the test for the hypothesis $H : \theta = \theta_0$ against $K : \theta \neq \theta_0$ on the significance level δ and two-sided confidence interval $[\pi_1, \pi_2]$ for the parameter θ on the confidence level $1-\delta$, where $\pi_1 = \pi_1(V_1, \ldots, V_n; \frac{\delta}{2})$ $\pi_2 = \pi_2(V_1, \ldots, V_n; \frac{\delta}{2})$, we can state the following proposition:

Proposition 3. *Let $X_1, \ldots, X_n$, denote a fuzzy random sample, where $X_i \in \mathcal{F}_c(\mathbb{R})$ for $i = 1, \ldots, n$, from the distribution with unknown real parameter θ and let $\xi \in [0,1]$. Let $\Lambda(\theta) \in \mathcal{F}_c(\mathbb{R})$ denote a fuzzy perception of θ and let $[\pi_1, \pi_2]$ be two-sided confidence interval for the parameter θ on the confidence level $1-\delta$. Then a function $\phi : (\mathcal{F}_c(\mathbb{R}))^n \to \{0,1\}$ such that*

$$\phi(X_1, \ldots, X_n) = \begin{cases} 1 & \text{if } \ (\Lambda_0)^U_{1-\xi} < \Pi^L_{1-\xi} \ \text{ or } \ (\Lambda_0)^L_{1-\xi} > \Pi^U_{1-\xi} \\ 0 & \text{otherwise,} \end{cases} \tag{22}$$

where

$$\begin{aligned} \Pi^L_{1-\xi} = \Pi^L_{1-\xi}(X_1, \ldots, X_n; \tfrac{\delta}{2}) = \inf \{ t \in \mathbb{R} : \forall i \in \{1, \ldots, n\} \\ \exists x_i \in (X_i)_{1-\xi} \ \text{ such that } \ \pi_1(x_1, \ldots, x_n) \leq t \}, \end{aligned} \tag{23}$$

$$\begin{aligned} \Pi^U_{1-\xi} = \Pi^U_{1-\xi}(X_1, \ldots, X_n; \tfrac{\delta}{2}) = \sup \{ t \in \mathbb{R} : \forall i \in \{1, \ldots, n\} \\ \exists x_i \in (X_i)_{1-\xi} \ \text{ such that } \ \pi_2(x_1, \ldots, x_n) \geq t \}, \end{aligned} \tag{24}$$

is a test for hypothesis $H : \Lambda(\theta) = \Lambda_0$ against one-sided alternative $K : Ness(\Lambda_0 \neq \Lambda(\theta)) \geq \xi$, on the significance level δ.

The proof is similar to that of Proposition 1.

4 Applications in statistical process control

Statistical process control (*SPC*) is a collection of methods for achieving continuous improvement in quality. This objective is accomplished by a continuous monitoring of the process under study in order to quickly detect the occurrence of assignable causes and undertake the necessary corrective actions. The most commonly used *SPC* tools are control charts.

The most popular $\overline{x}$ control chart for monitoring the process level contains three lines: a center line (CL) corresponding to the process level and two other horizontal lines, called the upper control limit (UCL) and the lower control limit (LCL), respectively. Suppose that the process under consideration is normally distributed. Let us first assume that we know the parameters of the process (i.e. its mean m_0 and standard deviation σ) when the process is thought to be in control. In such a case the traditional $\overline{x}$ control chart is given by lines

$$\begin{aligned} UCL &= m_0 + u_{1-\delta/2}\tfrac{\sigma}{\sqrt{n}}, \\ CL &= m_0, \\ LCL &= m_0 - u_{1-\delta/2}\tfrac{\sigma}{\sqrt{n}}, \end{aligned} \tag{25}$$

where $u_{1-\delta/2}$ is the $100(1-\delta/2)$ percentile of the standard normal distribution and δ is a significance level (traditionally $\delta = 0,0027$ is accepted). Of course, if the process parameters are not known we have to estimate them.

When applying this chart one draws samples of a fixed size n at specified time points, then he computes an arithmetical mean of each sample and plots it as a point on the chart. As long as the points lie within the control limits the process is assumed to be in control. However, if a point plots outside the control limits we are forced to assume that the process is no longer under control.

It is worth to notice that the control chart described above is equivalent to the following test

$$\phi(V_1, \ldots, V_n) = \begin{cases} 0, & \text{if } m_0 - u_{1-\delta/2}\frac{\sigma}{\sqrt{n}} < \overline{V} < m_0 + u_{1-\delta/2}\frac{\sigma}{\sqrt{n}}, \\ 1, & \text{otherwise}, \end{cases} \tag{26}$$

for the two-sided hypothesis testing problem $H : m = m_0$ against $K : m \neq m_0$.

The traditional *SPC* tools were constructed for precise data. However, sometimes we are not able to obtain exact numerical data but we deal with imprecise or even linguistic data. To use classical control charts in such situations we should compress these vague observations to exact data, but by doing this we often loose too much information. Thus it seems reasonable to use fuzzy sets for modelling vague or linguistic data and then to design control charts for these fuzzy data. Control charts for linguistic variables have been developed by Wang and Raz (1988, 1990), Raz and Wang (1990), and Kanagawa, Tamaki and Ohta (1993). Then Höppner (1994) and Höppner

and Wolff (1995) proposed a fuzzy-Shewhart control chart for monitoring the process level. Their charts are designed for very particular cases and have many drawbacks (see Grzegorzewski, 1997a), so cannot be recommended for applications.
Because of the correspondence between control charts and significance tests, it seems natural to use a general method for constructing fuzzy tests for fuzzy data to design control charts for fuzzy observations. This method was proposed by Grzegorzewski (2000, 2001). Control charts for monitoring the process level designed using this method, called *fuzzy control charts*, were also suggested by Grzegorzewski (1997b). Below we show how to construct a control chart based on NSD index.

Suppose our data are no longer precice but vague, i.e we observe fuzzy samples $X_1, \ldots, X_n$, where each X_i is a fuzzy number. It may happen that the target value m_0 is also not precise (e.g. expressed in a linguistic form). Moreover, if the true value of the process mean m_0 is not known, we have to estimate it. Since fuzzy data are used for the estimation, the estimate is also a fuzzy number.

As it was mentioned above, the traditional $\overline{x}$ control chart for crisp data is based on the two-sided test for the mean. Thus we can construct a control chart for fuzzy data corresponding to the test for the hypothesis $H : \Lambda(m_0) = \Lambda_0$ against two-sided alternative $K : Ness(\Lambda(m_0) \neq \Lambda_0) \geq \xi$, where ξ is a fixed number ($\xi \in [0,1]$) and $\Lambda(m_0) \in \mathcal{F}_c(\mathbb{R})$ denotes a fuzzy perception of the true mean m.

We can rewrite equation (22) as follows

$$\varphi'(X_1, \ldots, X_n) = \begin{cases} 1, & \text{if } \begin{cases} (\overline{X})^U_{1-\xi} < (\Lambda_0)^L_{1-\xi} - \zeta \text{ or} \\ (\overline{X})^L_{1-\xi} > (\Lambda_0)^U_{1-\xi} + \zeta, \end{cases} \\ 0, & \text{otherwise,} \end{cases} \tag{27}$$

where ζ is a constant depending on a sample size n, confidence level $1 - \delta$ and whether the true variance of the process is known.

Therefore, by the analogy to classical $\overline{x}$ control chart, the control lines of the new chart are

$$\begin{cases} LCL = (\Lambda_0)^L_{1-\xi} - \zeta, \\ UCL = (\Lambda_0)^U_{1-\xi} + \zeta. \end{cases} \tag{28}$$

However now, instead of the center line CL, we have a *center area* CA, where

$$CA = [(\Lambda_0)^L_{1-\xi}, (\Lambda_0)^U_{1-\xi}]. \tag{29}$$

The inspection with our new chart looks as follows: At the beginning one chooses a significance level δ and required value of the necessity index ξ. Then he draws a sample $X_1, \ldots, X_n$ of a fixed size n at specified time points, computes the arithmetical mean $\overline{X}$, determines interval I corresponding to $(1 - \xi)$th cut of $\overline{X}$, i.e.

$$I = [(\overline{X})^L_{1-\xi}, (\overline{X})^U_{1-\xi}], \tag{30}$$

and plots it on the chart. If the whole interval lies outside the control limits (i.e. below LCL or above UCL) it is interpreted that the process is no longer in control. If the interval intersects one of the control limits it is a warning. An example of the inspection with this control chart is given in Figure 1.

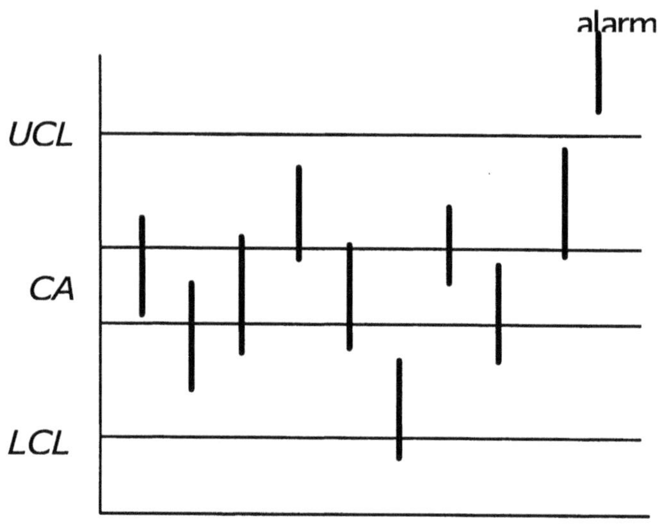

Fig. 1. A control chart based on NSD index

In practice we always estimate Λ_0 by the grand average of the means

$$\overline{\overline{X}} = \frac{1}{k}\sum_{j=1}^{k}(\frac{1}{n}\sum_{i=1}^{n} X_{ij}) \tag{31}$$

of the undisturbed prerun k samples (i.e. samples taken when the process is thought to be in control). Then assuming that the process is normally distributed with unknown standard deviation we get the following formulae for the chart

$$\begin{cases} LCL = (\overline{\overline{X}})^L_{1-\xi} - t^{[n-1]}_{1-\delta/2}\frac{1}{\sqrt{n}}(\overline{\overline{S}})^U_{1-\xi}, \\ UCL = (\overline{\overline{X}})^U_{1-\xi} + t^{[n-1]}_{1-\delta/2}\frac{1}{\sqrt{n}}(\overline{\overline{S}})^U_{1-\xi}, \end{cases} \tag{32}$$

$$CA = [(\overline{\overline{X}})^L_{1-\xi}, (\overline{\overline{X}})^U_{1-\xi}], \tag{33}$$

where $t^{[n-1]}_{1-\delta/2}$ is the $100(1-\delta/2)$ percentile of the t-distribution with $n-1$ degrees of freedom,

$$\begin{aligned} (\overline{\overline{X}})^L_{1-\xi} &= \tfrac{1}{kn}\textstyle\sum_{j=1}^{k}\sum_{j=1}^{n}(X_{ij})^L_{1-\xi}, \\ (\overline{\overline{X}})^U_{1-\xi} &= \tfrac{1}{kn}\textstyle\sum_{j=1}^{k}\sum_{j=1}^{n}(X_{ij})^U_{1-\xi}, \end{aligned} \tag{34}$$

$$(\overline{\overline{S}})_{1-\xi}^{U}=\frac{\Gamma(\frac{n-1}{2})}{\Gamma(\frac{n}{2})}\sqrt{\frac{n-1}{2}}\frac{1}{k}\sum_{j=1}^{k}\sqrt{\frac{1}{n-1}\sum_{i=1}^{n}\left((X_{ij})_{1-\xi}^{U}-(\overline{X}_{j})_{1-\xi}^{U}\right)^{2}}, \quad (35)$$

and Γ is the gamma function (see Höppner, 1994, or Höppner and Wolff, 1995).

5 Conclusions

The propositions given above show how to construct statistical tests for verifying fuzzy hypotheses with fuzzy data. In defining fuzzy alternatives we have used the Dubois-Prade necessity index of strict dominance. Of course, one may construct similar tests for other indices, like the possibility index of strict dominance, etc.

Our tests are well defined, because if we use crisp data instead of fuzzy observations and if we replace fuzzy hypotheses by crisp ones our tests reduce to the classical tests of significance. These tests are also very simple in use. Although we consider fuzziness both in data and in hypotheses the output of these test is crisp, i.e. our tests lead to precisely described decision: to rejection or to acceptance of the hypothesis under study. Thus they do not require any defuzzification method, which is also their advantage.

References

1. Arnold, B.F. (1995). Statistical Tests Optimally Meeting Certain Fuzzy Requirements on the Power Function and on the Sample Size, *Fuzzy Sets and Systems* **75**, 365–372.
2. Arnold, B.F. (1996). An Approach to Fuzzy Hypothesis Testing, *Metrika* **44**, 119–126.
3. Casals, M.R., Gil, M.A. and Gil, P. (1986a). On the Use of Zadeh's Probabilistic Definition for Testing Statistical Hypotheses from Fuzzy Information, *Fuzzy Sets and Systems* **20**, 175–190.
4. Casals, M.R., Gil, M.A. and Gil, P. (1986b). The Fuzzy Decision Problem: an Approach to the Problem of Testing Statistical Hypotheses with Fuzzy Information, *European J. Oper. Res.* **27**, 371–382.
5. Delgado, M., Verdegay, J.L. and Vila, M.A. (1985). Testing Fuzzy Hypotheses. A Bayesian Approach. In *Approximate Reasoning in Expert Systems* (Gupta, M.M., Kandel, A., Bandler, W. and Kiszka J.B., Eds.), Elsevier Science Publishers, 307–316.
6. Dubois, D. and Prade, H. (1980). *Fuzzy Sets and Systems: Theory and Applications.* Academic Press, New York.
7. Dubois, D. and Prade, H. (1983). Ranking Fuzzy Numbers in the Setting of Possibility Theory, *Inform. Sci.* **30**, 183–224.

8. Grzegorzewski, P. (1997a). *Statistical Decisions with Vague Data: Application in Statistical Quality Control.* PhD Thesis. Systems Research Institute, PAS (in Polish).
9. Grzegorzewski, P. (1997b). Control Charts for Fuzzy Data, *Proc. 5th European Congress Intelligent Techniques and Soft Comp. EUFIT'97*, Aachen, 1326–1330.
10. Grzegorzewski, P. (2000). Testing statistical hypotheses with vague data, *Fuzzy Sets and Systems* **112**, 501–510.
11. Grzegorzewski, P. (2001). Fuzzy tests - defuzzification and randomization, *Fuzzy Sets and Systems* **118**, 437-446.
12. Grzegorzewski, P. and Hryniewicz, O. (1997). Testing Hypotheses in Fuzzy Environment, *Mathware and Soft Computing* **4**), 203–217.
13. Höppner J. (1994). Statistiche Prozeßkontrolle mit Fuzzy-Daten. Ph.D. Dissertation, Ulm University.
14. Höppner J. and Wolff H. (1995). The Design of a Fuzzy-Shewhart Control Chart. Research Report, Würzburg University.
15. Hryniewicz, O. (1992). Statistical Acceptance Sampling with Uncertain Information from a Sample and Fuzzy Quality Criteria. Working Paper of SRI PAS, Warsaw, (in Polish).
16. Hryniewicz, O. (1994)., Statistical Decisions with Imprecise Data and Requirements. In *Systems Analysis and Decisions Support in Economics and Technology, Proc. 9th Polish-Italian and 6th Polish-Finnish Confer.* (R. Kulikowski, K. Szkatuła and J. Kacprzyk, Eds.). Omnitech Press, 135–143.
17. Kanagawa, A., Tamaki, F. and Ohta, H. (1993). Control Charts for Process Average and Variability Based on Linguistic Data, *Int. J. Prod. Res.* **31**, 913–922.
18. Kruse, R. (1982). The Strong Law of Large Numbers for Fuzzy Random Variables, *Inform. Sci.* **28**, 233–241.
19. Kruse, R. and Meyer, K.D. (1987). Statistics with Vague Data. D. Riedel Publishing Company.
20. Kwakernaak, H. (1978). Fuzzy Random Variables, Part I: Definitions and Theorems, *Inform. Sci.* **15**, 1–15.
21. Kwakernaak, H. (1978). Fuzzy Random Variables, Part II: Algorithms and Examples for the Discrete Case, *Inform. Sci.* **17**, 253–278.
22. Lehmann, E.L. (1986). *Testing Statistical Hypotheses.* 2nd ed., J. Wiley & Sons, New York.
23. Puri, M.L. and Ralescu, D.A. (1986). Fuzzy Random Variables, *J. Math. Anal. Appl.* **114**, 409–422.
24. Raz, T. and Wang, J.H. (1990). Probabilistic and Membership Approaches in Construction of Control Charts for Linguistic Data, *Production Planning & Control* **1**, 147–157.
25. Saade, J. (1994). Extension of Fuzzy Hypothesis Testing with Hybrid Data, *Fuzzy Sets and Systems* **63**, 57–71.
26. Saade, J. and Schwarzlander, H. (1990). Fuzzy Hypothesis Testing with Hybrid Data, *Fuzzy Sets and Systems* **35**, 197–212.
27. Wang, J.H. and Raz, T. (1988). Applying Fuzzy Set Theory in the Development of Quality Control Charts. International Industrial Engineering Conference Proceedings, Orlando, FL, 30–35.
28. Wang, J.H. and Raz, T. (1990). On the Construction of Control Charts Using Linguistic Variables, *Int. J. Prod. Res.* **28**, 477–487.

29. Watanabe, N. and Imaizumi, T. (1993). A Fuzzy Statistical Test of Fuzzy Hypotheses, *Fuzzy Sets and Systems* **53**, 167–178.

Possibilistic interpretation of fuzzy statistical tests

Olgierd Hryniewicz

Systems Research Institute, Newelska 6, 01-447 Warsaw, Poland

Abstract. A new possibilistic method for the interpretation of the results of fuzzy statistical tests has been proposed. The concept of the observed test size (p-value, significance) has been generalised to the case of fuzzy data. Indices of the possibility and necessity of dominance have been used for the comparison of null and alternative hypotheses. An example from statistical quality control is given.

1 Introduction

One of the most important application of statistical inference is to test certain hypotheses about phenomena which are random in nature. In such a case we formulate hypotheses about some finite or infinite population from which we draw one or more samples. The result of sampling provides statistical data for the verification of each stated hypothesis by a formalised procedure - a statistical test. When the result of a statistical test is used for making decisions we deal with the problem of statistical decisions.

In the classical statistics we assume that a phenomenon of interest is described by a random variable, say Z, distributed according to a certain probability distribution P_θ belonging to a family of probability distributions $\mathcal{P} = \{P_\theta : \theta \in \Theta\}$ indexed by a parameter θ (one- or multidimensional). We assume that decisions which are to be made upon the phenomenon of interest depend entirely on the value of the parameter θ. Usually, we formulate some hypotheses about the phenomenon of interest which are described by subsets of Θ, and the decision is related to the hypothesis which seems to be well confirmed by the existing data. If the value of θ were known we might take an appropriate decision without any problem finding a subset of Θ the known value of θ belongs to. However, the value of θ is not known, and we could only formulate some hypotheses about it. The basic hypothesis under test is called a *null hypothesis* $H : \theta \in \Theta_H$ (where $\Theta_H \subset \Theta$) upon a value of the parameter θ, and our original decision is equivalent to its acceptance or rejection. According to the Neyman-Pearson theory of statistical tests we should also formulate an alternative hypothesis $K : \theta \in \Theta_K$, where $\Theta_K \subset \Theta$, and $\Theta_H \cap \Theta_K = \emptyset$.

In the traditional statistics the theory of statistical test is well established, and all its notions are precise and well defined. However, when we have to apply this theory in practice we face certain problems with the interpretation of tests results. The simpliest and commonly used interpretation of statistical tests is provided in terms of frequencies of correct or incorrect decisions. It works perfectly in case of long series of similar decision problems such as problems encountered in statistical quality control of production processes. However, the interpretation in terms of frequencies does not seem to be sufficiently convincing for these practitioners who have to apply statistical tests for making single and unique decisions. This problem has been addressed in the paper by Hryniewicz (2000) who applied the notions from the theory of possibility in order to provide a decision maker with a simple interpretation of a test result.

Traditional statistical tests have been proposed for precisely defined crisp data. However, in many practical situations we face data which are not only random but vague as well. The intoduction of vagueness to the problem of statistical testing leads to a new class of statistical tests which have been proposed by many authors such as Arnold (1995), Casals *et al.* (1986) Delgado *et al.* (1985), Kruse and Meyer (1987), Saade (1994), Saade and Schwarzlander (1994), Son *et al.* (1992), and Watanabe and Imaizumi (1993). For a deeper discussion and a critical review of the problems cosidered there we refer the reader to the paper by Grzegorzewski and Hryniewicz (1997). Recently, Grzegorzewski (2000) has proposed a unified approach for testing statistical hypotheses with vague data which is a direct generalisation of the classical approach. Unfortunately, all these proposals do not address the problem mentioned in the previous paragraph, namely the problem of the interpratation of the test result that is used for making a single and unique decision. In this paper we extend the approach proposed in Hryniewicz (2000) to the case of statistical tests with vague data.

In the next section we present some basic notions and definitions used in the theory of statistical tests. Following Grzegorzewski (2000) we introduce the problem of testing statistical hypotheses with fuzzy data. Then, in the third section we present the possibilistic approach to the interpretation of the results of statistical tests that has been proposed in Hryniewicz (2000). In the fourth section we present the generalisation of these results to the case of vague data. Finally we present an example from the field of statistical quality control.

2 Statistical tests with fuzzy data

In statistical tests we assume that the observed random phenomenon is described by a crisp random variable Z. In such a case we observe a random sample $Z_1, Z_2, ..., Z_n$, and the following decisions are to be made: either to reject H (and to accept K) or not to reject H (usually identified with the acceptance of H). Let's denote by 0 the rejection, and by 1, the acceptance of H. Hence, the decision rule, called a statistical test, can be defined as a function $\varphi : \mathbb{R}^n \rightarrow \{0,1\}$. Each nonrandomised statistical test divides the whole space of possible observations of the random variable Z into two exclusive subspaces: $\{(z_1, z_2, ..., z_n) \in \mathbb{R}^n : \varphi(z_1, z_2, ..., z_n) = 0\}$, and $\{(z_1, z_2, ..., z_n) \in \mathbb{R}^n : \varphi(z_1, z_2, ..., z_n) = 1\}$. First of these subspaces is called *a critical region*, and the second is called *an acceptance region*. In the majority of practical cases we deal with a certain *test statistic* $T = T(Z_1, Z_2, ..., Z_n)$, and we reject the considered null hypothesis H when the value of T belongs to a certain critical region $\mathcal{K}$, i.e. if $T = T(Z_1, Z_2, ..., Z_n) \in \mathcal{K}$. In such a case the decision rule looks like this

$$\varphi(Z_1, Z_2, ..., Z_n) = \begin{cases} 0 & \text{if } T(Z_1, Z_2, ..., Z_n) \in \mathcal{K}, \\ 1 & \text{otherwise.} \end{cases} \tag{1}$$

To define the critical region we must set an upper value for the probability of a wrong rejection of the null hypothesis H (the so called probability of type I error). This probability, denoted by δ, is called *a significance level* of the test. Thus, we have

$$P(\varphi(Z_1, Z_2, ..., Z_n) = 0|H) \leq \delta. \tag{2}$$

In general, for a given sample number n there may exist many statistical tests which fulfil this condition. However, only some of them may have additional desirable properties, and only those are used in practice. For more detailed description of the problem we refer the reader to the excellent book by Lehmann (1986).

Suppose now, that instead of a crisp random variable Z we observe a fuzzy random variable $\widetilde{Z}$. The notion of a fuzzy random variable has been defined by many authors. In this paper we use the definition proposed in Grzegorzewski (2000).

Definition 1. . Let $(\Omega, \mathcal{A}, P)$ be a probability space, where Ω is a set of all possible outcomes of the random experiment, $\mathcal{A}$ is a σ-algebra of subsets of Ω (the set of all possible events), and P is a probability measure.

A mapping $\widetilde{Z} : \Omega \to \mathcal{F}_c(\mathbb{R})$, where $\mathcal{F}_c(\mathbb{R})$ is the space of all fuzzy numbers, is called a *fuzzy random variable* if it satisfies the following properties:

(a) $\left\{\widetilde{Z}_\alpha(\omega) : \alpha \in [0,1]\right\}$ is a set representation of $\widetilde{Z}(\omega)$ for all $\omega \in \Omega$,

(b) for each $\alpha \in [0,1]$ both $\widetilde{Z}_\alpha^L = \widetilde{Z}_\alpha^L(\omega) = \inf \widetilde{Z}_\alpha(\omega)$ and $\widetilde{Z}_\alpha^U = \widetilde{Z}_\alpha^U(\omega) = \sup \widetilde{Z}_\alpha(\omega)$, are usual real-valued random variables on $(\Omega, \mathcal{A}, P)$.

This definition is similar to the definitions proposed by Kwakernaak (1978) and Kruse (1982), and the random variable $\widetilde{Z}$ may be considered as a perception of an unknown usual random variable $Z : \Omega \to \mathbb{R}$, called an original of $\widetilde{Z}$.

Let $\widetilde{Z}_1, \ldots, \widetilde{Z}_n$ denote a fuzzy sample, i.e. a fuzzy perception of the usual random sample $Z_1, \ldots, Z_n$, from the population with the distribution P_Θ. Let

δ be a given number from the interval $(0,1)$. Grzegorzewski (2000) defined a fuzzy test as follows:

Definition 2. . A function $\varphi : (\mathcal{F}_c(\mathbb{R}))^n \to \widetilde{\mathcal{F}}(\{0,1\})$, where $\widetilde{\mathcal{F}}(\{0,1\})$ is the set of possible decisions, is called a *fuzzy test* for the hypothesis H, on the significance level δ, if

$$\sup_{\alpha\in[0,1]} P\left\{\omega\in\Omega : \varphi_\alpha\left(\widetilde{Z}_1(\omega),\ldots,\widetilde{Z}_n(\omega)\right)\subseteq\{0\}\,|H\right\}\le\delta$$

where φ_α is the α-level set (α-cut) of $\varphi\left(\widetilde{Z}_1,\ldots,\widetilde{Z}_n\right)$.

It is well-known that in the statistical testing with crisp data there is an equivalence between the set of values of the considered probability distribution parameter for which the null hypothesis is accepted and a certain confidence interval for this parameter. The same equivalence exists in the case of statistical tests with fuzzy data.

Let us consider, for example, a statistical test with the null hypothesis $H : \theta \le \theta_0$, and the alternative hypothesis $K : \theta > \theta_0$. In the case of crisp data there is one-to-one correspondence between the acceptance region for this test on the significance level δ and the one-sided confidence interval $[\underline{\pi}_l, +\infty)$for the parameter θ on the confidence level $1-\delta$, where $\underline{\pi}_l = \underline{\pi}_l(Z_1,\ldots,Z_n;\delta)$.

Kruse and Meyer (1987) introduced the notion of a fuzzy confidence interval for the unknown parameter θ, when the data are fuzzy. In the considered case, a *fuzzy equivalent* of $[\underline{\pi}_l, +\infty)$ can be defined by the following α-cuts (for all $\alpha\in(0,1]$):

$$\begin{aligned}\underline{\Pi}^L_\alpha &= \underline{\Pi}^L_\alpha\left(\widetilde{Z}_1,\ldots,\widetilde{Z}_n;\delta\right)\\ &= \inf\left\{t\in\mathbb{R} : \forall i\in\{1,\ldots n\}\,\exists z_i\in\left(\widetilde{Z}_i\right)_\alpha\right.\\ &\qquad \text{such that } \underline{\pi}_l(z_1,\ldots,z_n;\delta)\le t\}\end{aligned} \tag{3}$$

Similarily, we can define a *fuzzy equivalent* of the one-sided confidence interval $(-\infty, \overline{\pi}_u]$, as given in Grzegorzewski (2000):

$$\begin{aligned}\overline{\Pi}^U_\alpha &= \overline{\Pi}^U_\alpha\left(\widetilde{Z}_1,\ldots,\widetilde{Z}_n;\delta\right)\\ &= \sup\left\{t\in\mathbb{R} : \forall i\in\{1,\ldots n\}\,\exists z_i\in\left(\widetilde{Z}_i\right)_\alpha\right.\\ &\qquad \text{such that } \overline{\pi}_u(z_1,\ldots,z_n;\delta)\ge t\}\end{aligned} \tag{4}$$

where $\overline{\pi}_u(z_1,\ldots,z_n;\delta) = \underline{\pi}_l(z_1,\ldots,z_n;1-\delta)$.

The notion of the *one-sided fuzzy interval* can be used to define a fuzzy test. In the considered case, a function $\varphi : (\mathcal{F}_c(\mathbb{R}))^n \to \widetilde{\mathcal{F}}(\{0,1\})$ with the following α-cuts:

$$\varphi_\alpha\left(\widetilde{Z}_1,\ldots,\widetilde{Z}_n\right) = \begin{cases}\{1\} & \text{if } \theta_0\in\left(\underline{\Pi}_\alpha\backslash(\neg\underline{\Pi})_\alpha\right),\\ \{0\} & \text{if } \theta_0\in\left((\neg\underline{\Pi})_\alpha\backslash\underline{\Pi}_\alpha\right),\\ \{0,1\} & \text{if } \theta_0\in\left(\underline{\Pi}_\alpha\bigcap(\neg\underline{\Pi})_\alpha\right),\\ \emptyset & \text{if } \theta_0\notin\left(\underline{\Pi}_\alpha\bigcup(\neg\underline{\Pi})_\alpha\right)\end{cases} \tag{5}$$

is a fuzzy test for $H : \theta \leq \theta_0$, against $K : \theta > \theta_0$, on the significance level δ (Grzegorzewski, 2000). In a similar way, we can define fuzzy tests for testing, for example, $H : \theta \geq \theta_0$, against $K : \theta < \theta_0$.

3 Possibilistic interpretation of statistical tests

The interpretation of the results of statistical tests creates problems for many practitioners. The probem arises from the fact that the null and the alternative hypotheses are not symmetric. It means, that the result of the test depends upon which hypothesis is considered as the null hypothesis, and which as the alternative one. Thus, it may happen that on a given significance level δ we cannot neither reject $H : \theta \leq \theta_0$, against $K : \theta > \theta_0$ nor reject $H : \theta > \theta_0$, against $K : \theta \leq \theta_0$. This confuses a decision maker, especially when the decision has to be made only once.

To overcome this problem Hryniewicz (2000) proposed a new possibilistic interpretation of the results of statistical tests. For this interpretation a well known concept of the observed test size (also known as the p-value) has been used. The observed test size p is the minimum value of the significance level δ for which the null hypothesis H has to be rejected for the given observed value of the test statistics T. In other words, if the chosen value of the significance level δ is greater than the value of p, then the null hypothesis H is rejected. Otherwise, we do not have any important reason to reject it.

Let us assume that our statistical decision problem is described, as usually, by setting two alternative hypotheses $H : \theta \in \Theta_H$ and $K : \theta \in \Theta_K$. Hryniewicz (2000) proposes to consider these two hypotheses *separately*. First, he analyses only the null hypothesis H. For the observed value of the test statistics $t = T(z_1, z_2, ..., z_n)$ he finds the observed test size p_H for this hypothesis. The value of the observed test size p_H shows how the observed data support the null hypothesis. When this value is relatively large we may say that the observed data strongly support H. Otherwise, we should say that the data do not sufficiently support H. It is worthwhile to note that in the latter case we do not claim that the data support K. In the considered case of the null hypothesis $H : \theta \leq \theta_0$ the observed test size is defined as

$$p_H = p_H(\theta_0) = \inf_{\delta} \{\theta_0 \leq \underline{\pi}_l\} \tag{6}$$

The same can be done for the alternative hypothesis K, so we can find for this hypothesis the observed test size p_K. When $\Theta_H \cup \Theta_K = \Theta$ we have $p_K = 1 - p_H$. Thus, $p_K(\theta_1) = 1 - p_H(\theta_1)$.

Let us denote by 0 a situation when we decide that the data do not support the considered hypothesis, and by 1 a situation when we decide to accept the hypothesis. Hryniewicz (2000) proposes to evaluate the null hypothesis H by a fuzzy set $\widetilde{H}$ with the following membership function

$$\mu_H(x) = \begin{cases} \min[1, 2(1-p_H)] & \text{if } x = 0 \\ \min[1, 2p_H] & \text{if } x = 1 \end{cases} \tag{7}$$

which may be interpreted as a *possibility distribution* (see Dubois and Prade, 1997) of H. It is worthy to note that $\sup(\mu_H(0), \mu_H(1)) = 1$, and $\mu_H(0) = 1$ indicates that it is *plausible* that the hypothesis H is not true. On the other hand, when $\mu_H(1) = 1$ we wouldn't be surprised if H was true. It is necessary to stress here that the values of $\mu_H(x)$ do not represent the probability that H is true, but only a *possibility* of a correct decision.

The same can be done for the alternative hypothesis K. Hryniewicz (2000) proposes to evaluate the alternative hypothesis K by a fuzzy set $\widetilde{K}$ with the following membership function

$$\mu_K(x) = \begin{cases} \min[1, 2(1-p_K)] & \text{if } x = 0 \\ \min[1, 2p_K] & \text{if } x = 1 \end{cases} \tag{8}$$

which may be interpreted as a *possibility distribution* of K.

To choose an appropriate decision, i.e. to choose either H or K Hryniewicz (2000) proposes to use four measures of possibility defined by Dubois and Prade (1983). First measure is named the *Possibility of Dominance*, and for two fuzzy sets A and B is defined as

$$PD = Poss(A \succeq B) = \sup_{x,y:x \geq y} \min\{\mu_A(x), \mu_B(y)\}. \tag{9}$$

PD is the measure for a possibility that the set A is not dominated by the set B. In the considered problem of testing hypotheses we have

$$\begin{aligned} &PD = Poss\left(\widetilde{H} \succeq \widetilde{K}\right) = \\ &= \sup[\min\{\mu_H(0), \mu_K(0)\}, \min\{\mu_H(1), \mu_K(0)\}, \min\{\mu_H(1), \mu_K(1)\}] = \\ &= \max\{\mu_H(1), \mu_K(0)\}, \end{aligned} \tag{10}$$

and PD represents a *possibility* that choosing H over K is not a worse solution.

Second measure is named the *Possibility of Strict Dominance*, and for two fuzzy sets A and B is defined as

$$PSD = Poss(A \succ B) = \sup_x \inf_{y:y \geq x} \min\{\mu_A(x), 1 - \mu_B(y)\} \tag{11}$$

PSD is the measure for a possibility that the set A strictly dominates the set B. In the considered problem of testing hypotheses we have

$$\begin{aligned} &PSD = Poss\left(\widetilde{H} \succ \widetilde{K}\right) = \\ &= \sup[\inf[\min\{\mu_H(0), 1 - \mu_K(0)\}, \min\{\mu_H(0), 1 - \mu_K(1)\}]] = \\ &= \min\{\mu_H(1), 1 - \mu_K(1)\}, \end{aligned} \tag{12}$$

and PSD represents a *possibility* that choosing H over K is a correct decision.

Third measure is named the *Necessity of Dominance*, and for two fuzzy sets A and B is defined as

$$\begin{aligned} ND &= Ness\,(A \succeq B) = \\ &= \inf_{x} \sup_{y: y \le x} \max\{1 - \mu_A(x), \mu_B(y)\} \end{aligned} \tag{13}$$

and ND represents a *necessity* that the set A dominates the set B. In the considered problem of testing hypotheses we have

$$\begin{aligned} ND &= \inf\{\sup[\max(1 - \mu_H(1), \mu_K(0)), \\ &\max(1 - \mu_H(1), \mu_K(1)), \max(1 - \mu_H(0), \mu_K(0))]\} = \\ &= \max\{1 - \mu_H(0), \mu_K(0)\}, \end{aligned} \tag{14}$$

and ND represents a *necessity* of choosing H over K.

Fourth measure is named the *Necessity of Strict Dominance*, and for two fuzzy sets A and B is defined as

$$\begin{aligned} NSD &= Ness\,(A \succ B) = 1 - \sup_{x,y: x \le y} \min\{\mu_A(x), \mu_B(y)\} = \\ &= 1 - Poss\,(B \ge A) \end{aligned} \tag{15}$$

and NSD represents a *necessity* that the set A strictly dominates the set B. In the considered problem of testing hypotheses we have

$$\begin{aligned} NSD &= \\ &= 1 - \sup[\min\{\mu_H(0), \mu_K(0)\}, \min\{\mu_H(0), \mu_K(1)\}, \min\{\mu_H(1), \mu_K(1)\}] \\ &= 1 - \max\{\mu_H(0), \mu_K(1)\}, \end{aligned} \tag{16}$$

and NSD represents a strict *necessity* of choosing H over K.

Close examinations of the proposed measures reveals that in the considered case of statistical test

$$PD \ge ND \ge PSD \ge NSD. \tag{17}$$

It means that according to the practical situation we can choose the appropriate measure of the correctness of our decision. If the choice between H and K leads to serious consequences we should choose the NSD measure. In such a case $p_H > 0,5$ is required to have $NSD > 0$. When these consequences are not so serious we may choose the PSD measure. In that case $PSD > 0$ when $p_K < 0,5$, i.e. when there is no strong evidence that the alternative hypothesis is true. Finally, the PD measure, which is always positive, gives us the information of the possibility that choosing H over K is not a wrong decision. It is also possible to use the ND measure instead of PD, especially when we expect a slightly stronger evidence that choosing H is allowable.

The proposed method allows to assign possibility measures for the results of classical statistical tests. It assigns numerical values for grades of possibility or necessity of choosing null hypothesis against the alternative one.

4 Possibilistic interpretation in the case of fuzzy data

When we observe imprecise data the problem of the interpretation of statistical tests becomes more complicated than in the classical case of crisp data. In the case of crisp data, on the given significance level δ, we always obtain a crisp result of a statistical test: either to reject or to accept the null hypothesis. In the case of fuzzy data, the statistical tests described in the second section of this paper may indicate a fuzzy decision. In such a case, the statistical test does not provide a decision maker with a univocal solution. When we interchange the null and the alternative hypotheses, we also obtain a fuzzy result. In such a case, a correct decision might be difficult even for an experienced statistician. Therefore, there is a need to propose a relatively simple method for making an appropriate decision, i.e. for choosing this hypothesis that sems to be a correct one.

To deal with this problem we propose to extend the results presented in the third section of this paper to the case of fuzzy data. As previously, we assume that a fuzzy random sample $\widetilde{Z}_1, \ldots, \widetilde{Z}_n$ is observed. When the test statistics for the considered hypothesis is known for the case of crisp data, we can fuzzify it using the well known extension principle. Then, for this fuzzy test statistics we can define fuzzy confidence intervals applying the methodology sketched in the second section of this paper.

Without the loss of generality we assume that we test the null hypothesis $H : \theta \leq \theta_0$, against the alternative $K : \theta > \theta_0$. Now, we shall consider these two hypothesis separately. When the test data are fuzzy, for each α-cut ($\alpha \in [0,1]$) we can introduce the following quantity

$$p_{H,\alpha} = p_{H,\alpha}(\theta_0) = \inf_{\delta} \left\{ \theta_0 \leq \underline{\Pi}_{\alpha}^{L} \right\}, \tag{18}$$

where $\underline{\Pi}_{\alpha}^{L}$ is the lower bound of the α-cut of the one-sided fuzzy confidence interval given by (3) .

For the alternative hypothesis the analogous quantity is defined as

$$p_{K,\alpha} = p_{K,\alpha}(\theta_1) = \inf_{\delta} \left\{ \theta_0 > \overline{\Pi}_{\alpha}^{U} \right\}, \tag{19}$$

where $\overline{\Pi}_{\alpha}^{U}$ is the upper bound of the α-cut of the one-sided fuzzy confidence interval given by (4) . It is easy to show that for each $\alpha_1 \geq \alpha_2$ we have $\underline{\Pi}_{\alpha_1}^{L} \geq \underline{\Pi}_{\alpha_2}^{L}$, and $\overline{\Pi}_{\alpha_1}^{U} \leq \overline{\Pi}_{\alpha_2}^{U}$. Thus, we have $p_{H,\alpha_1} \geq p_{H,\alpha_2}$ and $p_{K,\alpha_1} \geq p_{K,\alpha_2}$. Taking this into account, we can define the *fuzzy observed test size* (*fuzzy p-value*) for the null hypothesis as

$$\widetilde{p}_H = \sup \left\{ \alpha I_{[p_{H,1}, p_{H,\alpha}]}; \alpha \in [0,1] \right\}. \tag{20}$$

and the *fuzzy observed test size* for the alternative hypothesis as

$$\widetilde{p}_K = \sup \left\{ \alpha I_{[p_{K,1}, p_{K,\alpha}]}; \alpha \in [0,1] \right\}. \tag{21}$$

Having the fuzzy versions of the observed test sizes for both hypotheses we can define the fuzzy equivalents of the possibility and necessity indices presented in the previous section. It is worth of noting, that each of these indices can be described as a certain function $IND(p_H, p_K)$. Using the extension principle, we can define the fuzzy counterparts of these indices $\widetilde{IND}$ using the following α-cuts:

$$(IND)_\alpha := \{IND(p_H, p_K) : p_H \in [p_{H,1}, p_{H,\alpha}],\ p_K \in [p_{K,1}, p_{K,\alpha}]\}. \quad (22)$$

When a decision maker needs an advice presented in a non-fuzzy form, we can defuzzify indices $\widetilde{IND}$ using, for example, Yager's F_1 index defined as

$$F_1 = \frac{\int_0^1 x_{IND}\mu(x_{IND})\, dx_{IND}}{\int_0^1 \mu(x_{IND})\, dx_{IND}} \quad (23)$$

where $\mu(x_{IND})$ is the the membership function of the considered possibility (or necessity) index. The values of the defuzzified possibility (or necessity) indices may be interpreted in exactly the same way as in the case of crisp data. Without the defuzzification, this interpretation is not, unfortunately, so straightforward.

5 Applications in statistical quality control

One of the most imortant problems of the statistical quality control is the inspection of lots with the aim to reject the lots with a large number of nonconforming items. A random sample of n items is taken from the inspected lot and the conformance to some technical specifications is determined for each sample item. When the number of nonconforming items in the sample is too large, the lot is rejected. Otherwise, the lot is accepted. Let θ be the number of nonconforming items in the whole lot, θ_0 be the highiest allowable number of such items in that lot, and $\theta_1 > \theta_0$ be the number of nonconforming items in the whole lot that is considered as totally unacceptable. Therefore, the sampling inspection may be vieved upon as the statistical test of the null hypothesis $H : \theta \leq \theta_0$, against the alternative $K : \theta > \theta_1$.

There exists a simple and convincing interpretation of the testing procedures when there is a series of inspected lots. However, in a contemporary production circumstances (production "on demand") such a situation becomes not so frequent. Very often, only unique lots have to be inspected, and the frequency interpretation of the test results is not convincing, at least for a majority of practitioners. To cope with this problem, Hryniewicz (2000) proposed a possibilistic interpretation for the results of the acceptance sampling when all the test data are crisp. However, there may exist situations when test data are presented in a vague form. In such a case, we can apply the methodology developed in the previous section of this paper.

Suppose that the quality of the inspected items is asessed by users, who may express their opinions imprecisely, for example as "quite good", "nearly bad", etc. Denote by x_i,where

$$x_i = \begin{cases} 0 & \text{if the item is conforming} \\ 1 & \text{if the item is nonconforming,} \end{cases} \tag{24}$$

where $i = 1, \ldots, n$, the result of the quality inspection of a sample item in a non-fuzzy case. In general, in the case of imprecise quality asessment, each item may be described by a fuzzy set $\widetilde{x}_i = (x_i, \mu(x_i))$, where $x_i \in [0,1], \mu(x_i) \geq 0$, and $\sup \mu(x_i) = 1$, for $i = 1, \ldots, n$. The description of this type has been proposed in Hryniewicz (1994).

Hryniewicz (1994) proposed the simplest possible membership function which is either

$$\mu(x_i) = \begin{cases} \mu_{0,i} & \text{for } x_i = 0 \\ 1 & \text{for } x_i = 1, \end{cases} \tag{25}$$

where $\mu_{0,i} \in [0,1], i = 1, \ldots, n$ for all items asessed as nonconforming (or "rather" nonconforming), or

$$\mu(x_i) = \begin{cases} 1 & \text{for } x_i = 0 \\ \mu_{1,i} & \text{for } x_i = 1, \end{cases} \tag{26}$$

where $\mu_{1,i} \in [0,1], i = 1, \ldots, n$ for all items asessed as conforming (or "rather" conforming). For a fully nonconforming item we use (5) with $\mu_{0,i} = 0$, and for a fully conforming item we use (5) with $\mu_{1,i} = 0$. It is worth noting, that this description is equivalent to the description of each item by a single number $a \in [0,1]$. If $a \geq 0,5$ we use (5) with $\mu_0 = 2(1-a)$, and for $a \leq 0,5$ we use (5) with $\mu_1 = 2a$.

Assume that in the inspected sample of n items there is n_1 items with quality described by (5) that are numbered in such a way that $0 \leq \mu_{0,1} \leq \mu_{0,2} \leq \cdots \leq \mu_{0,n_1} \leq 1$, and n_2 $(n_1 + n_2 = n)$ items with quality described by (5) that a numbered in such a way that $1 \geq \mu_{1,1} \geq \mu_{1,2} \geq \cdots \geq \mu_{1,n_2} \geq 0$.

The total fuzzy number of nonconforming items in the sample $\widetilde{d}$ can be found (see Hryniewicz (1994) as

$$\widetilde{d} = \mu_{0,1}|0 + \mu_{0,2}|1 + \cdots + \mu_{0,n_1}|(n_1 - 1) + 1|n_1 + \mu_{1,1}|(n_1 + 1) + \cdots + \mu_{1,n_2}|(n_1 + n_2). \tag{27}$$

The observed number of the nonconforming items in the sample is used as the test statistics in testing the null hypothesis $H : \theta \leq \theta_0$, against the alternative $K : \theta > \theta_1$. In the case of fuzzy data for that purpose we propose to use $\widetilde{d}$ given by (27).

Let's consider a practical example. Suppose that a lot consisting of $N = 1000$ items of a certain equipment is delivered to the market. The producer wants to know whether consumers are satisfied with this product. He assumes that at most 10 negative opinions are considered satisfactory. On the other hand, if there are more than 50 consumers that have negative opinions about the quality of that equipment the producer will consider the whole shipment as a failure. Thus, the producer has to test the null hypothesis $H: \theta \leq \theta_0 = 10$, against the alternative $K: \theta > \theta_1 = 50$.

In the international standard for acceptance sampling ISO 2859-2 (1985) that provides the user with acceptance sampling plans for quality inspection we can find the required sample number $n = 125$. Thus, it is necessary to ask 125 users of the considered equipment about their asessment of its quality. Suppose now, that 120 users described the equipment as fully conforming to their requirements, 3 users gave a totally negative opinion, one user described the equipment as "rather nonconforming", and one user asessed the quality of the equipment as "practically conforming". Assume that the asessment "rather nonconforming" can be described by a fuzzy set with the membership function

$$\mu(x_i) = \begin{cases} 0,5 & \text{for } x_i = 0 \\ 1 & \text{for } x_i = 1 \end{cases},$$

and the asessment "practically conforming" can be described by a fuzzy set with the membership function

$$\mu(x_i) = \begin{cases} 1 & \text{for } x_i = 0 \\ 0,2 & \text{for } x_i = 1 \end{cases}.$$

The observed fuzzy number of nonconforming items in the sample is, therefore, given by

$$\tilde{d} = 0,5|3 + 1|4 + 0,2|5.$$

The random number of nonconforming items in a lot is distributed according to the hypergeometric distribution. When exactly d nonconforming items are found in the sample of n items taken from a lot of N items the observed test size of the hypothesis $H: \theta \leq \theta_0$ is given by (see a similar result for the binomial distribution in Bickel and Doksum, 1977)

$$p_H(d, \theta_0) = \sum_{i=d}^{n} \frac{\binom{\theta_0}{i}\binom{N-\theta_0}{n-i}}{\binom{N}{n}}. \tag{28}$$

Similarily, we can find that the observed test size of the hypothesis $K: \theta > \theta_1$ is given by

$$p_K(d, \theta_1) = 1 - p_H(d, \theta_1). \tag{29}$$

Now, we can find that the fuzzy observed test sizes are expressed as $\tilde{p}_H = 1|0.027 + 0,5|0.119$ and $\tilde{p}_K = 1|0,108 + 0,2|0,228$, respectively. Having the fuzzy observed test sizes for both (the null, and the alternative) hypotheses we can find the fuzzy indices of possibility (necessity) that the hypothesis that the quality of the lot is satisfactory is better supported by data than the alternative one. Simple calculation show that the PD and ND indices are, in the considered case, crisp and equal to 1. Moreover, we can find that the NSD index is also a crisp, and is equal to 0. Only the *Possibility of Strict Dominance* index PSD is fuzzy, and can be expressed as $\widetilde{PSD} = 1|0,054 + 0,5|0,238$. If we use Yager's F_1 index to defuzzify it, we find that $PSD^* = 0,115$.

The results of these computations can be interpreted as follows:

1. It is absolutely possible that choosing the hypothesis $H : \theta \leq \theta_0 = 10$ instead of $K : \theta > \theta_1 = 50$ is not a worse solution;
2. There is a rather slight possibility $(0,115)$ that choosing the hypothesis $H : \theta \leq \theta_0 = 10$ instead of $K : \theta > \theta_1 = 50$ is a definitely better decision;
3. The observed data do not indicate that there is any necessity of choosing the hypothesis $H : \theta \leq \theta_0 = 10$ instead of $K : \theta > \theta_1 = 50$.

This interpretation of the test results does not provide the decision maker with the definite answer about the quality of products. However, it clearly indicates ($NSD = 0$) that the test data do not firmly support the claim that the quality of the tested equipment is satisfactory.

References

1. Arnold, B.F. (1995). Statistical tests optimally meeting certain fuzzy requirements on the power function and on the sample size, *Fuzzy Sets and Systems* **75**, 365–372.
2. Bickel, P.J. and Doksum, K.A. (1977). *Mathematical Statistics. Basic Ideas and Selected Topics.* Holden Day, Inc., San Francisco.
3. Casals, M.R., Gil, M.A. and Gil, P. (1986). The fuzzy decision problem: an approach to the problem of testing statistical hypotheses with fuzzy information, *Europ. Journ. of Oper. Res.* **27**, 371–382.
4. Delgado, M., Verdegay, J.L. and Vila, M.A. (1985). Testing fuzzy hypotheses. A Bayesian approach. In *Approximate Reasoning in Expert Systems* (M.M. Gupta, A. Kandel, W. Bandler and J.B. Kiszka, Eds.). Elsevier, Amsterdam, 307–316.
5. Dubois, D. and Prade, H. (1983). Ranking fuzzy numbers in the setting of possibility theory, *Inform. Sci.* **30**, 184–244.
6. Dubois, D. and Prade, H. (1997). Qualitative possibility theory and its applications to reasoning and decision under uncertainty, *Belgian Journal of Operations Research, Statistics and Computer Science* **37**, 5–28.
7. Grzegorzewski, P. (2000). Testing statistical hypotheses with vague data, *Fuzzy Sets and Systems* **112**, 501–510.
8. Grzegorzewski, P. and Hryniewicz, O. (1997). Testing hypotheses in fuzzy environment, *Mathware and Soft Computing* **4**, 203–217.
9. Hryniewicz, O. (1994). Statistical decisions with imprecise data and requirements. In *Systems Analysis and Decision Support in Economics and Technology* (R. Kulikowski, K. Szkatula, J. Kacprzyk, Eds.). Omnitech Press, Warsaw, 135–143.
10. Hryniewicz, O. (2000) Possibilistic interpretation of the results of statistical tests. *Proc. 8ht International Conference IPMU*, Madrid, vol.I,215–219.

11. ISO 2859-2 (1985). Sampling procedures for inspection by attributes - Part 2: Sampling plans indexed by limiting quality (LQ) for isolated lot inspection.
12. Kruse, R. (1982). The strong law of large numbers for fuzzy random variables, *Inform. Sci.* **28**, 233–241.
13. Kruse, R. and Meyer, K.D. (1987). *Statistics with Vague Data.* Riedel, Dodrecht.
14. Kwakernaak, H. (1978). Fuzzy random variables, part I: definitions and theorems, *Inform. Sci.* **15**, 1–15; Part II: algorithms and examples for the discrete case , *Inform. Sci.* **17**, 253–278.
15. Lehmann, E.L. (1986) . *Testing Statistical Hypotheses*, 2nd ed., J. Wiley & Sons, New York.
16. Saade, J. (1994). Extension of fuzzy hypothesis testing with hybrid data, *Fuzzy Sets and Systems* **63**, 57–71.
17. Saade, J. and Schwarzlander, H. (1990). Fuzzy hypothesis testing with hybrid data, *Fuzzy Sets and Systems* **35**, 197–212.
18. Son, J.Ch., Song, I. and Kim, H.Y. (1992). A fuzzy decision problem based on the generalized Neyman-Pearson criterion, *Fuzzy Sets and Systems* **47**, 65–75.
19. Watanabe, N. and Imaizumi, T. (1993). A fuzzy statistical test of fuzzy hypotheses, *Fuzzy Sets and Systems* **53**, 167–178.
20. Zadeh, L.A. (1978). Fuzzy sets as a basis for a theory of possibility, *Fuzzy Sets and Systems* **1**, 3–28.

Possibilistic regression analysis

Hideo Tanaka[1], Peijun Guo[2]

[1] Toyohashi Sozo College, 20-1 Matsusita, Ushikawacho, Toyohashi, 440-8511, Japan
[2] Faculty of Economics, Kagawa University, Takamatsu, Kagawa 760-8523, Japan

Abstract. In this paper two possibilitis regression methods are presented, namely, the linear programming (LP)-based method and the quadratic programming (QP) one. Both methods are illustrated by means of some examples.

1 Introduction

Regression analysis is a fundamental analytic tool in many disciplines. The method gives a crisp relationship between the dependent and independent variables based on the given data from the statistic viewpoint. If a phenomenon under consideration has not stochastic variability but is also uncertain in some sense, it is more natural to seek a fuzzy functional relationship for the given data that may be fuzzy or crisp. That is to say, a fuzzy phenomenon should be modeled by a fuzzy functional relationship. This is the prime motivation for fuzzy regression analysis.

Fuzzy regression analysis was first proposed by Tanaka *et al.* (1982) and Tanaka (1987), where a fuzzy linear system was used as a regression model. Since membership functions of fuzzy sets are often described as possibility distributions, this approach is usually called possibilistic regression analysis (cf. Tanaka, 1987, Tanaka *et al.*, 1987, 1989, Tanaka and Watada, 1988, Hayashi and Tanaka, 1990), where the fuzzy coefficients are assumed to be noninteractive, that is, each membership function is determined independently. Because these methods reduce possibilistic regression problems to linear programming (LP) problems, some coefficients can become crisp because of the LP solutions. In order to deal with interactive possibility distributions of coefficients, quadratic and exponential possibility distributions have been used to formulate possibilistic regression models (Tanaka and Ishibuchi, 1991, Tanaka *et al.*, 1995). In these approaches, inclusion relations between the given outputs and the estimated outputs play an important role because inclusion relations naturally arise in possibility theory (see Dubois and Prade, 1988). In the paper by Tanaka *et al.* (1987), both possibility and necessity measures are used to evaluate inclusion relations.

There are three cases for input-output data to be analyzed: 1) crisp input-output data, 2) crisp input data and fuzzy output data and 3) fuzzy input-output data. Corresponding to the type of data structures, we have studied several combinations of data structures and models such as (crisp data, fuzzy functional relation), (fuzzy data, crisp functional relation) and (fuzzy data, fuzzy functional relation). Therefore, it is rather hard to classify types of fuzzy regression analyses into clear divisions. Nevertheless, regression techniques are classified into two distinct areas (see Kacprzyk and Fedrizzi, 1992), this is, possibility approaches and least square approaches (cf. Celminš, 1987ab, Diamond, 1988). Here, we focus on only possibility approaches, which are based on possibility models. We introduce two kinds of possibilistic regression methods. One is based on linear programming (LP) and the other is based on quadratic programming (QP) (see Tanaka and Lee, 1998). It can be seen that in LP-based possibilistic regression analysis, some coefficients tend to become crisp because of the characteristic of LP. On the other hand, QP approaches give more diverse spread coefficients than linear programming ones. Another advantage of QP-based possibilistic regression analysis is the ability to integrate both the property of central tendency and the possibilistic property in fuzzy regression. By changing the weights of the quadratic function, we can analyze the given data from different viewpoints. This paper is organized as follows. In Section 2, the LP-based possibilistic regression analysis is addressed. In Section 3, the QP-based methods is proposed. In Section 4, numerical examples are given to explain our approaches.

2 Interval regression by linear programming problems

Given the crisp data $(y_j, \mathbf{x}_j)$, the *interval regression model* is expressed as

$$Y = A_1 x_1 + \cdots + A_n x_n = \mathbf{A}\mathbf{x}, \tag{1}$$

where x_i is an *input variable*, A_i is an interval denoted as $A_i = (a_i, c_i)_I$, Y is an *estimated interval*, $\mathbf{x} = [x_1, \cdots, x_n]^t$ is an *input vector* and $\mathbf{A} = [A_1, \cdots, A_n]^t$ is an *interval coefficient vector.*

The interval output in (1) can be obtained as follows:

$$Y(\mathbf{x}) = (\mathbf{a}^t\mathbf{x}, \mathbf{c}^t|\mathbf{x}|)_I, \tag{2}$$

where $\mathbf{a} = [a_1, \cdots, a_n]^t$, $\mathbf{c} = [c_1, \cdots, c_n]^t$, and $|\mathbf{x}| = [|x_1|, \cdots, |x_n|]^t$.

Similarly, the following assumptions are given in order to formulate interval regression for crisp data.

1) The data are given as $(y_j, \mathbf{x}_j)$, $j = 1, \ldots, m$.
2) The data can be represented by the interval model (1).
3) The given output y_j should be included in the estimated interval $Y(\mathbf{x}_j) = (\mathbf{a}^t\mathbf{x}_j, \mathbf{c}^t|\mathbf{x}_j|)_I$, that is,

$$\mathbf{a}^t\mathbf{x}_j - \mathbf{c}^t|\mathbf{x}_j| \leq y_j \leq \mathbf{a}^t\mathbf{x}_j + \mathbf{c}^t|\mathbf{x}_j|. \tag{3}$$

4) The sum of spreads of the interval model should be minimized as follows:

$$\min_{\mathbf{a},\mathbf{c}} J = \sum_{j=1}^{m} \mathbf{c}^t|\mathbf{x}_j|. \tag{4}$$

Interval regression analysis determines the interval coefficients A_i, $i = 1,\ldots,n$, which minimize J subject to (3). This leads to the following *LP problem*

$$\min_{\mathbf{a},\mathbf{c}} J = \sum_{j=1}^{m} \mathbf{c}^t|\mathbf{x}_j|$$

$$\text{s.t. } \mathbf{a}^t\mathbf{x}_j - \mathbf{c}^t|\mathbf{x}_j| \leq y_j \; (j = 1,\ldots,m), \qquad (5)$$
$$\mathbf{a}^t\mathbf{x}_j + \mathbf{c}^t|\mathbf{x}_j| \geq y_j \; (j = 1,\ldots,m),$$
$$\mathbf{c} \geq \mathbf{0}.$$

Since interval regression analysis can be reduced to the LP problem (5), other constraint conditions for the coefficients can be introduced. For instance, if an input variable, say x_i, has a positive correlation with the output variable, it is advantageous to constrain A_i to be positive. Generally speaking, by introducing expert knowledge suggesting that the interval coefficient A_i should lie in some interval $B_i = (b_i, d_i)_I$, the interval A_i can be estimated within the limit of that knowledge B_i. Thus, we can introduce the following constraint condition:

$$A_i \subset B_i \Leftrightarrow b_i - d_i \leq a_i - c_i \text{ and } b_i + d_i \geq a_i + c_i. \tag{6}$$

Since A_i is constrained by the expert knowledge B_i, the obtained linear interval regression model appears to be acceptable. When the given outputs are intervals but the given inputs are crisp, we can consider two regression models, namely, an upper estimation model and a lower estimation model. The given data are denoted as $(Y_j, x_{j1},\ldots,x_{jn}) = (Y_j, \mathbf{x}_j)$ where Y_j is an interval output denoted as $(y_j, e_j)_I$. The upper and lower estimation models are defined respectively as follows:

$$Y_j^* = A_1^* x_{j1} + \cdots + A_n^* x_{jn}, \text{ (Upper model)} \tag{7}$$
$$Y_{*j} = A_{*1} x_{j1} + \cdots + A_{*n} x_{jn}. \text{ (Lower model)} \tag{8}$$

Two regression models are described as follows:

Upper regression model: The problem here is to satisfy

$$Y_j \subseteq Y_j^*, \; j = 1,\ldots,m \tag{9}$$

and to find the interval coefficients $A_i^* = (a_i^*, c_i^*)_I$ that minimize the sum of the spreads of the estimation intervals, that is,

$$J^* = \sum_{j=1}^{m} \mathbf{c}^{*t}|\mathbf{x}_j|, \tag{10}$$

where the minimization stems from the inclusion relations (9). Since the constraint conditions $Y_j \subseteq Y_j^*$ can be written as

$$\begin{aligned} y_j - e_j &\geq \mathbf{a}^{*t}\mathbf{x}_j - \mathbf{c}^{*t}|\mathbf{x}_j|, \\ y_j + e_j &\leq \mathbf{a}^{*t}\mathbf{x}_j + \mathbf{c}^{*t}|\mathbf{x}_j|, \end{aligned} \tag{11}$$

the problem for obtaining the interval coefficients A_i^* can be described as the following LP problem:

$$\min_{\mathbf{a}^*,\mathbf{c}^*} \sum_{j=1}^{m} \mathbf{c}^{*t}|\mathbf{x}_j| \tag{12}$$

$$\begin{aligned} \text{s.t. } & y_j - e_j \geq \mathbf{a}^{*t}\mathbf{x}_j - \mathbf{c}^{*t}|\mathbf{x}_j|, \\ & y_j + e_j \leq \mathbf{a}^{*t}\mathbf{x}_j + \mathbf{c}^{*t}|\mathbf{x}_j|, \\ & \mathbf{c}^* \geq \mathbf{0}. \end{aligned}$$

Lower regression model: The problem here is to satisfy

$$Y_{*j} \subseteq Y_j,\ j = 1, \ldots, m \tag{13}$$

and to find the interval coefficients $A_{*i} = (a_{*i}, c_{*i})_I$ that maximize the sum of the spreads of the estimation intervals:

$$J_* = \sum_{j=1}^{m} \mathbf{c}_*^t|\mathbf{x}_j|, \tag{14}$$

where the maximization stems from the inclusion relations (13). Since the constraint conditions $Y_{*j} \subseteq Y_j$ can be written as

$$\begin{aligned} y_j - e_j &\leq \mathbf{a}_*^t\mathbf{x}_j - \mathbf{c}_*^t|\mathbf{x}_j|, \\ y_j + e_j &\geq \mathbf{a}_*^t\mathbf{x}_j + \mathbf{c}_*^t|\mathbf{x}_j|, \end{aligned} \tag{15}$$

the problem for obtaining the interval coefficients A_{*i} can be described as the following LP problem:

$$\max_{\mathbf{a}_*,\mathbf{c}_*} \sum_{j=1}^{m} \mathbf{c}_*^t|\mathbf{x}_j| \tag{16}$$

$$\begin{aligned} \text{s.t. } & y_j - e_j \leq \mathbf{a}_*^t\mathbf{x}_j - \mathbf{c}_*^t|\mathbf{x}_j|, \\ & y_j + e_j \geq \mathbf{a}_*^t\mathbf{x}_j + \mathbf{c}_*^t|\mathbf{x}_j|, \\ & \mathbf{c}_* \geq \mathbf{0}. \end{aligned}$$

The reason for maximizing J_* is to find the widest estimated intervals Y_{*j} among those satisfying the constraint condition of (15). The estimated intervals from upper and lower estimation models satisfy inclusion relations $Y_{*j} \subseteq Y_j \subseteq Y_j^*$, $j = 1, \ldots, m$.

In order to show the validity of the above formulations, assume that the given data $(Y_j^0, \mathbf{x}_j^0)$, $j = 1, \ldots, m$ satisfy the linear interval system

$$Y_j^0 = A_1^0 x_{1j}^0 + \ldots + A_n^0 x_{nj}^0 = \mathbf{A}^0 \mathbf{x}_j, \tag{17}$$

where $\mathbf{A}^0 = (\mathbf{A}^0, \mathbf{C}^0)_i$.

Theorem 1. *If the given data $(Y_j^0, \mathbf{x}_j^0)$, $j = 1, \ldots, m$ satisfy (17), the interval vector $\mathbf{A}^*$ and $\mathbf{A}_*$ obtained from (12) and (16), respectively, are the same as $\mathbf{A}^0$. Thus, we have*

$$\mathbf{A}^* = \mathbf{A}_* = \mathbf{A}^0, \; Y_j^* = Y_{*j} = Y_j^0, \; j = 1, \ldots, m. \tag{18}$$

Proof. Let us prove only $\mathbf{A}^* = \mathbf{A}^0$ in the upper regression model. Since $(Y_j^0, \mathbf{x}_j^0)$ satisfies (12), we have

$$e_j^0 = \mathbf{c}^{0t} |\mathbf{x}_j|, \tag{19}$$
$$y_j^0 = \mathbf{a}^{0t} \mathbf{x}_j. \tag{20}$$

Substituting (19) and (20) to the constraint conditions of (12) yields that

$$\begin{aligned} y_j^0 &\geq \mathbf{a}^{*t}\mathbf{x}_j^0 - \mathbf{c}^{*t}|\mathbf{x}_j^0| + \mathbf{c}^{0t}|\mathbf{x}_j^0|, \\ y_j^0 &\leq \mathbf{a}^{*t}\mathbf{x}_j^0 + \mathbf{c}^{*t}|\mathbf{x}_j^0| - \mathbf{c}^{0t}|\mathbf{x}_j^0|. \end{aligned} \tag{21}$$

Setting $\mathbf{a}^* = \mathbf{a}^0$ and $\mathbf{c}^* = \mathbf{c}^0$, $(\mathbf{a}^0, \mathbf{c}^0)_I$ is a feasible solution of the LP problem (12). If there is another solution $\mathbf{c}'$ such that

$$\sum_{j=1}^{m} \mathbf{c}'^t |\mathbf{x}_j^0| < \sum_{j=1}^{m} \mathbf{c}^{0t} |\mathbf{x}_j^0|. \tag{22}$$

Thus, for some i we have

$$\mathbf{c}'^t |\mathbf{x}_i^0| < \mathbf{c}^{0t} |\mathbf{x}_i^0|. \tag{23}$$

The ith constraint condition of (12) can be rewritten as

$$\begin{aligned} y_i^0 &\geq \mathbf{a}^{*t}\mathbf{x}_i^0 + k_i, \\ y_i^0 &\leq \mathbf{a}^{*t}\mathbf{x}_i^0 - k_i, \end{aligned} \tag{24}$$

where k_i is as follows:

$$k_i = (\mathbf{c}^0 - \mathbf{c}')^t |\mathbf{x}_i^0| > 0. \tag{25}$$

It is obvious from the contradiction of (24) that (23) can not hold. Thus, the optimal solution $\mathbf{c}^*$ should be $\mathbf{c}^0$. Moreover, it follows from (21) with $\mathbf{c}^* = \mathbf{c}^0$ that $y_j^0 = \mathbf{a}^{*t}\mathbf{x}_j^0$. Thus, $\mathbf{a}^*$ is equal to $\mathbf{a}^0$. ∎

Theorem 2. *There exists always an optimal solution in the upper regression model (12) while it is not assured that there is always an optimal solution in the lower regression model (16) for interval linear systems.*

Proof. In the upper regression model, there is an admissible set of the constraint conditions (11) if a sufficient large positive vector is taken for $\mathbf{c}^*$. On the contrary, there is a case where there is no admissible set of (15) even if a zero vector is taken for $\mathbf{c}^*$. ∎

When there is no solution for the assumed linear model in the lower regression model, it is needed that we introduce the other input variables to the linear model or assume a non-linear interval model. Thus, if the assumed model is appropriate for the given data, we have an optimal solution in lower regression models (cf. Tanaka *et al.*, 1998).

The linear interval regression model formulated above includes the following features.

1) The upper and lower regression models for the interval data are similar to the concepts of upper and lower approximation of rough sets (Pawlak, 1984). When the output data are intervals, the upper and lower intervals are obtained in the context of incompleteness of data.
2) The range of the estimated interval widens as the number of data increases. This is due to the fact that the increased analytical data result in more information and wider possibilities for decision-making. In contrast, an estimation interval in conventional regression analysis diminishes as the number of data increases. Since conventional regression analysis is based on a probability model, objective analysis is done for a large number of data, but interval regression analysis is based on a possibility model and is useful for the problem of deciding what is possible.
3) Since interval regression models can be reduced to LP problems, constraint conditions for the coefficients can be introduced easily. For instance, it is advantageous to constrain coefficients to be positive if the variables corresponding to those coefficients have a positive correlation with the output. In addition, generally speaking, by introducing expert knowledge, interval coefficients can be estimated within the limits of that knowledge. Since it is constrained by expert knowledge, the obtained linear interval regression model appears to be acceptable.
4) In general, since the interval represents partial ignorance, this should also be reflected in the analytical results.

3 Interval regression analysis by quadratic programming approach

3.1 Basic model by quadratic programming

Here we introduce a basic model by QP in interval regression analysis corresponding to the former LP-based interval regression model. A QP approach is an optimization problem which involves minimizing a quadratic objective function subject to linear constraints.

To formulate interval regression by QP, the following assumptions are given.

(1) The input-output data are given as $(y_j, \mathbf{x}_j)$ $(j = 1, \ldots, p)$, where

$$\mathbf{x}_j = [1, x_{j1}, \ldots, x_{jn}]^t.$$

(2) The data can be represented by the interval linear model (1).
(3) The given output y_j should be included in the estimated output $Y(\mathbf{x}_j)$, that is, $y_j \in Y(\mathbf{x}_j)$ $(j = 1, \ldots, p)$.
(4) The objective function is defined by

$$J = \sum_{j=1}^{p} (\mathbf{c}^t |\mathbf{x}_j|)^2 = \mathbf{c}^t \left(\sum_{j=1}^{p} |\mathbf{x}_j||\mathbf{x}_j|^t \right) \mathbf{c}, \tag{26}$$

which is the sum of squared spreads of the estimated outputs and matrix $\sum_{j=1}^{p} |\mathbf{x}_j||\mathbf{x}_j|^t$ is an $(n+1) \times (n+1)$ positive definite one.

Based on the above assumptions, the *QP-based interval regression* is to determine the optimal interval coefficients $\mathbf{A}_i = (\mathbf{a}_i, \mathbf{c}_i)_I$, $i = 0, \ldots, p$, that minimize the objective function (26) subject to the linear constraints (3). Thus, the basic QP-based model can be expressed as the following QP problem:

$$\min_{\mathbf{a},\mathbf{c}} J = \sum_{j=1}^{p} (\mathbf{c}^t |\mathbf{x}_j|)^2 = \mathbf{c}^t \left(\sum_{j=1}^{p} |\mathbf{x}_j||\mathbf{x}_j|^t \right) \mathbf{c} + \xi \mathbf{a}^t \mathbf{a} \tag{27}$$

$$\begin{aligned} \text{s.t. } & \mathbf{a}^t \mathbf{x}_j - \mathbf{c}^t |\mathbf{x}_j| \leq y_j \ (j = 1, \ldots, p), \\ & \mathbf{a}^t \mathbf{x}_j + \mathbf{c}^t |\mathbf{x}_j| \geq y_j \ (j = 1, \ldots, p), \\ & \mathbf{c} \geq \mathbf{0}. \end{aligned}$$

where ξ is a small positive number. $\xi \mathbf{a}^t \mathbf{a}$ is added to the objective function (26) so that (27) becomes a strictly convex quadratic programming because of positive definite matrix respect to decision variables $\mathbf{a}$ and $\mathbf{c}$.

3.2 Model integrating central tendency and possibilistic property

In this section, a new objective function is introduced with considering minimizing the sum of squared spreads of the estimated outputs and the sum of squared distances between the estimated output centers and the observed outputs as follows:

$$J = k_1 \sum_{j=1}^{p} (y_j - \mathbf{a}^t \mathbf{x}_j)^2 + k_2 \mathbf{c}^t \left(\sum_{j=1}^{p} |\mathbf{x}_j||\mathbf{x}_j|^t \right) \mathbf{c}$$

$$= k_1 \left(\mathbf{a}^t \left(\sum_{j=1}^{p} \mathbf{x}_j \mathbf{x}_j^t \right) \mathbf{a} - 2 \sum_{j=1}^{p} y_j \mathbf{x}_j^t y_j \mathbf{a} + \sum_{j=1}^{p} y_j^2 \right) + k_2 \mathbf{c}^t \left(\sum_{j=1}^{p} |\mathbf{x}_j||\mathbf{x}_j|^t \right) \mathbf{c}, \tag{28}$$

where $\sum_{j=1}^{p}(y_j - \mathbf{a}^t\mathbf{x}_j)^2$ corresponds to the least squares estimation, k_1 and k_2 are positive weight coefficients.

Using this new objective function (28), interval regression analysis is to determine the interval coefficients $\mathbf{A}_i = (\mathbf{a}_i, \mathbf{c}_i)_I$, $i = 0, \ldots, p$, that minimize the objective function (28) subject to the linear constraints (3) which can be expressed as the following QP problem

$$\min_{\mathbf{a},\mathbf{c}} k_1 \sum_{j=1}^{p} (y_j - \mathbf{a}^t\mathbf{x}_j)^2 + k_2 \mathbf{c}^t \left(\sum_{j=1}^{p} |\mathbf{x}_j||\mathbf{x}_j|^t \right) \mathbf{c} \tag{29}$$

$$\begin{aligned} \text{s.t. } & \mathbf{a}^t\mathbf{x}_j - \mathbf{c}^t|\mathbf{x}_j| \le y_j \ (j = 1, \ldots, p), \\ & \mathbf{a}^t\mathbf{x}_j + \mathbf{c}^t|\mathbf{x}_j| \ge y_j \ (j = 1, \ldots, p), \\ & \mathbf{c} \ge \mathbf{0}. \end{aligned}$$

Likewise, when a data set with crisp inputs and interval outputs is given, we can consider two interval regression models, i.e., the upper and the lower models (7) and (8) by QP problems. In order to guarantee that $Y_*(\mathbf{x}_j) \subseteq Y_j \subseteq Y^*(\mathbf{x}_j)$ for any arbitrary $\mathbf{x}_i$, it is convenient to assume that

$$A_i^* = (a_i, c_i + d_i)_I, \ A_{*i} = (a_i, c_i)_I. \tag{30}$$

The objective function is introduced as the following quadratic function:

$$J = \sum_{j=1}^{p} (\mathbf{d}^t|\mathbf{x}_j|^t)^2 = \mathbf{d}^t \left(\sum_{j=1}^{p} |\mathbf{x}_j||\mathbf{x}_j|^t \right) \mathbf{d}, \tag{31}$$

which is the sum of squared spread differences between upper and lower regression model. Therefore, interval regression analysis is to determine the interval coefficients A_i^* and A_{*i} $(i = 0, 1, \ldots, n)$ that minimize the objective function (31) and satisfy inclusion relations $Y_*(\mathbf{x}_j) \subseteq Y_j \subseteq Y^*(\mathbf{x}_j)$ $(j = 1, \ldots, p)$, which can be described as the following QP problem:

$$\min_{\mathbf{a},\mathbf{c},\mathbf{d}} J = \mathbf{d}^t \left(\sum_{j=1}^{p} |\mathbf{x}_j||\mathbf{x}_j|^t \right) \mathbf{d} + \xi(\mathbf{a}^t\mathbf{a} + \mathbf{c}^t\mathbf{c}) \tag{32}$$

$$\begin{aligned} \text{s.t. } & \mathbf{a}^t\mathbf{x}_j + \mathbf{c}^t|\mathbf{x}_j| + \mathbf{d}^t|\mathbf{x}_j| \ge y_j + e_j, \\ & \mathbf{a}^t\mathbf{x}_j - \mathbf{c}^t|\mathbf{x}_j| - \mathbf{d}^t|\mathbf{x}_j| \ge y_j - e_j, \\ & \mathbf{a}^t\mathbf{x}_j + \mathbf{c}^t|\mathbf{x}_j| \le y_j + e_j, \\ & \mathbf{a}^t\mathbf{x}_j - \mathbf{c}^t|\mathbf{x}_j| \le y_j - e_j, \\ & \mathbf{c} \ge \mathbf{0}. \end{aligned}$$

where ξ is a small positive number. Likewise, $\xi(\mathbf{a}^t\mathbf{a} + \mathbf{c}^t\mathbf{c})$ is added to the objective function (31) so that (32) becomes a strictly convex quadratic programming because of positive definite matrix respect to decision variables $\mathbf{a}$, $\mathbf{c}$ and $\mathbf{d}$. This approach for obtaining the upper and lower regression models is called a unified approach (Ishibuchi and Tanaka, 1993).

4 Numerical examples

In this section, we first compare the QP-based basic regression model with the corresponding LP method by a house price example. Then the QP formulation integrating the central tendency and possibilistic properties is used for the same data set. After that, the upper and lower regression models are obtained by QP for the given crisp input and output data set.

4.1 Comparison between LP-based and QP-based interval regression models

$No(j)$	y	x_1	x_2	x_3
1	606	1	38.09	36.43
2	710	1	62.10	26.50
3	808	1	63.76	44.71
4	826	1	74.52	38.09
5	865	1	75.38	41.10
6	852	2	52.99	26.49
7	917	2	62.93	26.49
8	1031	2	72.04	33.12
9	1092	2	76.12	43.06
10	1203	2	90.26	342.64
11	1394	3	85.70	31.33
12	1420	3	95.27	27.64
13	1601	3	105.98	27.64
14	1632	3	79.25	66.81
15	1699	3	120.50	32.25

Table 1. House price data

The house price data are shown in Table 1 where x_1 is the goodness of the material, x_2 is the area of the first floor (m^2), x_3 is the area of the second floor (m^2), and y is a sale price (10,000 yen). Note that x_1 can have three values, namely, 1, 2, and 3 representing low, medium, and high grades of materials, respectively.

The interval regression model of the house price data is assumed as

$$Y(\mathbf{x}) = A_0 + A_1 x_1 + A_2 x_2 + A_3 x_3. \tag{33}$$

In this model it is assumed that the coefficients are positive since it is reasonable that as x_i $(i = 1, 2, 3)$ increases, the price of house y increases. Thus, we let $\mathbf{a} \geq \mathbf{0}$. Using the QP-based method (27), we obtained the following optimal interval regression model:

$$Y^*(\mathbf{x}) = (0, 8.813)_I + (234.151, 24.814)x_1 + (6.137, 0)_I + (4.9, 0.487)x_3. \tag{34}$$

Next, using the LP-based method (5), we obtained the following optimal interval regression model:

$$Y^*(\mathbf{x}) = (0, 0) + (245.167, 37.674)x_1 + (5.853, 0)x_2 + (4.786, 0)x_3. \tag{35}$$

$No(j)$	y	$Y^*(\mathbf{x}_j)$
1	606	$[595.048, 697.785]$
2	710	$[698.576, 791.641]$
3	808	$[789.124, 899.926]$
4	826	$[825.944, 930.298]$
5	865	$[844.505, 951.791]$
6	852	$[851.961, 994.644]$
7	917	$[912.963, 1055.650]$
8	1031	$[998.129, 1147.270]$
9	1092	$[1067.030, 1225.860]$
10	1203	$[1151.960, 1310.370]$
11	1394	$[1283.400, 1480.420]$
12	1420	$[1325.850, 1519.280]$
13	1601	$[1391.570, 1601.000]$
14	1632	$[1400.390, 1632.000]$
15	1699	$[1501.030, 1699.000]$

Table 2. Obtained intervals by (29) for house price

From (34) and (35), it can be noticed that quadratic programming approaches give more diverse spread coefficients than linear programming ones, i.e., in the QP-based model only $c_2 = 0$ whereas in the LP-based model $c_0 = c_2 = c_3 = 0$. The difference comes from the characteristics of quadratic and linear functions in the optimization problems. Therefore, in general, interval regression by QP is preferable to the one by LP in the sense that the regression coefficients obtained by QP tend to become more non-crisp than those by LP. The more non-crisp coefficients we have, the better result we have since the spread of data can be interpreted by many coefficients. The obtained intervals by (27) for house price is listed in Table 2. The obtained interval outputs in Table 2 indicate that sale prices for expensive houses are located at the upper ends of the estimation intervals. This implies that the prices of high-grade products are set higher to reflect their additional values.

	weights		optimal coefficients vectors	
case	k_1	k_2	$\mathbf{a}^*$	$\mathbf{c}^*$
(a)	1	.0001	$(0, 270.380, 5.686, 3.862)$	$(52.665, 19.853, 0, 0)$
(b)	1	.5	$(0, 245.989, 5.804, 4.810)$	$(8.951, 30.623, 0, 0.179)$
(c)	1	1	$(0, 239.047, 5.969, 4.893)$	$(1.916, 31.170, 0, 0.292)$
(d)	.5	1	$(0, 237.033, 6.040, 4.885)$	$(3.68, 28.806, 0, 0.386)$
(e)	.0001	1	$(0, 234.152, 6.137, 4.900)$	$(8.811, 24.816, 0, 0.487)$

Table 3. Optimal coefficients by QP (11)
with variable weights for house price data

Let us apply again the house price data in Table 1 for the proposed QP method (29). Here we examine the influence of weight coefficients k_1 and k_2. Using (29), we obtained five sets of optimal coefficients by changing weights k_1 and k_2, shown in Table 3. It is noticed in Table 3 that the center regression line in the case of $k_1 >> k_2$, i.e. the case (a), tends to be the regression line obtained by conventional regression analysis. Furthermore, the case (e) is considering mainly to reduce the fuzziness of the model which is one of possibilistic properties of fuzzy regression. Thus, our proposed QP approach is combining the above mentioned two properties in one optimization problem where weight coefficients and are determined by considering the trade-off between $\sum_{j=1}^{p}(y_j - \mathbf{a}^t\mathbf{x}_j)^2$ and $\mathbf{c}^t \left(\sum_{j=1}^{p} |\mathbf{x}_j||\mathbf{x}_j|^t\right) \mathbf{c}$. From the obtained optimal coefficients for different combinations of weights, we calculated $\sum_{j=1}^{p}(y_j - \mathbf{a}^t\mathbf{x}_j)^2$ and $\mathbf{c}^t \left(\sum_{j=1}^{p} |\mathbf{x}_j||\mathbf{x}_j|^t\right) \mathbf{c}$ shown in Table 4. It is noticed that $\sum_{j=1}^{p}(y_j - \mathbf{a}^t\mathbf{x}_j)^2$ is increasing gradually in the order of (a), (b), (c), (d), (e) in Table 4 which is indicative of the fact that the sum of squared deviations between the given outputs and the estimation centers increases as we put more weight on k_2 than k_1. On the contrary, values of

$\mathbf{c}^t \left(\sum_{j=1}^{p} |\mathbf{x}_j||\mathbf{x}_j|^t\right) \mathbf{c}$ show a decreasing trend in the order of (a), (b), (c), (d), (e) in Table 4 which tells that the sum of squared spreads decreases as we put more weight on k_2 than k_1. Figure 1 shows the regression models for (a), (c), and (e). It can be noticed that Figure 1 (a) has more central tendency than Figure 1 (c) and Figure 1 (e) because the number of observations being on the edge of the estimated intervals in Figure 1 (a) is fewer than those in Figure 1 (c) and Figure 1 (e).

case	k_1	k_2	$\sum_{j=1}^{p}(y_j - \mathbf{a}^t\mathbf{x}_j)^2$	%	$\sum_{j=1}^{p} \mathbf{c}^t|\mathbf{x}_j||\mathbf{x}_j|^t\mathbf{c}$	%
(a)	1	.0001	48748.0	84.2	131926.0	140.5
(b)	1	.5	55696.9	96.2	97601.2	104.0
(c)	1	1	57910.6	100.0	93870.9	100.0
(d)	.5	1	58330.6	100.7	93563.7	99.7
(e)	.0001	1	59197.7	102.2	93375.8	99.5

Table 4. Comparison of results with variable weights for price data

4.2 Upper and lower regression models for crisp input and fuzzy output data

The data set of crisp inputs and interval outputs is shown in Table 5.

$No(j)$	1	2	3	4
Input (x)	1	2	3	4
Output (Y)	$[15, 30]$	$[20, 37.5]$	$[15, 35]$	$[25, 60]$
$No(j)$	5	6	7	8
Input (x)	5	6	7	8
Output (Y)	$[25, 55]$	$[40, 65]$	$[55, 95]$	$[70, 100]$

Table 5. Data sets with crisp inputs and interval output

The linear interval model for the given input and output is assumed as

$$Y(x) = A_0 + A_1 x. \tag{36}$$

Using (32), we obtained the upper model $Y^*(x)$ and the lower model $Y_*(x)$ as

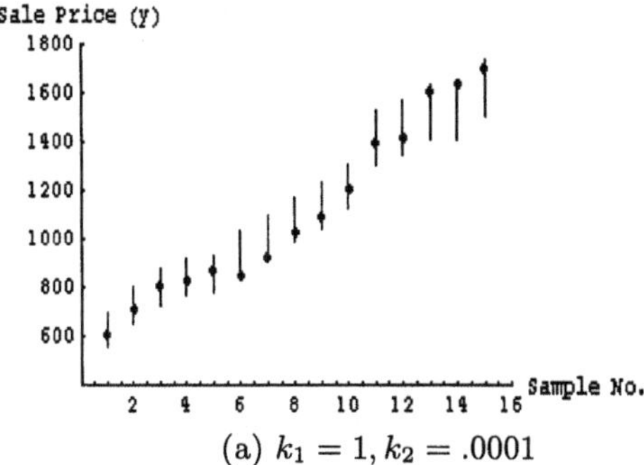

(a) $k_1 = 1, k_2 = .0001$

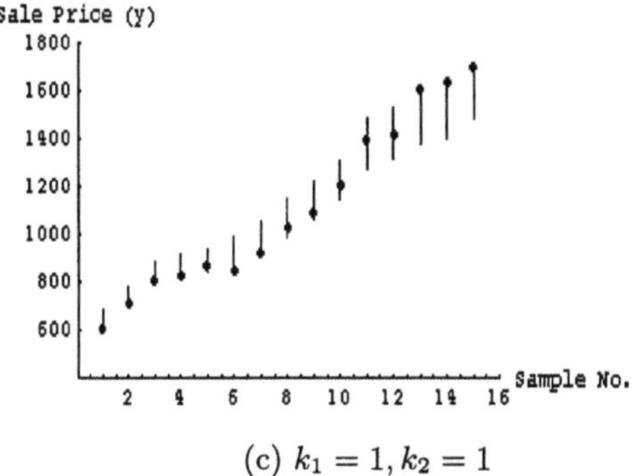

(c) $k_1 = 1, k_2 = 1$

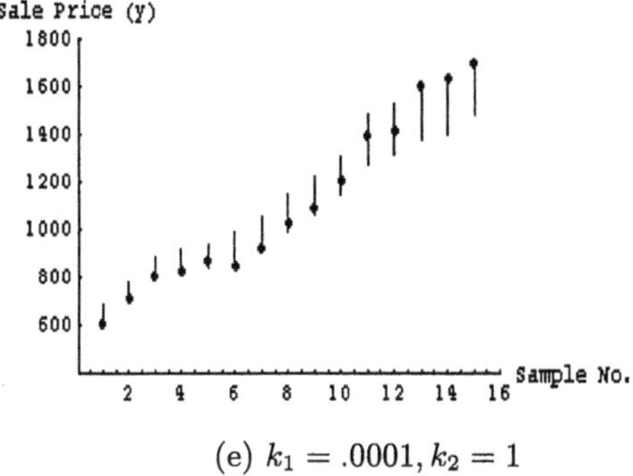

(e) $k_1 = .0001, k_2 = 1$

Fig. 1. Fuzzy regression models by (29) for house price data

$$Y^*(x) = (7.311, 11.856) + (8.371, 2.462)x, \tag{37}$$

$$Y_*(x) = (7.311, 0.168) + (8.371, 0.514)x, \tag{38}$$

which are depicted in Figure 2 where the outer 2 lines represent the upper model $Y^*(x)$ and the inner 2 lines represent the lower model $Y_*(x)$. Let us change the linear model (36) into the following non-linear model:

$$Y(x) = A_0 + A_1 x + A_2 x^2. \tag{39}$$

Using (32) again, the upper model $Y^*(x)$ and the lower model $Y_*(x)$ are obtained as

$$Y^*(x) = (10.463, 13.241) + (5.648, 1.944)x + (0.370, 0)x^2, \tag{40}$$

$$Y_*(x) = (10.463, 1.204) + (5.648, 1.019)x + (0.370, 0)x^2, \tag{41}$$

which are depicted in Figure 3 where the outer 2 lines represent the upper model $Y^*(x)$ and the inner 2 lines represent the lower model $Y_*(x)$. It is obviously known that from Figure 2 and Figure 3 that polynomial (39) can make the upper and lower regression model closer than the linear function (36).

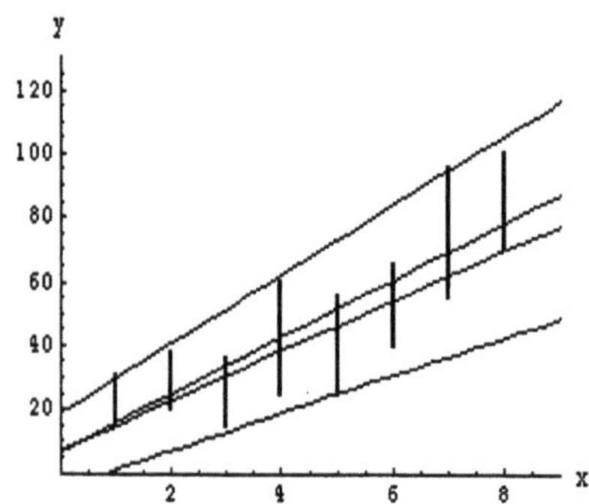

Fig. 2. The upper and lower models based on (35)

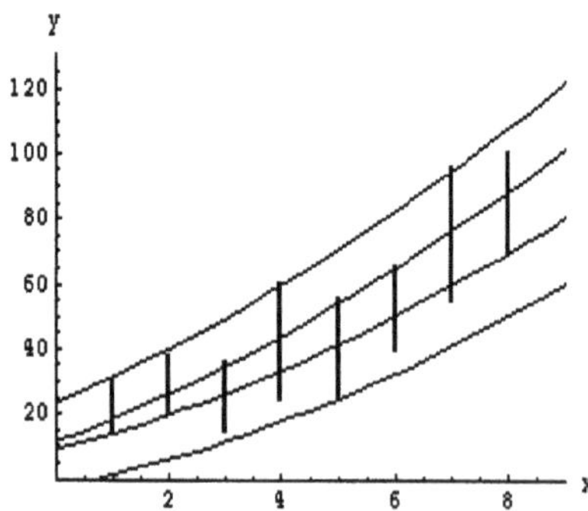

Fig. 3. The upper and lower models based on (38)

5 Conclusions

In this paper we introduced possibilistic regression analyses by LP and QP approaches. Even though QP-based models are somewhat more complicated to be solved than LP-based ones, interval regression by QP gives more diverse spread coefficients than by LP. The proposed interval regression by QP can effectively reflect the property of central tendency in least squares and the possibilistic property in fuzzy regression. From our proposed approach, by changing the weight coefficients of the objective function in QP, the given phenomenon can be explained well from different viewpoints. Moreover, we can obtain the upper and lower regression models based on QP to reflect two kind of possibilistic relations between the given and estimated interval outputs. Intervals can be extended to fuzzy intervals in possibility regression which is shown in Tanaka and Guo (1999) in detail.

References

1. Celminš, A. (1987a). Least squares model fitting to fuzzy vector data, *Fuzzy Sets and Systems* **22**, 245-269.
2. Celminš, A. (1987b). Multidimensional fitting of fuzzy model, *Mathematical Modeling* **9**, 669-690.
3. Diamond, P. (1988). Fuzzy least squares, *Inform. Sci.* **46**, 141-157.
4. Dubois, D. and Prade, H. (1988). *Possibility Theory.* Plenum Press, New York.
5. Hayashi, I. and Tanaka, H. (1990). The fuzzy GMDH algorithm by possibility models and its application, *Fuzzy Sets and Systems* **36**, 245-258.
6. Ishibuchi, H. and Tanaka, H. (1993). A unified approach to possibility and necessity analysis with interval regression models, *Proc. 5th IFSA World Congress*, 501-504.
7. Kacprzyk, J. and Fedrizzi, M. (Eds.) (1992), *Fuzzy Regression Analysis*, Physica-Verlag, Heidelberg.
8. Pawlak, Z. (1984). Rough classification, *Intern. J. Man-Mach. Stud.* **20**, 469-485.
9. Tanaka, H. (1987). Fuzzy data analysis by possibilistic linear models, *Fuzzy Sets and Systems* **24**, 363-375.
10. Tanaka, H. and Guo, P. (1999). *Possibilistic Data Analysis for Operations Research.* Physica-Verlag, Heidelberg.
11. Tanaka, H., Hayashi, I. and Watada, J. (1987). Possibilistic linear regression analysis based on possibility measure, *Proc. 2nd IFSA World Congress*, 317-320.
12. Tanaka, H., Hayashi, I. and Watada, J. (1989). Possibilistic linear regression analysis for fuzzy data, *European J. Oper. Res.* **40** 389-396.
13. Tanaka, H. and Ishibuchi, H. (1991). Identification of possibilistic linear system by quadratic membership functions, *Fuzzy Sets and Systems* **36**, 145-160.
14. Tanaka, H., Ishibuchi, H. and Yoshikawa, S. (1995). Exponential possibility regression analysis, *Fuzzy Sets and Systems* **69**, 305-318.
15. Tanaka, H. and Lee, H. (1998). Interval regression analysis by quadratic programming approach, *IEEE Trans. Fuzzy Systems* **6**, 473-481.

16. Tanaka, H., Lee, H. and Guo, P. (1998). Possibility data analysis with rough set concept, *Proc. 6th IEEE Intern. Confer. Fuzzy Syst.*, 117-122.
17. Tanaka, H., Uejima, S. and Asai, K. (1982). Linear regression analysis with fuzzy model, *IEEE Trans. Syst., Man Cyber.* **12** 903-907.
18. Tanaka, H. and Watada, J. (1988). Possibilistic linear systems and their application to the linear regression model, *Fuzzy Sets and Systems* **27**, 275-289.

Linear regression in a fuzzy context. The least square method

Antonia Salas[1], Norberto Corral[1] and Carlo Bertoluzza[2]

[1] Departamento de Estadística e I.O. y D.M., Universidad de Oviedo, Facultad de Ciencias, C/ Calvo Sotelo s/n, 33007 Oviedo (Spain)
[2] Dipartimento di Informatica e Sistemistica, University of Pavia, Pavia (Italy)

Abstract. This paper deals with the statistical linear regression. Its aim is to extend the classical least square method to the case where the observations are not crisp, but fuzzy numbers. In order to attain our purpose, we introduce a suitable and very general squared distance between fuzzy numbers which substitute the classical $(a-b)^2$ on the real line. Then we use a well known theorem of functional theory to prove that, under suitable and reasonable conditions, there exists a unique set of fuzzy coefficients which minimize the sum of the squared distances between the previsions and the observations. Nevertheless the classical variational methods cannot be used to find the regression coefficient due to the particular structure of the domain of the functional to be minimized (its interior is empty). So we complete the paper by describing a numerical solution based on active constraint method.

1 Introduction and basic definitions

Let (X,Y) be a pair of random variables and let us suppose that there exists a functional dependence between X and Y. We try to approximate this dependence by means of a suitable polinomial of degree k. In order to achieve our aim, we have at our disposal a set of couples (X_i, Y_i) of experimental observations. If X and Y are real variables, and X_i, Y_i are real numbers, then the classical least square method furnishes the coefficients of the best polinomial which approximate the functional dependence.

However in many cases the random variables, as well as the observations, are fuzzy quantities, and the above mentioned method cannot be applied. We propose here a generalization, which reduces to the classical least square when the variables and the observations are real.

In order to attain our purpose we remember some definitions regarding the fuzzy sets and the fuzzy numbers, as well as some basic results of the fuzzy theory.

Definition 1. (Zadeh, 1965) A *fuzzy subset* of a given universe Ω is a map from Ω into the interval $[0,1]$ (*membership function*).

If the range of the map reduces to the set $\{0,1\}$, then the map itself reduces to a characteristic function, and the fuzzy set reduces to a crisp one.

From now on we will represent fuzzy sets and fuzzy quantities by means of an overlining tilde. So a fuzzy set $\widetilde{A}$ is identified by a map

$$\widetilde{A} : \Omega \longrightarrow [0,1].$$

Definition 2. For each $\alpha \in]0,1]$ the *α-cut* A_α of the set $\widetilde{A}$ is the crisp subset of Ω defined by

$$A_\alpha = \{\omega \in \Omega \mid \widetilde{A}(\omega) \geq \alpha\}$$

(the set $A_1 = \{\omega \in \Omega \mid \widetilde{A}(\omega) = 1\}$ is the *core* of the fuzzy set). Moreover we define the 0-*cut* by means of

$$A_0 = \text{supp}(\widetilde{A}) = \overline{\{\omega \in \Omega \mid \widetilde{A}(\omega) > 0\}}.$$

Definition 3. A fuzzy subset is said to be

- *normal* if $A_1 \neq \emptyset$,
- *convex* if all its α-cuts are convex subsets of Ω [1].

Definition 4. A *fuzzy number* is a normal and convex subset of the real line $\mathbb{R}$ with bounded support[2]. We will denote by $\mathcal{F}_c(\mathbb{R})$ the space of the fuzzy numbers.

The α-cuts A_α of a fuzzy number $\widetilde{A}$ are intervals, whose first and second endpoints are denoted by $a^{(1)}(\alpha), a^{(2)}(\alpha)$ or $a_\alpha^{(1)}, a_\alpha^{(2)}$. These endpoints can or cannot belong to the α-cuts; however this fact does not have a great importance in our approach to the regression problem. As the α-cuts determine the fuzzy number, we can identify it by means of the family of couples $[a^{(1)}(\alpha), a^{(2)}(\alpha)]$. So we may write

$$\widetilde{A} = \{A_\alpha = [a^{(1)}(\alpha), a^{(2)}(\alpha)] \mid \alpha \in [0,1]\}. \tag{1}$$

The two functions $a^{(i)}$, $i = 1,2$ appearing in (1) have the following properties

$$\begin{aligned}
\text{Dom}(a^{(i)}) &= [0,1] \\
a^{(1)} &\text{ is not decreasing} \\
a^{(2)} &\text{ is not increasing} \\
a^{(1)}(1) &\leq a^{(2)}(1) \\
a^{(i)} &\in L^2([0,1])
\end{aligned}$$

[1] Obviously this definition holds iff Ω is a linear space.

[2] The support to be bounded is a condition which is not requested in the usual definitions; however, we think that fuzzy numbers with unbounded support do not present any interest either from theoretical or practical points of view.

Definition 5. The *sum* and *product* of two fuzzy numbers are defined, using the extension principle (Nguyen, 1978). Their α-cuts are:

$$(\widetilde{A}+\widetilde{B})_\alpha = \left[a_\alpha^{(1)}+b_\alpha^{(1)}\ ,a_\alpha^{(2)}+b_\alpha^{(2)}\right] \tag{2}$$

$$\begin{aligned}(\widetilde{A}\cdot\widetilde{B})_\alpha = & \left[\min\{a_\alpha^{(1)}\cdot b_\alpha^{(1)},a_\alpha^{(1)}\cdot b_\alpha^{(2)},a_\alpha^{(2)}\cdot b_\alpha^{(1)},a_\alpha^{(2)}\cdot b_\alpha^{(2)}\},\right.\\ & \left.\max\{a_\alpha^{(1)}\cdot b_\alpha^{(1)},a_\alpha^{(1)}\cdot b_\alpha^{(2)},a_\alpha^{(2)}\cdot b_\alpha^{(1)},a_\alpha^{(2)}\cdot b_\alpha^{(2)}\}\right]\end{aligned} \tag{3}$$

It is easy to recognize that the subset of $\mathcal{F}_c(\mathbb{R})$ composed by the functions

$$\widetilde{\delta}_a(x) = \begin{cases}1 \text{ if } x=a\\ 0 \text{ if } x\neq a.\end{cases}$$

is isomorph to the algebra of the real numbers, so we can identify the real line with this particular subset.

2 The distance

In order to define the distance $D(\widetilde{A},\widetilde{B})$ between two fuzzy numbers $\widetilde{A}$ and $\widetilde{B}$, we begin by introducing a distance $d(A_\alpha,B_\alpha)$ between their correspondent α-cuts; then $D(\widetilde{A},\widetilde{B})$ is defined as a suitable weighted mean of the values $d(A_\alpha,B_\alpha)$. The crucial point of this process consists in the definition of $d(A_\alpha,B_\alpha)$. So our first task consists of defining a measure of the distance between two intervals; we propose the following one

Definition 6. Let g be a probability measure on $([0,1],\mathcal{B}([0,1]))$. The *distance d* [3] *between two intervals* $A=[a^{(1)},a^{(2)}]$, $B=[b^{(1)},b^{(2)}]$ is given by

$$d(A,B) = \sqrt{\int_0^1 \left[t\cdot|a^{(1)}-b^{(1)}|+(1-t)|a^{(2)}-b^{(2)}|\right]^2 dg(t)}. \tag{4}$$

Some comments are needed in order to justify why we do not accept the two best known interval distances used in fuzzy number theory.

1. The Hausdorff distance $d_H(A,B)$ is defined by $d_H(A,B) = \max[\delta(A,B),\ \delta(B,A)]$, where $\delta(X,Y)=\sup_{x\in X}\{\inf_{y\in Y}|x-y|\}$

   It assigns the same distance to the two pairs of intervals A,B and $\widehat{A},\widehat{B}$ shown in Figure 1, whereas it seems to be obvious that the distance between the second pair has to be greater than the distance between the first one.

[3] As d depends on the choice of the measure g, the appropriate notation for these distances would be d_g. We will omit the subindex g as long as no equivocations appear.

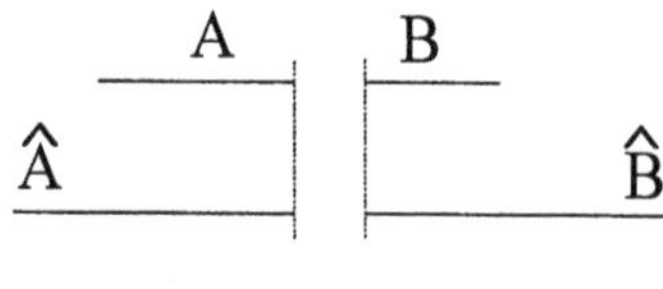

Fig. 1.

2. The second distance $d^*(A,B)$ is a suitable "combination" of the two differences $|a^{(1)} - b^{(1)}|$ and $|a^{(2)} - b^{(2)}|$. The best known are

$$d_1^*(A,B) = \tfrac{1}{2}[|a^{(1)} - b^{(1)}| + |a^{(2)} - b^{(2)}|]$$

$$d_2^*(A,B) = \sqrt{\tfrac{1}{2}[(a^{(1)} - b^{(1)})^2 + (a^{(2)} - b^{(2)})^2]}$$

Both these expressions, as well as all the "combinations" d_1^* and d_2^* attribute the same distance to the two pairs A, B and $\widehat{A}, \widehat{B}$ of Figure 2, but we think that the distance between the second pair has to be greater than the distance between the first one.

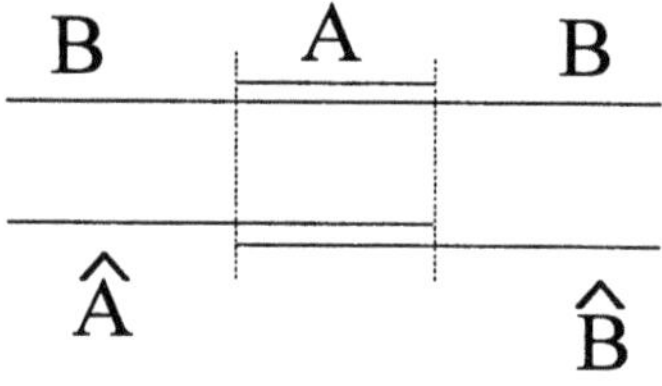

Fig. 2.

Our definition does not have these disadvantages. It is easy to recognize that the distance d between the second pair is greater than that between the first one, unless the probability measure g is concentrated at the points 0 and 1. In this latter case d reduces to d_2^* . Other particular cases of (4) are proposed in Bertoluzza *et al.* (1991) and Diamond (1991).

In this paper we restrict ourselves to probability measures which are the sum of a term continuous with respect to the Lebesgue measure and of a finite probability distribution placed at S points $t_1 \ldots t_S$, that is

$$dg = \gamma(t)dt \quad \text{with}$$

$$\gamma(t) = \overline{\gamma}(t) + \sum_{s=1}^{S} k_s \delta(t - t_s) \tag{5}$$

where $\overline{\gamma}$ is a Lebesgue measurable function, and δ is the Dirac distribution $\delta(t - t_s) = 0 \ \forall t \neq t_s$ and $\int \delta(t - t_s)\, dt = 1$ in all the intervals containing

t_s. In this case (4) becomes

$$d^2(A,B) = \int_0^1 \overline{\gamma}(t)\left[t\cdot(a^{(1)}-b^{(1)})+(1-t)(a^{(2)}-b^{(2)})\right]^2 dt + \sum_{s=1}^{S} k_s[a_s-b_s]^2 \tag{6}$$

where $a_s = t_s a^{(1)} + (1-t_s)a^{(2)}$, $b_s = t_s b^{(1)} + (1-t_s)b^{(2)}$

Since γ defines a probability, the following conditions must hold

$$\gamma(t) \geq 0, \; \int_0^1 \gamma(t)\, dt = 1$$

However, we think that some other properties have to be imposed if (4) has to represent a reasonable way of measuring a distance. In particular: (*i*) the endpoints of the intervals are of great importance in the definition of the distance (conditions (7A) and (7C) below), (*ii*) no reason exists to prefer the left or the right half side of the interval (condition (7B)). Therefore we propose the following complementary properties

$$\begin{array}{lll} \gamma(0) > 0 \;,\; \gamma(1) > 0 & & [A] \\ \gamma(t) = \gamma(1-t) & & [B] \\ t_1 = 0 \;,\; t_S = 1 & \text{if } S > 1 & [C] \end{array} \tag{7}$$

It is easy to prove, using the Minkowski inequalities

$$\sqrt{\textstyle\sum_i (a_i+b_i+\ldots+c_i)^2} \leq \sqrt{\textstyle\sum_i a_i^2} + \sqrt{\textstyle\sum_i b_i^2} + \ldots + \sqrt{\textstyle\sum_i b_i^2},$$

$$\sqrt{\textstyle\int (u+v+\ldots+w)^2\, dt} \leq \sqrt{\textstyle\int u^2\, dt} + \sqrt{\textstyle\int v^2\, dt} + \ldots + \sqrt{\textstyle\int w^2\, dt},$$

that the positive function, defined on the family of the intervals $A = [a^{(1)}, a^{(2)}]$ by means of

$$\nu(A) = \nu([a^{(1)}, a^{(2)}]) = \sqrt{\int_0^1 \gamma(t)[ta^{(1)} + (1-t)a^{(2)}]^2\, dt}$$

is a norm, provided that γ is a function of type (5); therefore the function d defined by (4) really represents a distance.

The general form (4) is too complicated from the operational point of view. We think that we obtain a sufficiently good measure if we choose in (6) $\overline{\gamma}(t) = 0$, $S = 3$ (that is if we choose a probability measure concentrated in the points 0, 0.5, 1). In this case (6) reduces to

$$d^2(A,B) = k[a^{(1)} - b^{(1)}]^2 + h[a^{(G)} - b^{(G)}]^2 + k[a^{(2)} - b^{(2)}]^2$$

where $a^{(G)} = \frac{a^{(1)}+a^{(2)}}{2}$ and $2k + h = 1$.

Starting from $d(A_\alpha, B_\alpha)$ we construct $D(\widetilde{A}, \widetilde{B})$ as follows. Let f be another probability measure on $([0,1], \mathcal{B}([0,1]))$ generated by a density function $\varphi(\alpha)$ of the form

$$\varphi(\alpha) = \widetilde{\varphi}(\alpha) + \sum_{l=1}^{L} h_l \delta(\alpha - \alpha_l)$$

where, as function γ, $\overline{\varphi}$ is a Lebesgue measurable function and δ is the Dirac distribution.

Definition 7. The *distance* D *between two fuzzy numbers* $\widetilde{A}$ and $\widetilde{B}$ is defined by

$$D(\widetilde{A}, \widetilde{B}) = \sqrt{\int_0^1 [d(A_\alpha, B_\alpha)]^2 \varphi(\alpha)\, d\alpha}. \tag{8}$$

$D(\widetilde{A}, \widetilde{B})$ reduces to the usual distance $|b - a|$ on the real line when $\widetilde{A} = a$ and $\widetilde{B} = b$ are real numbers.

It is easy to prove that the function D has all the properties of a distance in $\mathbb{R}$, provided that the functions $a^{(i)}, b^{(i)}$ belong to $L^2([0,1])$ and $\alpha > 0 \Rightarrow \varphi(\alpha) > 0$.

Remark 1. In some cases it is useful to choose $\varphi(\alpha) = 0$ in a suitable interval $[0, \overline{\alpha}]$. This happens when we establish that the α-cuts corresponding to low degrees of membership $(\alpha \leq \overline{\alpha})$ do not affect the measure of the distance. In this case (8) does not represent a true distance because $D(\widetilde{A}, \widetilde{B}) = 0$ does not imply $\widetilde{A} = \widetilde{B}$. However, we may introduce a suitable equivalence relation (and its corresponding equivalence classes) by posing $\widetilde{A} \cong \widetilde{B} \Leftrightarrow D(\widetilde{A}, \widetilde{B}) = 0$; the function D defines a true distance on the space of the equivalence classes.

The supplementary condition to be imposed on the probability "density" φ arises from the fact that in computing the distance the intervals with greater membership degree count more than the lower ones. Thus the condition we impose is

$$\alpha_1 < \alpha_2 \implies \varphi(\alpha_1) \leq \varphi(\alpha_2) \tag{9}$$

The practical choice of the function φ depends on the structure of the fuzzy numbers we deal with. Thus if the membership function takes only a finite or countable set of values, then it seems to be natural to select φ in such a way that the probability f is concentrated at these points. On the contrary, if all the values are allowed, then we think that a measurable function has to be selected, that is $\varphi = \overline{\varphi}$. In particular the two following functions seem to be quite useful

$$\begin{array}{lllll} \varphi(\alpha) & = & \overline{\varphi}(\alpha) & = & 1 & & [A] \\ \varphi(\alpha) & = & \overline{\varphi}(\alpha) & = & e(\alpha - \frac{1}{2}) + 1 & \text{with } 0 < e \leq 2 & [B] \end{array} \tag{10}$$

It has been proved that two distances D' and D'' of the type (8), are generate topologies T', T'' which are equivalent if the following conditions

$$\begin{array}{lllllll} p' & = & \varphi'(0) > 0 & & p'' & = & \varphi''(0) > 0 \\ q' & = & \varphi'(1) < +\infty & & q'' & = & \varphi''(1) < +\infty \end{array}$$

hold. The proof can be found in Bertoluzza *et al.* (1995).

3 The main problem

Let $\mathcal{X}$ and $\mathcal{Y}$ be two fuzzy random variables and let us suppose that there exists a *linear dependence* between $\mathcal{X}$ and $\mathcal{Y}$, that is

$$\mathcal{Y} = \alpha\mathcal{X} + \beta$$

where α and β are real or fuzzy numbers. Our primary task consists in determining the regression coefficients α, β . In order to reach this goal we have at our disposal a finite family of experimental observations represented by a sequence $\{(\widetilde{X}_i, \widetilde{Y}_i) \mid i = 1 \dots n\}$ of couples of fuzzy numbers.

If the observations as well as the coefficients were real valued numbers, then the classical least square method would give the solution to our problem. In fact in this case the regression coefficients are the values which minimize the function

$$f(\alpha, \beta) = \sum_{i=1}^{n} (y_i - \alpha x_i - \beta)^2$$

But in the actual case at least the observations are fuzzy numbers, and this method cannot be applied because the square of a fuzzy number is also a fuzzy value and no total ordering can be usefully defined on $\mathcal{F}_c(\mathbb{R})$. In order to avoid this difficulty we propose to substitute the function to be minimized by the sum of the squared distance between the observations $\widetilde{Y}_i$ and the predictions $\alpha\widetilde{X}_i + \beta$.

If we consider the most general case, then we do not impose any limitation to the regression coefficients and therefore we have to suppose that $\alpha = \widetilde{A}$ and $\beta = \widetilde{B}$ are fuzzy numbers. So what we have to do is to find the values $\widetilde{A}, \widetilde{B}$ which minimize the functional

$$F^*(\widetilde{A}, \widetilde{B}) = \sum_{i=1}^{n} D^2(\widetilde{Y}_i, \widetilde{A}\widetilde{X}_i + \widetilde{B}) \tag{11}$$

Nevertheless let us consider the linear fuzzy function $\widetilde{Y} = \widetilde{A}\widetilde{X} + \widetilde{B}$; as in the real case, we can try to interpret geometrically this function by representing a fuzzy straight line in the fuzzy two-dimensional plane. But a fuzzy straight line is, in our pictorial vision, a fuzzy subset $\widetilde{R}$ of the plane whose support R_0 lies between two parallel real straight lines (see Figure 3).

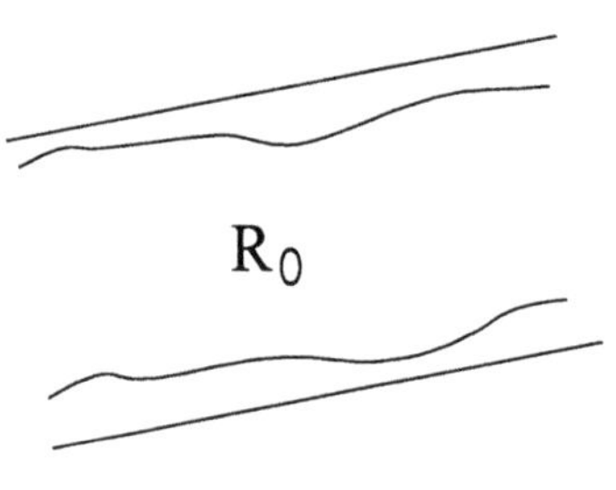

Fig. 3.

This means that both the $\widetilde{X}$ and $\widetilde{Y}$ coordinates of the fuzzy points on the fuzzy line have necessarily supports with an upper bounded measure. In order for this condition to hold, it is necessary for the first coefficient $\widetilde{A} = a$ to be a real number (see e.g. Ciliberti, 1990-91, Ciampa, 1991-92). In this case the generalized least square method leads to minimizing the functional

$$F^*(a, \widetilde{B}) = \sum_{i=1}^{n} D^2(\widetilde{Y}_i,\ a\widetilde{X}_i + \widetilde{B}) \tag{12}$$

Thus we can generalize the least square method in two different ways: (i) interpolate the experimental data (X_i, Y_i) by means of a fuzzy linear function, that is minimize (11), or (ii) interpolate the experimental data by means of a fuzzy straight line, that is minimize (12). Although the first case is more general, we think that the second one is more meaningful, and therefore we will develop in detail this latter problem and we will give only some suggestions of how to study the first one.

In this paper we have identified the fuzzy numbers with two functions defined on $[0,1]$, which characterize respectively the non-decreasing and the non-increasing part of the membership function. Therefore, in order to complete the description of our problem we have to specify the form of the functional (11) in terms of the functions which describe the fuzzy numbers $\widetilde{A}, \widetilde{B}$, or the form of the functional (12) in terms of a and of the functions $b^{(1)}, b^{(2)}$.

$$\begin{aligned} F^*(\widetilde{A},\widetilde{B}) &= F[a^{(1)}(\alpha), a^{(2)}(\alpha), b^{(1)}(\alpha), b^{(2)}(\alpha)] && [A] \\ F^*(a,\widetilde{B}) &= F[a, b^{(1)}(\alpha), b^{(2)}(\alpha)] && [B] \end{aligned} \qquad (13)$$

where the arguments of the functions F are subject to the conditions

$$\begin{array}{ll} a^{(1)}, b^{(1)} & \text{non} - \text{decreasing} \\ a^{(2)}, b^{(2)} & \text{non} - \text{increasing} \\ a^{(i)}, b^{(i)} & \in L^2([0,1]) \\ a^{(1)}(1) & \leq a^{(2)}(1) \\ b^{(1)}(1) & \leq b^{(2)}(1) \\ a & \in \mathbb{R} \end{array}$$

The expressions (13A,B) can be obtained explicitly by using the definition of product and sum of fuzzy numbers and the definition of distance. They are quite complicated because the endpoints of the α-cuts of the product have different expressions according to the signs of the end point of the α-cuts of the factors.

Nevertheless if the slope of the fuzzy interpolating linear funtion is real, then we obtain easily the explicit form of the functional (6). In fact let us observe firstly that we do not lose in generality if we suppose that all the observations $\widetilde{X}_i$ are positive, namely that $\inf\{x_i^{(1)}(0)|i = 1 \dots n\} \geq 0$ (on the contrary it sufficies to operate a suitable coordinate translation on the x axis). In this case it is easy to verify that

$$\begin{aligned} &F[a, b^{(1)}(\alpha), b^{(2)}(\alpha)] \\ &= \textstyle\sum_{i=1}^n \int_0^1 \varphi(\alpha) \Big[\int_0^1 \gamma(t) \{t[y_i^{(1)}(\alpha) - a x_i^{(1)}(\alpha) - b^{(1)}(\alpha)] \\ &+ (1-t)[y_i^{(2)}(\alpha) - a x_i^{(2)}(\alpha) - b^{(2)}(\alpha)]\}^2 \, dt \Big] \, d\alpha \qquad \text{if } a \geq 0 \end{aligned} \qquad (14)$$

$$\begin{aligned} &F[a, b^{(1)}(\alpha), b^{(2)}(\alpha)] \\ &= \textstyle\sum_{i=1}^n \int_0^1 \varphi(\alpha) \Big[\int_0^1 \gamma(t) \{t[y_i^{(1)}(\alpha) - a x_i^{(2)}(\alpha) - b^{(1)}(\alpha)] \\ &+ (1-t)[y_i^{(2)}(\alpha) - a x_i^{(1)}(\alpha) - b^{(2)}(\alpha)]\}^2 \, dt \Big] \, d\alpha \qquad \text{if } a \leq 0 \end{aligned} \qquad (15)$$

4 Existence and uniqueness of the best linear approximation

We will prove in this section that the problem of the minimization of the functional (13B) has a unique solution, provided that some conditions are satisfied, which cover all the cases of statistical interest. We will also give a partial answer to the problem of minimizing the functional (13A).

The proof of this result is based on the following well-known theorem of functional analysis (see e.g Baiocchi and Capelo, 1984).

Theorem 1. *Let Γ be a closed and convex subset of a reflexive Banach space Ω. If f is a real valued, lower semicontinuous, convex and cohercive map defined on Ω, then there exists a minimum of f in the subset Γ. Moreover if f is strictly convex, then the minimum is unique.*

We know that the functional (13B) may be written in the form (14,15). Then we proceed by showing that the functional (14) has a unique minimum M^+ in the subset where $a \geq 0$ and the functional (15) has a unique minimum M^- in the subset where $a \leq 0$. The minimum between M^+ and M^- corresponds to the point of the absolute minimum of the functional (13B).

$$\Omega = \mathbb{R}^+ \times L^2([0,1]) \times L^2([0,1]) \tag{16}$$

$$\begin{array}{lr} \Gamma^+ = \{(a, b^{(1)}, b^{(2)}) \mid a \geq 0\} \quad \text{with} & \\ a \geq 0 & [A] \\ \quad b^{(1)}(\alpha) \ \text{non-decreasing} & [B] \\ \quad b^{(2)}(\alpha) \ \text{non-increasing} & [C] \\ \quad b^{(1)}(1) \leq b^{(2)}(1) & [D] \end{array} \tag{17}$$

$$\begin{aligned} &F^+[a, b^{(1)}(\alpha), b^{(2)}(\alpha)] \\ &= \sum_{i=1}^n \int_0^1 \varphi(\alpha)\Big[\int_0^1 \gamma(t)\{t[y_i^{(1)}(\alpha) - ax_i^{(1)}(\alpha) - b^{(1)}(\alpha)] \\ &+(1-t)[y_i^{(2)}(\alpha) - ax_i^{(2)}(\alpha) - b^{(2)}(\alpha)]\}^2 dt\Big]\, d\alpha \end{aligned} \tag{18}$$

It is easy to prove that: (i) Ω is a reflexive Banach space (because it is the cartesian product of three Hilbert spaces), (ii) Γ^+ is a convex subset of Ω (obvious), (iii) f is lower semicontinuous (because it is continuous except at the boundary points) (see e.g. Salas, 1991).

In order to apply Theorem 1, we have to prove that Γ^+ is closed and F^+ is (strictly) convex and cohercive.

(iv) Γ^+ is closed. Let $\omega = (a, b^{(1)}, b^{(2)}) \in \overline{\Gamma^+}$, and let us suppose that $(a, b^{(1)}, b^{(2)}) \in \overline{\Gamma^+} - \Gamma^+$. Then there exists a sequence of points of

Γ^+ which converges to ω in $L^2([0,1])$ and consequently there exists a subsequence $\omega_r = (a_r, b_r^{(1)}, b_r^{(2)}) \in \Gamma^+$ which converges to ω a.e. This means that $a = \lim a_r \geq 0$, $b^{(1)} = \lim b_r^{(1)}$ is almost everywhere non-decreasing, and $b^{(2)} = \lim b_r^{(2)}$ is a.e. non-increasing. Moreover we have $b^{(1)} \leq b^{(2)}$ a.e. in $[0,1]$. This means that in the space $L^2([0,1])$ the point ω fulfil the conditions (17) and therefore belongs to Γ. So we proved that the set Γ^+, considered as subset of $\Omega = L^2([0,1])$, is closed.

(v) In order to prove that F^+ is convex let us pose $\omega_1 = (a_1, b_1^{(1)}, b_1^{(2)})$, $\omega_2 = (a_2, b_2^{(1)}, b_2^{(2)})$. We have to prove that

$$F^+(\lambda\omega_1 + (1-\lambda)\omega_2) \leq \lambda F^+(\omega_1) + (1-\lambda)F^+(\omega_2) \tag{19}$$

By posing

$$\begin{aligned} y_i^{(t)}(\alpha) &= t y_i^{(1)}(\alpha) + (1-t) y_i^{(2)}(\alpha) \\ x_i^{(t)}(\alpha) &= t x_i^{(1)}(\alpha) + (1-t) x_i^{(2)}(\alpha) \\ b^{(t)}(\alpha) &= t b^{(1)}(\alpha) + (1-t) b^{(2)}(\alpha) \end{aligned}$$

we obtain immediately

$$\begin{aligned} &\Big\{t\big[y_i^{(1)}(\alpha) - [\lambda a_1 + (1-\lambda)a_2]x_i^{(1)}(\alpha) - [\lambda b_1^{(1)}(\alpha) + (1-\lambda)b_2^{(1)}(\alpha)]\big] \\ &+(1-t)\big[y_i^{(2)}(\alpha) - [\lambda a_1 + (1-\lambda a_2]x_i^{(2)}(\alpha) - [\lambda b_1^{(2)}(\alpha) + (1-\lambda)b_2^{(2)}(\alpha]\big]\Big\}^2 \\ &= \Big\{y_i^{(t)}(\alpha) - [\lambda a_1 + (1-\lambda)a_2]x^{(t)}(\alpha) - [\lambda b_1^{(t)} + (1-\lambda)b_2^{(t)}]\Big\}^2 \\ &= \Big\{\lambda[y_i^{(t)}(\alpha) - a_1 x_i^{(t)}(\alpha) - b_1^{(t)}(\alpha)] + (1-\lambda)[y_i^{(t)}(\alpha) - a_2 x_i^{(t)}(\alpha) - b_2^{(t)}(\alpha)]\Big\}^2 \\ &\leq \lambda[y_i^{(t)}(\alpha) - a_1 x_i^{(t)}(\alpha) - b_1^{(t)}(\alpha)]^2 + (1-\lambda)[y_i^{(t)}(\alpha) - a_2 x_i^{(t)}(\alpha) - b_2^{(t)}(\alpha)]^2 \end{aligned} \tag{20}$$

If we multiply first and second sides of (20) by the non negative term $\gamma(t)\varphi(\alpha)$, integrate between 0 and 1 with respect to t and α and add with respect to the index i, then we recognize that inequality (19) holds, and therefore the functional (14) is convex.

(vi) In order to prove that F^+ is cohercive, let us consider, for each $\alpha \in [0,1]$, the function

$$f_\alpha(z_1, z_2, z_3) = \sum_{i=1}^{n} d^2[(\widetilde{Y}_i)_\alpha \ , \ (a\widetilde{X}_i + \widetilde{B})_\alpha]$$

where $z_1 = a$, $z_2 = b^{(1)}(\alpha)$, $z_3 = b^{(2)}(\alpha)$. It is easy to recognize, using (4), that the function f_α may be written in the form

$$f_\alpha(z_1,\ z_2,\ z_3) \ = \ \sum_{r=1}^{3}\sum_{s=1}^{3} g_{rs}(\alpha) z_r z_s \ + \ \sum_{r=1}^{3} e_r(\alpha) z_r \ + \ d(\alpha) \qquad (21)$$

where

$$\begin{aligned}
g_{11}(\alpha) &= \textstyle\sum_{i=1}^{n} \int_0^1 \gamma(t)[x_i^{(t)}(\alpha)]^2\, dt \\
g_{22}(\alpha) &= n \textstyle\int_0^1 \gamma(t) t^2\, dt \\
g_{33}(\alpha) &= n \textstyle\int_0^1 \gamma(t)(1-t)^2\, dt \\
g_{12}(\alpha) &= g_{21}(\alpha) \ = \ \textstyle\sum_{i=1}^{n} \int_0^1 \gamma(t) t [x_i^{(t)}(\alpha)]\, dt \\
g_{13}(\alpha) &= g_{31}(\alpha) \ = \ \textstyle\sum_{i=1}^{n} \int_0^1 \gamma(t)(1-t)[x_i^{(t)}(\alpha)]\, dt \\
g_{23}(\alpha) &= g_{32}(\alpha) \ = \ n \textstyle\int_0^1 \gamma(t) t(1-t))\, dt \\
e_1(\alpha) &= -2 \textstyle\sum_{i=1}^{n} \int_0^1 \gamma(t)[x_i^{(t)}(\alpha)][y_i^{(t)}(\alpha)]\, dt \\
e_2(\alpha) &= -2 \textstyle\sum_{i=1}^{n} \int_0^1 \gamma(t) t [y_i^{(t)}(\alpha)]\, dt \\
e_3(\alpha) &= -2 \textstyle\sum_{i=1}^{n} \int_0^1 \gamma(t)(1-t)[y_i^{(t)}(\alpha)]\, dt \\
d(\alpha) &= \textstyle\sum_{i=1}^{n} \int_0^1 \gamma(t)[y_i^{(t)}(\alpha)]^2\, dt
\end{aligned}$$

Since the function (21) represents a quadric surface, there exists a linear orthogonal coordinate transform

$$\overline{z}_r = \sum_{s=1}^{3} \eta_{rs}(z_s - \zeta_s) \qquad (22)$$

which reduces f_α to its normal form

$$f_\alpha(z_1, z_2, z_3) \ = \ \overline{f}_\alpha(\overline{z}_1, \overline{z}_2, \overline{z}_3) \ = \ \sum_{r=1}^{3} \overline{g}_r(\alpha) \overline{z}_r^2 + \overline{d}(\alpha)$$

All the values $\overline{g}_r(\alpha)$ as well as $\overline{d}(\alpha)$ are non-negative, because f_α is a sum of squares. Moreover (22) preserves the norm, that is

$$z_1^2 + z_2^2 + z_3^2 \ = \ \|(z_1, z_2, z_3)\|^2 \ = \ \|(\overline{z}_1, \overline{z}_2, \overline{z}_3)\|^2 \ = \ \overline{z}_1^2 + \overline{z}_2^2 + \overline{z}_3^2 \qquad (23)$$

Now let us pose

$$\overline{g}(\alpha) \ = \ \inf\{\overline{g}_1(\alpha), \overline{g}_2(\alpha), \overline{g}_3(\alpha)\}$$

It is evident that

$$f_\alpha(z_1, z_2, z_3) \ = \ \overline{f}_\alpha(\overline{z}_1, \overline{z}_2, \overline{z}_3) \ \geq \overline{g}(\alpha)[\ \overline{z}_1^2 + \overline{z}_2^2 + \overline{z}_3^2] \qquad (24)$$

If we remember that $z_1 = a,\ z_2 = b^{(1)}(\alpha),\ z_3 = b^{(2)}(\alpha)$, then we have from (24)(23)

$$
\begin{aligned}
F^+(a,\widetilde{B}) &= \int_0^1 \varphi(\alpha) f_\alpha[a, b^{(1)}(\alpha), b^{(2)}(\alpha)]\, d\alpha \\
&\geq \int_0^1 \varphi(\alpha)\overline{g}(\alpha)\{a^2 + [b^{(1)}(\alpha)]^2 + [b^{(2)}(\alpha)]^2\}\, d\alpha \\
&\geq \overline{G}\|(a, b^{(1)}(\alpha), b^{(2)}(\alpha))\|
\end{aligned}
\tag{25}
$$

with $\overline{G} \in [\min_{\alpha\in[0,1]} \varphi(\alpha)\overline{g}(\alpha)\ ,\ \max_{\alpha\in[0,1]} \varphi(\alpha)\overline{g}(\alpha)]$; the last equality holds because of the law of the integral mean. If both the conditions

$$
\begin{array}{lll}
\varphi(\alpha) > 0 \quad \text{a.e.} & \text{and} & [A] \\
\overline{g}(\alpha) > 0 \quad \text{a.e.} & & [B]
\end{array}
\tag{26}
$$

hold, then we have also $\overline{G} > 0$ and, consequently, equation (25) assures that F^+ is cohercive. Moreover if $\overline{g}(\alpha) > 0$, then the function f_α represents a paraboloid, that is a strictly convex function. Thus if (26B) holds, then the functional F^+ is strictly convex and therefore this theorem (16) assures that F has a unique minimum M^+ in Γ^+.

In the same way we can prove that, under the same conditions of the previous case, the functional F has a unique minimum M^- in the subset

$$\Gamma^- = \{(a, b^{(1)}, b^{(2)}) \mid a \leq 0\}$$

By comparing the two minimum values which have been found in Γ^+ and in Γ^- respectively, we are able to determine the absolute minimum of F in the whole space $\mathbb{R} \times L^2([0,1]) \times L^2([0,1])$, provided that $M^+ \neq M^-$. So we can enunciate the following

Theorem 2. *If conditions (26) hold, then the functional F has one or exceptionally two absolute minimum points in the space $\Gamma = \mathbb{R} \times L^2([0,1]) \times L^2([0,1])$.*

We will complete this paragraph by analyzing the cases where conditions (26) do not hold.

CASE A. Let us remember that φ is non-negative and non-decreasing. Consequently if $\varphi(\alpha^*) = 0$, then $\varphi(\alpha) = 0\ \forall \alpha \in [0, \alpha^*]$. Thus the condition (26A) does not hold if and only if there exists a value $\underline{\alpha} > 0$ such that

$$
\varphi(\alpha) \begin{cases} = 0 & \text{if } \alpha \leq \underline{\alpha} \\ > 0 & \text{if } \alpha > \underline{\alpha} \end{cases}
$$

But if we choose $\varphi = 0$ in $[0, \underline{\alpha}]$, then we consider that the α-cuts corresponding to degrees of membership lower than $\underline{\alpha}$ do not play any role in the computation of the distance, and consequently in the determination of the functional F. This means that we do not lose any generality if we restrict the functions involved in our problem to the class $L^2([\underline{\alpha}, 1])$, and substitute the interval $[0,1]$ by the interval $[\underline{\alpha}, 1]$ in all our considerations. In this interval, condition (26A) holds.

CASE B We discuss the case where condition (26B) does not hold under some simplifying hypothesis, and more precisely we suppose that the measure g is concentrated in S point $0 = t_1 \dots t_S = 1$ of the interval $[0,1]$, so that the distance between two intervals $A = [a^{(1)}, a^{(2)}]$, $B = [b^{(1)}, b^{(2)}]$ takes the form

$$d^2(A,B) = \sum_{s=1}^{S} k_s [a_s - b_s]^2$$

where $a_s = (1-t_s)a^{(1)} + t_s a^{(2)}$, $b_s = (1-t_s)b^{(1)} + t_s b^{(2)}$ (clearly $a_1 = a^{(1)}, b_1 = b^{(1)}$ and $a_S = a^{(2)}, b_S = b^{(2)}$). The functional to be minimized is

$$F^+ = \sum_{i=1}^{n} \int_0^1 \varphi(\alpha) \sum_{s=1}^{S} [y_{i,s}(\alpha) - a x_{i,s}(\alpha) - b_s(\alpha)]^2 \, d\alpha \tag{27}$$

We can reformulate our problem by supposing that F is defined on the Banach space

$$\Omega = \mathbb{R} \times \left[\mathrm{L}^2([0,1])\right]^{\mathrm{S}}$$

Then we can recognize that our main problem is equivalent to the following one:

"Find the minimum of the functional (27) in the subset $\Gamma \subset \Omega$ defined by

$$\begin{aligned} \Gamma^+ = \{ & [a, b_1(\alpha), \dots, b_S(\alpha)] \mid \\ & a \in \mathbb{R}^+ \\ & b_1 \in L^2([0,1]) \\ & b_S \in L^2([0,1]) \\ & b_1 \text{ is non-decreasing} \\ & b_S \text{ is non-increasing} \\ & b_1(1) \leq b_S(1) \\ & b_s(\alpha) = t_s b_1(\alpha) + (1-t_s) b_S(\alpha) \}" \end{aligned}$$

If condition (26B) does not hold, let us consider a point $\overline{\alpha}$ where $\overline{g}(\overline{\alpha}) = 0$,where $\overline{g}_s(\overline{\alpha})$ are the eigenvalues of the matrix $\{g_{rs}(\overline{\alpha})\}$. Then we have $Det[\{g_{rs}(\overline{\alpha})\}] = \prod_{s=1}^{S} \overline{g}_s(\overline{\alpha}) = 0$. It is easy to verify that the matrix $\{g_{rs}\}$ has the form [4]

[4] For the sake of simplicity we omit to indicate explicitly the dependence of the variable $\overline{\alpha}$

$$\begin{bmatrix} nk_1 & 0 & \dots & 0 & k_1 \sum_{i=1}^n x_{i,1} \\ 0 & nk_2 & \dots & 0 & k_2 \sum_{i=1}^n x_{i,2} \\ \dots & \dots & \dots & \dots & \dots \\ 0 & 0 & \dots & nk_S & k_S \sum_{i=1}^n x_{i,S} \\ k_1 \sum_{i=1}^n x_{i,1} & k_2 \sum_{i=1}^n x_{i,2} & \dots & k_S \sum_{i=1}^n x_{i,S} & \sum_{s=1}^S \sum_{i=1}^n x_{i,s}^2 \end{bmatrix}$$

Its determinant may be reduced, by means of straightforward computations, to the form

$$n^{S-1} \prod_{s=1}^{S} k_s \Big\{ \sum_{s=1}^{S} \big[nk_s \sum_{i=1}^{n} x_{i,s}^2 - k_s (\sum_{i=1}^{n} x_{i,s})^2 \big] \Big\}$$

It is easy to recognize that each of the addendum of the sum in (35) is non-negative, and it is zero if and only if

$$x_{1,s} = x_{2,s} = \dots x_{n,s} \quad \forall s = 1, S$$

that is if and only if the α-cuts of all the observations $\widetilde{X}_i$ are identical. Thus, for $\overline{g}(\alpha)$ to be greater than 0, it sufficies that at least two observations $\widetilde{X}'$ and $\widetilde{X}''$ have different α-cuts. Thus condition (26B) leads to the following corollary of Theorem 2:

Corollary 1. *The functional attains a unique minimum if and only if for almost every value* α *in* $[0, 1]$ *, the* α*-cuts of the observations* $\widetilde{X}_i$ *are not equal.*

5 Regression with fuzzy slope

We examine in this paragraph what happens when both regression coefficients are fuzzy numbers. In this case some difficulties arise because the functional to be minimized takes different forms according to the sign of $a_\alpha^{(1)}, a_\alpha^{(2)}, x_\alpha^{(1)}, x_\alpha^{(2)}$. So we can give only a partial answer to the main problem. More precisely we have to suppose that

(1) all the oservations $\widetilde{X}_i$ are positive,
(2) either $a^{(2)}(1) \geq 0$ or
(3) $a^{(2)}(0) \leq 0$.

When conditions (1)(2) holds, the functional to be minimized takes the form

$$F^+[a^{(1)}, a^{(2)}, b^{(1)}(\alpha), b^{(2)}(\alpha)]$$
$$= \sum_{i=1}^n \int_0^1 \varphi(\alpha) \Big[\int_0^1 \gamma(t)\{t[y_i^{(1)}(\alpha) - a^{(1)}(\alpha)x_i^{(1)}(\alpha) - b^{(1)}(\alpha)]$$
$$+(1-t)[y_i^{(2)}(\alpha) - a^{(2)}(\alpha)x_i^{(2)}(\alpha) - b^{(2)}(\alpha)]\}^2 \, dt \Big] \, d\alpha$$

From now on we can follow the development of paragraph 4 with the following changes:

(1) $\Omega = \mathbb{R}^+ \times L^2([0,1]) \times L^2([0,1])$ has to be changed to $L^2([0,1]) \times L^2([0,1]) \times L^2([0,1]) \times L^2([0,1])$

(2) $\Gamma^+ = (a, b^{(1)}, b^{(2)})$ has to be changed to $\Gamma^+ = \{(a^{(1)}, a^{(2)}, b^{(1)}, b^{(2)})\}$ with

$$\begin{array}{l} a^{(1)}(\alpha) \ \text{non} - \text{decreasing} \\ a^{(2)}(\alpha) \ \text{non} - \text{increasing} \\ a^{(1)}(1) \leq a^{(2)}(1) \\ a^{(2)}(1) \geq 0 \\ b^{(1)}(\alpha) \ \text{non} - \text{decreasing} \\ b^{(2)}(\alpha) \ \text{non} - \text{increasing} \\ b^{(1)}(1) \leq b^{(2)}(1) \end{array}$$

Following the same steps of Paragraph 4, we can easyly recognize that the functional F^+ has a unique minimum in Γ^+, provided that at least two $\widetilde{X}_i$ are different. So $\sum d^2(\widetilde{Y}_i, \widetilde{A}\widetilde{X}_i + \widetilde{B})$ has a unique minimum in Γ^+. In the same way we recognize that if conditions (2) and (3) hold, then $\sum d^2(\widetilde{Y}_i, \widetilde{A}\widetilde{X}_i + \widetilde{B})$ has a unique minimum in Γ^-, where Γ^- is Γ^+ with condition $a^{(2)}(1) \geq 0$ replaced by $a^{(1)}(0) \leq 0$.

There are many situations where the first condition holds: for example all the cases where the observations are measures or distances. The second and third ones implies that our result has not yet been proved when $\widetilde{A} = \widetilde{0}^-$ where $\widetilde{0}^-$ indicates the fuzzy numbers with negative core and 0 in the support, that is numbers as showed in Figure 4.

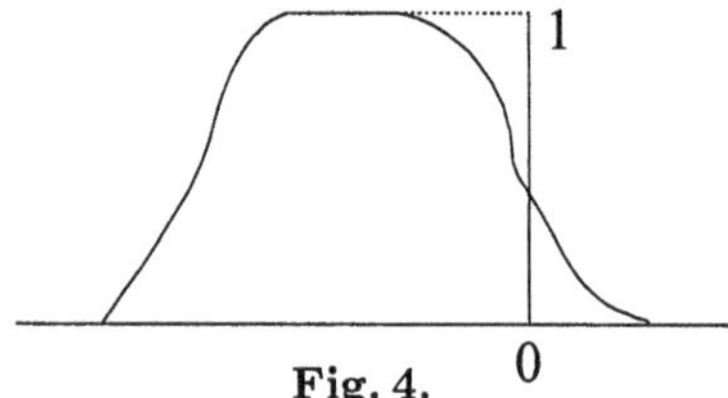

Fig. 4.

So the result of this paragraph can be applied in two cases.

(a) When we have some information (of physical or technical nature) which tell us that the slope cannot be of type $\widetilde{0}^-$.
(b) In the case where this kind of slope does not present any interest.

6 Numerical evaluation of the fuzzy coefficients

The classical variational methods cannot be employed in order to determine analitically the minimum of the functional F, because in every single neighbourhood of all the points in Γ, there exist points which do not belong to Γ. Then, in order to obtain the regression coefficients, we have to use numerical methods. Here we propose to approximate the functions $b^{(1)}$ and $b^{(2)}$ (which appear as second and third coordinates of the points of Γ^+) by means of suitable piecewise linear functions: afterwards we minimize F in the set of these approximations. We treat in detail the case of the functional F^+, that is we try to determine the minimum point in the set where the condition

$$a \geq 0 \tag{28}$$

holds, but what we develop still holds, with suitable modifications, when $a \leq 0$. In particular we proceed as follows. We fix $M+1$ values

$$0 = \alpha_0 < \alpha_1 < \ldots < \alpha_M = 1$$

of the parameter α. Then we approximate the functions $b^{(1)}, b^{(2)}$ which define a fuzzy number $\widetilde{B}$ by means of pieces of straight lines passing respectively through the points

$$(\alpha_i, b_i^{(1)}),\ (\alpha_{i+1}, b_{i+1}^{(1)}) \quad i = 0 \ldots M$$

$$(\alpha_i, b_i^{(2)}),\ (\alpha_{i+1}, b_{i+1}^{(2)}) \quad i = 0 \ldots M$$

where $b_i^{(1)} = b^{(1)}(\alpha_i)$, $b_i^{(2)} = b^{(2)}(\alpha_i)$.

The approximation of the fuzzy number $\widetilde{B}$ is completely determined by the $2M+2$ values $\{b_i^{(1)}, b_i^{(2)} \mid i = 0 \ldots M\}$. Since the functions $b^{(1)}$ and $b^{(2)}$ are respectively non-decreasing and non-increasing, the values $b_i^{(1)}$ and $b_i^{(2)}$ fulfil the following conditions

$$\begin{array}{ll} b^{(1)}(\alpha_0) \leq b^{(1)}(\alpha_1) \leq \ldots b^{(1)}(\alpha_M) & [A] \\ b^{(1)}(\alpha_M) \leq b^{(2)}(\alpha_M) & [B] \\ b^{(2)}(\alpha_0) \geq b^{(2)}(\alpha_1) \geq \ldots b^{(2)}(\alpha_M) & [C] \end{array} \tag{29}$$

Now our problem is reduced to determine the slope a and the values $b_i^{(1)}, b_i^{(2)}$ which individuate the best piecewise linear approximation of the minimum point of the functional F^+. The following hypotheses simplify the description of the solution method.

- the measure g is concentrated at the points $0, 0.5, 1$ (we can assume, without any loss of generality, $k_1 = k_3 = 1, k_2 = k$); consequently the distance between two intervals assumes the form

$$d^2(A,B) = k[a^{(1)} - b^{(1)}]^2 + h[a^{(G)} - b^{(G)}]^2 + k[a^{(2)} - b^{(2)}]^2,$$

- the measure f is uniformly distributed over the interval $[0,1]$, that is $\varphi(\alpha)=1$,
- the values α_i are uniformly distributed in the interval $[0,1]$, that is $|\alpha_{i+1}-\alpha_i| = h = \frac{1}{M}$ and $\alpha_{i+1} = i \cdot h$,
- the observations $(\widetilde{X}_j, \widetilde{Y}_j)$ are also represented by piecewise linear functions passing through the (known) points $(\alpha_i, x_i^{(1)}), (\alpha_i, x_i^{(2)})$ and $(\alpha_i, y_i^{(1)})$, $(\alpha_i, y_i^{(2)})$.

Nevertheless the results we will establish still hold in the general case; the assumptions we made allow us to simplify some tedious, but straightforward computations.

Under these hypotheses the fuzzy number $\widetilde{B}$ is approximated by the two following families of piecewise linear functions

$$\begin{aligned} b^{(1)}(\alpha) &= \{r_i^{(1)}(\alpha) = b_i^{(1)} + h(b_{i+1}^{(1)} - b(1)_i)(\alpha - \alpha_i) \mid i = 0 \dots M-1\} \quad [A] \\ b^{(2)}(\alpha) &= \{r_i^{(2)}(\alpha) = b_i^{(2)} + h(b_{i+1}^{(2)} - b(2)_i)(\alpha - \alpha_i) \mid i = 0 \dots M-1\} \quad [B] \end{aligned} \tag{30}$$

The form of the functional to be minimized may be obtained, in terms of the discretized unknowns $a, b_i^{(1)}, b_i^{(2)}$, by substituting (30) in the expression (14). The necessary computations are boring but straightforward; they may be found in Bollani (2001). The result is

$$\begin{aligned} F^+(a, b_i^{(1)}, b_i^{(2)}) &= \textstyle\sum_{j=1}^n D^2(\widetilde{Y}_j, a\widetilde{X}_j + \widetilde{B}) \\ = \tfrac{h}{3(2+k)} \textstyle\sum_{j=1}^n \sum_{i=0}^{M-1} &\Big\{ \big[(y_{j,i+1}^{(1)} - ax_{j,i+1}^{(1)} - b_{i+1}^{(1)})^2 \\ &+ (y_{j,i}^{(1)} - ax_{j,i}^{(1)} - b_i^{(1)})^2 \\ &+ (y_{j,i+1}^{(1)} - ax_{j,i+1}^{(1)} - b_{i+1}^{(1)})(y_{j,i}^{(1)} - ax_{j,i}^{(1)} - b_i^{(1)})\big] \\ &+ \big[(y_{j,i+1}^{(2)} - ax_{j,i+1}^{(2)} - b_{i+1}^{(2)})^2 \\ &+(y_{j,i}^{(2)} - ax_{j,i}^{(2)} - b_i^{(2)})^2 \\ &+ (y_{j,i+1}^{(2)} - ax_{j,i+1}^{(2)} - b_{i+1}^{(2)})(y_{j,i}^{(2)} - ax_{j,i}^{(2)} - b_i^{(2)})\big] \\ &+k \cdot \big[(y_{j,i+1}^{(G)} - ax_{j,i+1}^{(G)} - b_{i+1}^{(G)})^2 \\ &+(y_{j,i}^{(G)} - ax_{j,i}^{(G)} - b_i^{(G)})^2 \\ &+ (y_{j,i+1}^{(G)} - ax_{j,i+1}^{(G)} - b_{i+1}^{(G)})(y_{j,i}^{(G)} - ax_{j,i}^{(G)} - b_i^{(G)})\big]\Big\} \end{aligned} \tag{31}$$

where the quantities $\xi^{(G)}$ have to be expressed in terms of $\xi^{(1)}, \xi^{(2)}$ by means of $\xi^{(G)} = \frac{1}{2}(\xi^{(1)} + \xi^{(2)})$. Thus we recognize that F^+ is the sum of a quadratic and a linear form in its variables.

Since the constraints are inequalities, it is natural to use the active constraint technique. So let us rename our variables by letting

$$\begin{aligned} z_1 &= a & & [A] \\ z_{i+1} &= b^{(1)}_{i-1} \qquad i = 1 \dots M & & [B] \\ z_{M+i+2} &= b^{(2)}_{M-i} & & [C] \end{aligned} \tag{32}$$

The numerical solution of our minimum problem involves mainly the quadratic part of the functional F^+ and the constraints (29). Then it is important to analyze its structure. They appear as depicted in the figures below

*	*	*	*	*	*	*	*	*	*	*	*	*
*	*	*					*	*				
*	*	*	*				*	*	*			
*		*	*	*				*	*	*		
*			*	*	*				*	*	*	
*				*	*	*				*	*	*
*					*	*					*	*
*	*	*					*	*				
*	*	*	*				*	*	*			
*		*	*	*				*	*	*		
*			*	*	*				*	*	*	
*				*	*	*				*	*	*
*					*	*					*	*

Structure of the quadratic form

1	0	0	0	0	0	0	0	0	0	0	0	0
0	−1	1	0	0	0	0	0	0	0	0	0	0
0	0	−1	1	0	0	0	0	0	0	0	0	0
0	0	0	−1	1	0	0	0	0	0	0	0	0
0	0	0	0	−1	1	0	0	0	0	0	0	0
0	0	0	0	0	−1	1	0	0	0	0	0	0
0	0	0	0	0	0	0	−1	1	0	0	0	0
0	0	0	0	0	0	0	0	−1	1	0	0	0
0	0	0	0	0	0	0	0	0	−1	1	0	0
0	0	0	0	0	0	0	0	0	0	−1	1	0
0	0	0	0	0	0	0	0	0	0	0	−1	1

Constraint matrix
(when active)

Really the last matrix is modified after the remotion or the addition of an active constraint (in the sense that changes the number of the rows and columns), but it has always this form with or without first row and first column.

If we indicate by M the matrix of the coefficients of the quadratic form, by A the matrix which represent the constraints (the matrix of figure 5.2) and by $\boldsymbol{q}$ the coefficients of the linear form, then our problem may be reformulated as follows: find the minimum of the functional

$$F^{+}(\boldsymbol{z}) = \boldsymbol{z}^{\top} M \boldsymbol{z} + \boldsymbol{q}^{\top} \boldsymbol{z} \tag{33}$$

subject to the constraint

$$A\boldsymbol{z} \geq \boldsymbol{0} \tag{34}$$

The active constraints algorithm starts by examining an arbitrary admissible point $\boldsymbol{z}_{o}$ and creating a sequence of vectors $\boldsymbol{z}_{k}$ converging to the solution of the problem in a finite number of iterations. It is not an approximation method, but an exact one. It examines a lot of subsystems corresponding to a suitable subset of active constraints till the exact solution is reached. Since the number of all the possible combinations of the active constraints is finite, the steps of the procedure are finite. The details of the algorithm may be found in Bollani (2001), see also Gill *et al.* (1981). Here we give a short guideline. Let A^{*} be the completely active constraint matrix.

Step 1. Set $A_{0} = A^{*}$ and $k = 0$.

Step 2. Find the minimum of the functional $F^{+}(\boldsymbol{z})$ over the set $A_{k}\boldsymbol{z} = 0$; call $\boldsymbol{z}_{k}$ the minimum point.

Step 3. Compute the vector $\boldsymbol{\lambda}_{k} = (\lambda_{k,1} \dots \lambda_{k,n^{k}})$ of the lagrangean parameters of the actual problem.

Step 4. If $\boldsymbol{z}_{k}$ is not admissible ($A_{k}\boldsymbol{z}_{k} < 0$), then go to *Step 6*

else (if $\boldsymbol{z}_{k}$ is admissible).

Step 5. If $\lambda_{k,i} \geq 0 \; \forall i$ then stop
else
⋆ erase all the rows of A_{k} corresponding to $\lambda_{k,i} < 0$
set $k = k + 1$ and go to *Step 2.*

Step 6. ⋆ Insert the constraint (a row) corresponding to the actual search line set $k = k + 1$ and go to it Step 2.

In order to illustrate the proposed algorithm we have to remark that:

- The points marked by $\star$ (it Steps 5 and 6, erase or insert constraints) correspond to the modification of the matrix A_k, thus passing from A_k to A_{k+1}.
- If $\lambda_{k,i} < 0$, then the functional F^+ decreases along the normal to the ith constraint in the direction compatible with the constraint inequality; then the corresponding constraint have to be erased (*Step* 5).
- If $\boldsymbol{z}_k$ is not admissible (*Step* 4), then at least one of the constraints previously erased has to be re-inserted (*Step* 6). In particular we re-insert the ones which correspond to the components of $\boldsymbol{z}_k$ which do not agree with the constraint matrix.
- If $\boldsymbol{z}_k$ is not admissible, then some of the values of the vector $A\boldsymbol{z}_k$ are negative. Thus the constraint that we insert at *Step 6* are the rows corresponding to these values.

APPENDIX A: Coefficients of the polinomial F^+

We recognized, in Paragraph 6, that the discretized functional F^+ to be minimized, given by the formula (31), is a polinomial of degree 2 in the variables a, $b_i^{(1)}$, $b_i^{(2)}$. Then we will indicate by

A_{00} the coefficient of a^2,
A_{0i}^1 the coefficient of $ab_i^{(1)}$
A_{0i}^2 the coefficient of $ab_i^{(2)}$
B_{il}^1 the coefficient of $b_i^{(1)}b_l^{(1)}$
B_{il}^2 the coefficient of $b_i^{(2)}b_l^{(2)}$
B_{il}^{12} the coefficient of $b_i^{(1)}b_l^{(2)}$
A^* the coefficient of a
B_i^{*1} the coefficient of $b_i^{(1)}$
B_i^{*2} the coefficient of $b_i^{(2)}$
C the constant term of the linear form.

We can obtain the expressions of these coefficients by developing explicitly the expression (31). As final result we have (see Bollani, 2001):

A.1 - Coefficients of the quadratic form

$$\begin{aligned}
A_{00} &= \textstyle\sum_{j=1}^{n}\sum_{i=0}^{M-1}\big[(x_{j,i}^{(1)2}+x_{j,i+1}^{(1)2}+x_{j,i}^{(1)}x_{j,i+1}^{(1)}) \\
&\quad +(x_{j,i}^{(2)2}+x_{j,i+1}^{(2)2}+x_{j,i}^{(2)}x_{j,i+1}^{(2)}) \\
&\quad +k(x_{j,i}^{(G)2}+x_{j,i+1}^{(G)2}+x_{j,i}^{(G)}x_{j,i+1}^{(G1)})\big]
\end{aligned}$$

$$\begin{aligned}
A_{00}^{1} &= \textstyle\sum_{j=1}^{n}[2x_{j,0}^{(1)}+x_{j,1}^{(1)}] \\
&\quad +\tfrac{k}{2}\textstyle\sum_{j=1}^{n}[2x_{j,0}^{(G)}+x_{j,1}^{(G)}]
\end{aligned}$$

$$\begin{aligned}
A_{0M}^{1} &= \textstyle\sum_{j=1}^{n}[2x_{j,M}^{(1)}+x_{j,M-1}^{(1)}] \\
&\quad +\tfrac{k}{2}\textstyle\sum_{j=1}^{n}[2x_{j,M}^{(G)}+x_{j,M-1}^{(G)}]
\end{aligned}$$

$$\begin{aligned}
A_{0i}^{1} &= \textstyle\sum_{j=1}^{n}[4x_{j,i}^{(1)}+x_{j,i+1}^{(1)}+x_{j,i-1}^{(1)}] \\
&\quad +\tfrac{k}{2}\textstyle\sum_{j=1}^{n}[4x_{j,i}^{(G)}+x_{j,i+1}^{(G)}+x_{j,i-1}^{(G)}]
\end{aligned}$$

$$\begin{aligned}
A_{00}^{2} &= \textstyle\sum_{j=1}^{n}[2x_{j,0}^{(2)}+x_{j,1}^{(2)}] \\
&\quad +\tfrac{k}{2}\textstyle\sum_{j=1}^{n}[2x_{j,0}^{(G)}+x_{j,1}^{(G)}]
\end{aligned}$$

$$\begin{aligned}
A_{0M}^{2} &= \textstyle\sum_{j=1}^{n}[2x_{j,M}^{(2)}+x_{j,M-1}^{(2)}] \\
&\quad +\tfrac{k}{2}\textstyle\sum_{j=1}^{n}[2x_{j,M}^{(G)}+x_{j,M-1}^{(G)}]
\end{aligned}$$

$$\begin{aligned}
A_{0i}^{2} &= \textstyle\sum_{j=1}^{n}[4x_{j,i}^{(2)}+x_{j,i+1}^{(2)}+x_{j,i-1}^{(2)}] \\
&\quad +\tfrac{k}{2}\textstyle\sum_{j=1}^{n}[4x_{j,i}^{(G)}+x_{j,i+1}^{(G)}+x_{j,i-1}^{(G)}]
\end{aligned}$$

$$B_{00}^{1} = B_{MM}^{1} = B_{00}^{2} = B_{MM}^{2} = n+\tfrac{k}{4}n$$

$$B_{ii}^{1} = B_{ii}^{2} = 2n+\tfrac{k}{2}n \quad i\neq 0,M$$

$$B_{i,i+1}^{1} = B_{i,i+1}^{2} = n+\tfrac{k}{4}n$$

$$B_{ij}^{1} = B_{ij}^{2} = 0 \qquad j\neq i+1$$

$$B_{00}^{12} = B_{MM}^{12} = \tfrac{k}{2}n$$

$$B_{ii}^{12} = kn \quad i\neq 0,M$$

$$B_{i,i+1}^{12} = B_{i+1,i}^{12} = \tfrac{k}{4}n$$

$$B_{ij}^{12} = 0 \quad j\neq i-1,i+1$$

A.2 - COEFFICIENTS OF THE LINEAR FORM

$$\begin{aligned}
A^* = & -\sum_{j=1}^{n}\sum_{i=1}^{M-1}(4x_{j,i}^{(1)}y_{j,i}^{(1)} + x_{j,i}^{(1)}y_{j,i+1}^{(1)} + x_{j,i}^{(1)}y_{j,i-1}^{(1)}) \\
& - \sum_{j=1}^{n}(2x_{j,0}^{(1)}y_{j,0}^{(1)} + x_{j,0}^{(1)}y_{j,0}^{(1)}) - \sum_{j=1}^{n}(2x_{j,M}^{(1)}y_{j,M}^{(1)} + x_{j,M}^{(1)}y_{j,M}^{(1)}) \\
& -\sum_{j=1}^{n}\sum_{i=1}^{M-1}(4x_{j,i}^{(2)}y_{j,i}^{(2)} + x_{j,i}^{(2)}y_{j,i+1}^{(2)} + x_{j,i}^{(2)}y_{j,i-1}^{(2)}) \\
& - \sum_{j=1}^{n}(2x_{j,0}^{(2)}y_{j,0}^{(2)} + x_{j,0}^{(2)}y_{j,0}^{(2)}) - \sum_{j=1}^{n}(2x_{j,M}^{(2)}y_{j,M}^{(2)} + x_{j,M}^{(2)}y_{j,M}^{(2)}) \\
& -k\sum_{j=1}^{n}\sum_{i=1}^{M-1}(4x_{j,i}^{(G)}y_{j,i}^{(G)} + x_{j,i}^{(G)}y_{j,i+1}^{(G)} + x_{j,i}^{(G)}y_{j,i-1}^{(G)}) \\
& - k\sum_{j=1}^{n}(2x_{j,0}^{(G)}y_{j,0}^{(G)} + x_{j,0}^{(G)}y_{j,0}^{(G)}) - k\sum_{j=1}^{n}(2x_{j,M}^{(G)}y_{j,M}^{(G)} + x_{j,M}^{(G)}y_{j,M}^{(G)})
\end{aligned}$$

$$B_0^{*1} = -\sum_{j=1}^{n}(2y_{j,0}^{(1)} + y_{j,1}^{(1)}) - \tfrac{k}{2}\sum_{j=1}^{n}(2y_{j,0}^{(G)} + y_{j,1}^{(G)})$$

$$B_M^{*1} = -\sum_{j=1}^{n}(2y_{j,M}^{(1)} + y_{j,M-1}^{(1)}) - \tfrac{k}{2}\sum_{j=1}^{n}(2y_{j,M}^{(G)} + y_{j,M-1}^{(G)})$$

$$B_i^{*1} = -\sum_{j=1}^{n}(2y_{j,i}^{(1)} + y_{j,i+1}^{(1)} + y_{j,i-1}^{(1)} - \tfrac{k}{2}\sum_{j=1}^{n}(2y_{j,i}^{(G)} + y_{j,i+1}^{(G)} + y_{j,i-1}^{(G)})$$

$$B_0^{*2} = -\sum_{j=1}^{n}(2y_{j,0}^{(2)} + y_{j,1}^{(2)}) - \tfrac{k}{2}\sum_{j=1}^{n}(2y_{j,0}^{(G)} + y_{j,1}^{(G)})$$

$$B_M^{*1} = -\sum_{j=1}^{n}(2y_{j,M}^{(2)} + y_{j,M-1}^{(2)}) - \tfrac{k}{2}\sum_{j=1}^{n}(2y_{j,M}^{(G)} + y_{j,M-1}^{(G)})$$

$$B_i^{*1} = -\sum_{j=1}^{n}(2y_{j,i}^{(2)} + y_{j,i+1}^{(2)} + y_{j,i-1}^{(2)} - \tfrac{k}{2}\sum_{j=1}^{n}(2y_{j,i}^{(G)} + y_{j,i+1}^{(G)} + y_{j,i-1}^{(G)})$$

A.3 - CONSTANT TERM

$$\begin{aligned}
C = & \sum_{j=1}^{n}\sum_{i=0}^{M-1}(y_{j,i+1}^{(1)2} + y_{j,i}^{(1)2} + y_{j,i+1}^{(1)}y_{j,i}^{(1)}) \\
& + \sum_{j=1}^{n}\sum_{i=0}^{M-1}(y_{j,i+1}^{(2)2} + y_{j,i}^{(2)2} + y_{j,i+1}^{(2)}y_{j,i}^{(2)}) \\
& + k\sum_{j=1}^{n}\sum_{i=0}^{M-1}(y_{j,i+1}^{(G)2} + y_{j,i}^{(G)2} + y_{j,i+1}^{(G)}y_{j,i}^{(G)})
\end{aligned}$$

APPENDIX B: The algorithm detailed

As described in the shortened algorithm of paragraph 6, our method consists of minimizing at each step k the functional

$$F^+(\boldsymbol{z}) = \boldsymbol{z}^\top M\boldsymbol{z} + \boldsymbol{q}^\top\boldsymbol{z} \tag{35}$$

subject to the constraint

$$A_k\boldsymbol{z} = \boldsymbol{0} \tag{36}$$

The classical gradient method leaves to the linear system

$$\begin{aligned} M\boldsymbol{z} + \boldsymbol{q} &= -A_k \boldsymbol{\lambda}_k \\ A_k^T \boldsymbol{z} &= 0 \end{aligned}$$

that is

$$\begin{pmatrix} M & A_k \\ & \\ A_k^T & 0 \end{pmatrix} \begin{pmatrix} \boldsymbol{y}_k \\ \\ \boldsymbol{\lambda}_k \end{pmatrix} = \begin{pmatrix} -\boldsymbol{q} \\ \\ \boldsymbol{0} \end{pmatrix} \tag{36}$$

It is easy to recognize that the remotion or the insertion of some constraint equations does not change the fundamental structure of the constraint matrix. In fact, the non null part of A_k may be decomposed in blocks of dimension either $r \times (r-1)$ or $r \times r$ respectively of the form

−1	1	0	0	0	0
0	−1	1	0	0	0
0	0	−1	1	0	0
0	0	0	−1	1	0
0	0	0	0	−1	1

Fig. A

1	0	0	0	0	0
−1	1	0	0	0	0
0	−1	1	0	0	0
0	0	−1	1	0	0
0	0	0	−1	1	0
0	0	0	0	−1	1

Fig. B

Moreover each of these blocks is placed in such a way that all the elements of the matrix A_k situated above, below, on the right and on the left of the block are zero.

The structure of both kinds of matrices shown in the Figures A and B allows us to easily invert the matrix appearing in (37) (its inverse appears as a matrix composed by "disjoint" upper triangular matrices). Consequently the minimization problem (35)(36) can be solved without great difficulties.

We begin by rearranging the variables y_i in such a way that the constraint matrix A would be of the form

$$A = [A_1 \; N]$$

where A_1 is a matrix containing only squared blocks of the type

$$\begin{array}{|ccccc|}
\hline
-1 & 1 & 0 & 0 & 0 \\
0 & -1 & 1 & 0 & 0 \\
0 & 0 & -1 & 1 & 0 \\
0 & 0 & 0 & -1 & 1 \\
0 & 0 & 0 & 0 & -1 \\
\hline
\end{array}$$

and the columns of the matrix N contain at most one "1" in a suitable position. Then the equation (37) assumes the form

$$\begin{pmatrix} M_{11} & A_{k1} & M_{12} \\ A_{k1}^T & 0 & N \\ M21 & N^T & M_{22} \end{pmatrix} \begin{pmatrix} \boldsymbol{z}_{k1} \\ \boldsymbol{\lambda}_k \\ \boldsymbol{z}_{k2} \end{pmatrix} = \begin{pmatrix} -\boldsymbol{c}_1 \\ \boldsymbol{0} \\ \boldsymbol{c}_2 \end{pmatrix} \tag{37}$$

where $[\boldsymbol{z}_{k1}, \boldsymbol{z}_{k2}]$ and $[c_2, c_2]$ are suitable permutations of $\boldsymbol{y}_k$ and $-\boldsymbol{q}_k$.

In order to simplify the numerical solution we can observe that the matrix of the system (38) may be written as the product of three simpler matrices in the following form

$$\begin{pmatrix} I_1 & 0 \\ X_k^T & I_1 \end{pmatrix} \begin{pmatrix} K & 0 \\ 0 & S \end{pmatrix} \begin{pmatrix} I_1 & X \\ 0 & I_2 \end{pmatrix}$$

where

$$K = \begin{pmatrix} M_{11} & A_1 \\ A_1^T & 0 \end{pmatrix}$$

$$X^T = \begin{pmatrix} M_{21} & N^T \end{pmatrix} K^{-1}$$

$$S = M_{22} - X^T K^{-1} \begin{pmatrix} M_{12} \\ N \end{pmatrix}$$

The dimensions of the matrices K, X, S change at each step of the procedure according to the number of inserted or erased constraints.

So our problem reduces to solve successively the following three linear system

$$\begin{pmatrix} I & 0 & 0 \\ 0 & I & 0 \\ X_1^T & X_2^T & I \end{pmatrix} \begin{pmatrix} \boldsymbol{v}_1 \\ \boldsymbol{v}_2 \\ \boldsymbol{v}_3 \end{pmatrix} = \begin{pmatrix} -\boldsymbol{c}_1 \\ \boldsymbol{0} \\ \boldsymbol{c}_1 \end{pmatrix} \tag{38}$$

$$\begin{pmatrix} M_{11} & A_1 & 0 \\ A_1^T & 0 & 0 \\ 0 & 0 & S \end{pmatrix} \begin{pmatrix} u_1 \\ u_2 \\ u_3 \end{pmatrix} = \begin{pmatrix} -v_1 \\ v_2 \\ v_3 \end{pmatrix} \tag{39}$$

$$\begin{pmatrix} I & 0 & X_1 \\ 0 & I & X_2 \\ 0 & 0 & I \end{pmatrix} \begin{pmatrix} z_{k1} \\ \lambda_k \\ z_{k2} \end{pmatrix} = \begin{pmatrix} -u_1 \\ u_2 \\ u_3 \end{pmatrix} \tag{40}$$

where X_1^T and X_2^T are the two parts of the matrix $X^T = [M_{21}\ N^T]K^{-1}$ and X_1, X_2 the corresponding two parts of its transposed. Each of the three systems (39,40,41) can be solved in a much shorter computational time than the one needed for (38). This is why we use this factorization instead of solving the system (38) directly.

Acknowledgements

The research in this paper has been partially supported by DGESIC Grant No. DGE-99-PB98-1534. This financial support is deeply acknowledged.

References

1. Baiocchi, C. and Capelo, A. (1984). *Variational and Quasivariational inequalities*, John Wiley and Sons, New York.
2. Bertoluzza, C., Corral, N. and Salas, A. (1991). Fuzzy Linear Regression: Existence of Solution for a generalized Least Square Method, *Proc. IFSA'91*, Bruxelles.
3. Bertoluzza, C., Corral, N. and Salas, A. (1995). On a new class of distances between fuzzy sets, *Mathware & Soft Computing* **2**, 71-84.
4. Bollani, C. (2001). *Determinazione approssimata dei coefficienti di regressione in ambiente sfumato* (temptative title). Tesi di Laurea, Universitá di Pavia (to be discussed).
5. Ciampa, S. (1991-92). *Problemi connessi con la definizione di retta sfumata*, Tesi di Laurea, Universitá di Pavia.
6. Ciliberti, M.G. (1990-91). *Una possibile definizione di distanza per punti sfumati*, Tesi di Laurea, Universitá di Pavia.
7. Diamond, P. (1991). Least Square Method in Fuzzy Data Analysis, *Proc. IFSA'91*, Bruxelles.
8. Gill, P.E., Murray, W. and Wright, M.H. (1981). *Practical Optimization*, Academic Press, London.
9. Kaufmann, A. and Gupta, M.M. (1985). *Introduction to Fuzzy Arithmetics; Theory and Applications*, Van Nostrand Reinhold, New York.
10. Mainardi, S. (1991-92). *Regressione lineare sfumata: un'analisi dei parametri che influenzano i coefficienti di interpolazione*, Tesi di Laurea, Universitá di Pavia.

11. Nguyen, H.T. (1978). A Note on the Extension Principle for Fuzzy Sets, *J. Math. Anal. Appl.* **64**, 369-380.
12. Salas, A. (1991). *Regresión lineal con observaciones difusas*, PhD Thesis, Universidad de Oviedo.
13. Zadeh, L.A. (1965). Fuzzy sets, *Information and Control* **8**, 338-353.

Linear regression with random fuzzy observations

Wolfgang Näther and Ralf Körner

Faculty of Mathematics and Computer Sciences, Freiberg University of Mining and Technology, 09596 Freiberg, Germany

Abstract. In this paper, on the basis of the concept of fuzzy random variable a kind of (though not strict) linear estimation theory is developed. Modified linear estimators are presented and discussed, and the least squares approximation principle is used for constructing estimators.

1 Introduction, preliminaries and overview

Let us denote a classical crisp linear regression model by a random variable y which depends on the regressor $x \in \mathbb{R}^k$ by

$$\mathbb{E}y = f_1(x)\beta_1 + \ldots + f_m(x)\beta_m = f(x)^t\beta \tag{1}$$

where $f : \mathbb{R}^k \to \mathbb{R}^m$ is a known setup-function and $\beta \in \mathbb{R}^m$ is an unknown regression parameter. Given observations $y_1, \ldots, y_n$ of y at the design points $x_1, \ldots, x_n \in \mathbb{R}^k$, the parameter β is to be estimated, e.g. by the classical least-squares estimator

$$\check{\beta} = (F^tF)^{-1}F^t\boldsymbol{y} \tag{2}$$

where $\boldsymbol{y} = (y_1, \ldots, y_n)^t$ and $F = (f(x_1), \ldots, f(x_n))^t$ is supposed to have full rank.

The problem now is that only fuzzy observations $Y_1, \ldots, Y_n$ are available, for example the clouding y for given atmospheric pressure x is reported by linguistic expressions like: Cloudless, Clear, Fair, Cloudy, Overcast. Thus (1) has to be generalized by

$$\mathbb{E}Y = f(x)^t\boldsymbol{B}, \tag{3}$$

where now Y is a random fuzzy variable and $\boldsymbol{B}$ is a fuzzy parameter vector. The question is: How to estimate $\boldsymbol{B}$?

In what follows, a fuzzy subset A of $\mathbb{R}^n$ is identified with its membership function $m_A(z)$. For any $\alpha \in (0,1]$ the crisp set $A_\alpha = \{z \in \mathbb{R}^n : m_A(z) \geq \alpha\}$ is called the α-cut of A. A is called convex iff all α-cuts are convex and normal if A_1 is not empty.

We will model fuzzy observations mainly by fuzzy numbers. A fuzzy number A is a special kind of a normal and convex fuzzy set on $\mathbb{R}^1$ which is characterized by the fact that A_1 is a one-point-set, say $A_1 = \{\mu\}$. Often we will use the following parametric class of fuzzy numbers, the so called LR-fuzzy numbers:

$$m_A(z) = \begin{cases} L(\frac{\mu - z}{l}) & \text{if } z \leq \mu \\ R(\frac{z-\mu}{r}) & \text{if } z > \mu \end{cases} \tag{4}$$

Here $L : \mathbb{R}^+ \to [0,1]$ and $R : \mathbb{R}^+ \to [0,1]$ are fixed left-continuous non-increasing functions with $L(0) = R(0) = 1$. L and R are called left and right shape functions, μ the modal point of A and $l, r > 0$ the left/right spread of A. We abbreviate an LR-fuzzy number by

$$A = (\mu, l, r)_{LR}. \tag{5}$$

Sometimes only special cases of (5) are considered, e.g. shape-symmetric fuzzy numbers with $L = R$, or totally symmetric fuzzy numbers with $L = R$ and $l = r =: \Delta$, i.e. $A = (\mu, \Delta)_L$. If $L(x) = R(x) = [1-x]^+$ than A is called a triangular fuzzy number.

Basically in fuzzy set theory is *Zadeh's extension principle*

$$m_{g(A_1,\ldots,A_r)}(t) = \sup_{g(z_1,\ldots,z_r)=t} \min\{m_{A_1}(z_1), \ldots, m_{A_r}(z_r)\}. \tag{6}$$

This provides a general method how classical functions $g(z_1, \ldots, z_r)$ on $\mathbb{R}^r$ can be extended for fuzzy input $A_1, \ldots, A_r$. In particular, addition $\oplus$ and scalar multiplication $\odot$ of fuzzy sets are the extensions of proper addition $g(z_1, z_2) = z_1 + z_2$ and scalar multiplication $g(z) = \lambda z$ of real numbers. For notational simplicity, we will suppress the sign $\odot$ and only write λA.

For linear regression and linear estimation theory, we will need these operations. The advantage of LR-fuzzy numbers is that $\oplus$ and $\cdot$ can be expressed by simple operations w.r.t. the parameters μ, l and r:

$$(\mu_1, l_1, r_1)_{LR} \oplus (\mu_2, l_2, r_2)_{LR} = (\mu_1 + \mu_2, l_1 + l_2, r_1 + r_2)_{LR} \tag{7}$$

$$\lambda(\mu, l, r)_{LR} = \begin{cases} (\lambda\mu, \lambda l, \lambda r)_{LR} & \text{if } \lambda > 0 \\ (\lambda\mu, -\lambda r, -\lambda l)_{RL} & \text{if } \lambda < 0 \\ 1_{\{0\}} & \text{if } \lambda = 0 \end{cases} \tag{8}$$

Here 1_A is the indicator function of a set A.

Because of (8), LR-fuzzy numbers are not closed under $\cdot$ and $\oplus$. They only build a convex cone w.r.t $\cdot$ and $\oplus$. Shape-symmetric fuzzy numbers, however, are closed under $\cdot$ and $\oplus$. Especially for totally symmetric fuzzy numbers we have

$$\lambda_1 (\mu_1, \Delta_1)_L \oplus \lambda_2 (\mu_2, \Delta_2)_L = (\lambda_1\mu_1 + \lambda_2\mu_2, |\lambda_1|\Delta_1 + |\lambda_2|\Delta_2)_L. \tag{9}$$

Now the paper is organized as follows:

In Section 2, we sketch some data-analytic approaches from literature, i.e. approaches where it is not necessary to have a stochastic model like (3). At first, we mention the approach by Tanaka and his school where the regression problem is discussed from a possibilistic point of view (see, e.g. Tanaka, 1987). Second, we briefly discuss an approach by Bandemer where the fuzzy observations are used to evaluate possible functional relationships (see e.g. Bandemer, 1985). A third approach consists in a straightforward application of Zadeh's extension principle to well justified classical crisp estimates. e.g. to the least-squares estimator (2).

In Section 3, we use the concept of fuzzy random variables and try to develop some kind of linear estimation theory. Unfortunately, since fuzzy sets fail to constitute a linear speace w.r.t. $\cdot$ and $\oplus$ we cannot establish a strict linear theory. Moreover, strict linear unbiased estimators in general do not exist. Nevertheless, modified linear estimators often lead to better estimates than by use of the extension principle. In the case that the observations are LR fuzzy numbers relatively detailed results are presented.

In Section 4, the least squares approximation principle is used for construction of estimators, allowing general LR-fuzzy-number-data. Least squares approaches for triangular fuzzy data go back to Diamond (1988, 1992), applications can be found in Kacprzyk and Fedrizzi (1992) by Bardossy *et al.* (1992). Our approach in Section 4 is more general and not restricted to triangular fuzzy data. Also a new approximation approach is discussed which uses the so called Hukuhara difference between fuzzy sets.

Sections 3 and 4 are strongly related to Körner and Näther (1998) and Körner (1997a).

2 Some data-analytic approaches to the regression problem

2.1 The Tanaka-approach

For brief description of this approach, we do not choose the most general setting. We follow the starting paper by Tanaka *et al.* (1982).

The output-data $Y_1, \ldots, Y_n$ taken at n inputs $x_1, \ldots, x_n \in \mathbb{R}^k$ are assumed to be symmetric fuzzy numbers, i.e.

$$Y_i = (y_i, \Delta_i)_L \, ; \; i = 1, \ldots, n \tag{10}$$

and the objective is to fit these data to a fuzzy functional relationship. The core of this relationship is assumed to have the form (see (1))

$$\beta_1 x_1 + \cdots + \beta_k x_k = \beta^t x \tag{11}$$

and fuzziness is introduced by assuming the parameters β_j to be the core of symmetrical fuzzy numbers

$$B_j = (\beta_j, \delta_j)_L \, ; \; j = 1, \cdots, k \tag{12}$$

with the same reference function L as in (10). Hence, using (9) the fuzzy functional relationship has the form

$$F(x, \boldsymbol{B}) = \left(\beta^t x,\ \sum_{j=1}^{k} \delta_j |x_j| \right)_L . \tag{13}$$

Now, such parameter $\hat{\boldsymbol{B}} = (\hat{B}_1, \dots, \hat{B}_k)^t$ is to be found that $F(x, \hat{\boldsymbol{B}})$ appears as the best covering function of the data $Y_1, \dots, Y_n$ in the sense that

i) all Y_i are covered bei $F(x, \hat{\boldsymbol{B}})$ at least with a given degree $\alpha^* \leq 1$, i.e. for $\alpha \leq \alpha^*$ the α-cuts of Y_i are contained in the α-cuts of $F(x, \hat{\boldsymbol{B}})$ and
ii) the spread of $F(x, \hat{\boldsymbol{B}})$ is as small as possible. More specific: The sum of the spreads in (13) taken for the inputs x_i; $i = 1, \dots, n$; is to be minimized.

This leads to the optimization problem

$$\min_{\beta_j, \delta_j} \sum_{i=1}^{n} \sum_{j=1}^{k} \delta_j |x_{ij}| \tag{14}$$

with the side conditions

$$\begin{aligned} y_i &\leq L^{-1}(\alpha^*) \sum_j \delta_j |x_{ij}| - L^{-1}(\alpha^*) \Delta_i + \beta^t x_i \\ -y_i &\leq L^{-1}(\alpha^*) \sum_j \delta_j |x_{ij}| - L^{-1}(\alpha^*) \Delta_i - \beta^t x_i \end{aligned} \quad i = 1, \dots, n. \tag{15}$$

Note that (14) with (15) is a standard problem in linear programming. For generalizations (e.g. to nonlinear relationships or to the fuzzy-input-fuzzy output-case) and for a possiblistic interpretation see Bardossy (1990), Sakawa and Yano (1992), Tanaka (1987), Tanaka and Watada (1988).

2.2 The Bandemer-approach

Let be given fuzzy data $Y_1, \dots, Y_n$ which are fuzzy sets on $\mathbb{R}^{k+1}, k \geq 1$, i.e. the Y_i are not only fuzzy in y-direction but also in x-direction. Find a function $f(x, \beta^*) : \mathbb{R}^k \to \mathbb{R}^1$ from a given setup $\{f(x, \beta) : \beta \in \mathbb{R}^m\}$ which describes the data in a best possible way, at least for $x \in Q \subset \mathbb{R}^k$. Note that the setup functions are not assumed to be linear in β.

Following Bandemer (1985) and Bandemer and Näther (1992), the procedure consists in the following steps:

i) Aggregate $Y_1, \dots, Y_n$ to one fuzzy set on $\mathbb{R}^{k+1}$, e.g. by the union of the Y_i, i.e. for $x \in \mathbb{R}^k,\ y \in \mathbb{R}^1$

$$Y = \bigcup_i Y_i; \qquad \text{with} \qquad m_Y(x, y) = \max_i \{m_{Y_i}(x, y)\} . \tag{16}$$

ii) Evaluate the function $f(x,\beta)$ by the degree to which the graph $\{(x, f(x,\beta)\}_{x\in Q}$ hits the aggregated datum Y, e.g. by

$$m(\beta) := \int_Q m_Y(x, f(x,\beta))\; w(x)dx; \qquad \int_Q w(x)dx = 1. \tag{17}$$

$m(\beta)$ can be interpreted as membership function of a fuzzy parameter $\boldsymbol{B}$: $m(\beta) =: m_{\boldsymbol{B}}(\beta)$.

iii) β^* with e.g.

$$m_{\boldsymbol{B}}(\beta^*) = \max_{\beta} m_{\boldsymbol{B}}(\beta) \tag{18}$$

identifies the function $f(x,\beta^*)$ which "hits" the data in a best possible way.

Clearly, one can choose other aggregation operations in (16) (e.g. some kind of averages), other evaluation operations in (17) (e.g. $\sup_{x\in Q} m_Y(x, f(x,\beta))$) and other defuzzifications in (18) (e.g. the well known centroid method). In all cases, the fuzziness of the data is transfered into fuzziness of the parameter β.

This procedure works well if the Y_i more or less cover the interesting region in $\mathbb{R}^{k+1}$ - this happens if either n or the fuzziness of the Y_i is large enough. It works bad for small n or relatively crisp Y_i since then it only seldom happens that $f(x,\beta)$ hits Y and since in the sketched approach no idea of approximation is included.

2.3 Extended classical estimators

Let $T(y_1,\ldots,y_n)$ be a classical sample function, e.g. a point estimator of a certain parameter. If only fuzzy date $Y_1,\ldots,Y_n$ are available, via extension principle (6) we find the fuzzified version of T by

$$m_{T(Y_1,\ldots,Y_n)}(t) = \sup_{T(y_1,\ldots,y_n)=t} \min\{m_{Y_1}(y_1),\ldots,m_{Y_n}(y_n)\} \tag{19}$$

which will be called extended sample function, e.g. extended estimator. For linear sample functions like $\check{\beta}$ from (19) and shape-symmetric LR-fuzzy-number-data $Y_1,\ldots,Y_n$, say with shape function L, the extended estimators can be computed easily by application of (7) and (8), with the result that the components of the fuzzified $\check{\beta}$, say $\check{\boldsymbol{B}} = (F^tF)^{-1}F^t\boldsymbol{Y}$, appear as fuzzy numbers with the same L. For more details see Körner and Näther (1998).

However, there are some critical comments:

i) Application of the extension principle (19) leads to ambiguous results. Consider the following simple example: In real analysis clearly holds the identity $2y = 3y - y$. Now assume $Y = (\mu, \Delta)_L$. Then, following (8), $2y$ extends to $2Y = (2\mu, 2\Delta)_L$, but according to (9) $3y - y$ extends to $3Y \oplus (-1)Y = (2\mu, 4\Delta)_L$ which is different from $2Y$.

The reason is that fuzzy sets with $\oplus$ do not constitute a linear space, i.e. there is no inverse element $-A$ with $A \oplus (-A) = 0$.

ii) This ambiguity has consequences also for regression problems. Consider classical simple linear regression $Ey = ax + b$ with the least squares estimators

$$\hat{a} = \frac{\sum (x_i - \overline{x})(y_i - \overline{y})}{\sum (x_i - \overline{x})^2}; \qquad \hat{b} = \overline{y} - \frac{\sum (x_i - \overline{x})(y_i - \overline{y})}{\sum (x_i - \overline{x})^2}\overline{x}. \tag{20}$$

If we apply (19) to $\hat{b}$ from (20) we find the result $\hat{B}_1$. We find another result if we first extend $\hat{a}$ to $\hat{A}$ and then find $\hat{B}_2 = \overline{Y} \oplus (-1)\hat{A}\overline{x}$. The modal values of $\hat{B}_1$ and $\hat{B}_2$ coincide, but $\hat{B}_2$ has a larger spread, see also the illustrating Example 1.

Example 1. There are given eight triangular fuzzy data:

$$\begin{array}{ll}
x_1 = 6.15,\ Y_1 = (2.5, 0.07, 0.50)_{LL} & x_2 = 9.00,\ Y_2 = (2.0, 0.12, 0.45)_{LL} \\
x_3 = 12.00,\ Y_3 = (3.0, 0.20, 0.50)_{LL} & x_4 = 12.15,\ Y_4 = (2.5, 0.15, 0.43)_{LL} \\
x_5 = 15.00,\ Y_5 = (3.0, 0.23, 0.36)_{LL} & x_6 = 15.19,\ Y_6 = (3.5, 0.35, 0.23)_{LL} \\
x_7 = 17.55,\ Y_7 = (3.5, 0.32, 0.17)_{LL} & x_8 = 20.48,\ Y_8 = (4.0, 0.42, 0.21)_{LL}
\end{array}$$

Then we calculate $\hat{A} = (0.1259, 0.0819, 0.0314)_{LL}$,

$$\hat{B}_1 = (1.3079, 0.3685, 1.1705)_{LL}, \qquad \hat{B}_2 = (1.3079, 0.6547, 1.4567)_{LL},$$

which leads to $\hat{Y}_1 = \hat{A}x \oplus \hat{B}_1$ and $\hat{Y}_2 = \hat{A}x \oplus \hat{B}_2$.

In Figure 1 the triangles represent the data, the solid line the 1-level-cuts of $\hat{Y}_1$ and $\hat{Y}_2$, the dashed lines the 0-level-cut of $\hat{Y}_1$ and the dashed-dotted lines the 0-level-cut of $\hat{Y}_2$.

As a further critical comment, Figure 1 shows that estimations based on the extension principle seem to be too pessimistic, i.e. the spread of the estimated straight line seems to be too large in comparison with the relatively regular fuzzy data.

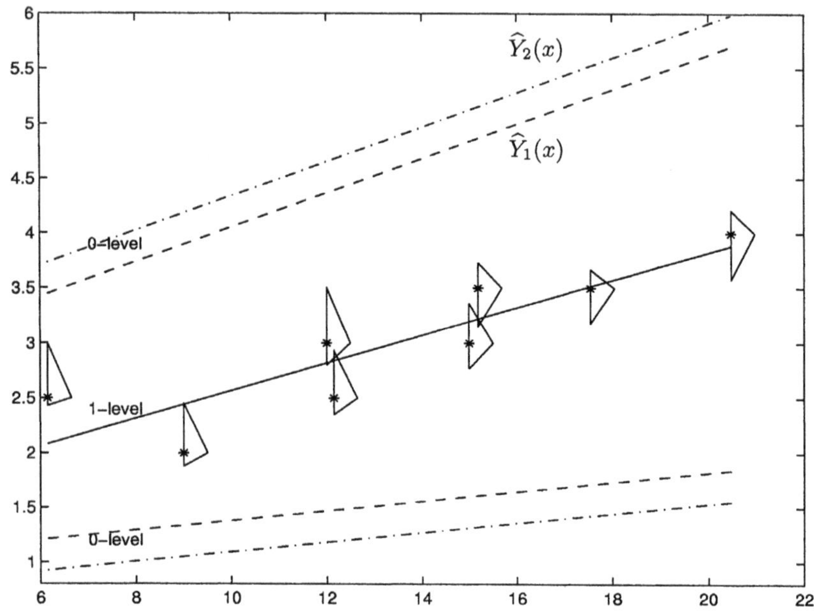

Fig. 1. Linear regression by the extension principle

3 Best linear unbiased estimation (BLUE)

3.1 Random fuzzy numbers and their expectation and variance

In this section we emphasize the stochastic background of the data, i.e. we assume that the data $Y_1, \ldots, Y_n$ are realizations of a random fuzzy variable Y, especially of a random fuzzy number. Since we are interested in BLUE procedures we have to work with a convenient expectation and variance of Y and we refer to Körner and Näther in Part 2 in this book.

Especially for a random LR-fuzzy number $Y = (\mu, l, r)_{LR}$ with random μ, l, r and fixed L, R we have

$$\begin{aligned} \mathbb{E}(\mu, l, r)_{LR} &= (\mathbb{E}\mu, \mathbb{E}l, \mathbb{E}r)_{LR} \\ \operatorname{Var}(\mu, l, r)_{LR} &= \operatorname{Var}\mu + L_2 \operatorname{Var} l + R_2 \operatorname{Var} r \\ &\quad -2L_1 \operatorname{Cov}(m, l) + 2R_1 \operatorname{Cov}(m, r) \end{aligned} \tag{21}$$

with

$$L_1 = \tfrac{1}{2} \int_0^1 L^{-1}(\alpha)\, d\alpha, \qquad R_1 = \tfrac{1}{2} \int_0^1 R^{-1}(\alpha)\, d\alpha$$

$$L_2 = \tfrac{1}{2} \int_0^1 (L^{-1}(\alpha))^2 d\alpha, \qquad R_2 = \tfrac{1}{2} \int_0^1 (R^{-1}(\alpha))^2 d\alpha.$$

For a random symmetric fuzzy number $Y = (\mu, \Delta)_L$ this reduces to

$$\begin{aligned} \mathbb{E}\,(\mu, \Delta)_L &= (\mathbb{E}\mu, \mathbb{E}\Delta)_L \\ \operatorname{Var}(\mu, \Delta)_L &= \operatorname{Var}\mu + 2L_2 \operatorname{Var}\Delta. \end{aligned} \tag{22}$$

Note that for two independent random fuzzy numbers it holds (see Körner, 1997b):

$$\operatorname{Var}(\lambda_1 Y_1 \oplus \lambda_2 Y_2) = \lambda_1^2 \operatorname{Var} Y_1 + \lambda_2^2 \operatorname{Var} Y_2. \tag{23}$$

3.2 BLUE

Now the starting point is the *fuzzified regression model* (3), i.e. more exactly

$$\mathbb{E}Y(x) = f(x)^t \boldsymbol{B} = f_1(x)B_1 \oplus \cdots \oplus f_m(x)B_m. \tag{24}$$

Let be given n fuzzy number data, say

$$Y_i = (y_i, \Delta_i)_L\,; \qquad i = 1, \ldots, n. \tag{25}$$

The aim is to estimate the $B_j; j = 1, \ldots, m$; by a linear estimation

$$\hat{B}_j = \lambda_{1j} Y_1 \oplus \cdots \oplus \lambda_{nj} Y_n =: \lambda_j^t \boldsymbol{Y} \tag{26}$$

which is unbiased in the sense that

$$\mathbb{E}\hat{B}_j = B_j. \tag{27}$$

Taking into account (9), for unbiasedness it is necessary that B_j is modeled as symmetric fuzzy number with the same shape function L, say

$$B_j = (\beta_j, \delta_j)_L\,; \qquad j = 1, \ldots, m. \tag{28}$$

Then (24) writes

$$\begin{aligned} \mathbb{E}Y(x) &= (f_1(x)\beta_1 + \cdots + f_m(x)\beta_m,\ |f_1(x)|\delta_1 + \cdots + |f_m(x)|\delta_m)_L \\ &=: \left(f(x)^t\beta,\ |f(x)^t|\delta\right)_L \end{aligned}$$

and, since the Y_i are assumed to be a realization of $Y(x_i)$,

$$\mathbb{E}\hat{B}_j =: \left(\lambda_j^t F\beta,\ |\lambda_j^t||F|\delta\right)_L \tag{29}$$

with $\lambda_j^t = (\lambda_{1j}, \ldots, \lambda_{nj})$, $|\lambda_j^t| = (|\lambda_{1j}|, \ldots, |\lambda_{nj}|)$; $j = 1, \ldots, m$; $F = (f(x_1), \ldots, f(x_n))^t, |F| = (|f(x_1)|, \ldots, |f(x_n)|)^t$. More condensed, with $\Lambda := (\lambda_1, \ldots, \lambda_m)^t$ and $\boldsymbol{B} = (B_1, \ldots, B_m)^t =: (\beta, \delta)_L$, an estimator $\hat{\boldsymbol{B}}$ is unbiased iff

$$\mathbb{E}\hat{\boldsymbol{B}} = (\Lambda F\beta, |\Lambda||F|\delta)_L = (\beta, \delta)_L \tag{30}$$

which is satisfied iff simultaneously

$$\Lambda F = I_m, \ |\Lambda||F| = I_m. \tag{31}$$

The first equation ensures unbiasedness of the centre, the second unbiasedness of the spreads. Unfortunately, in general it is not possible to obtain unbiasedness of the spreads. This can be seen already in the simple linear regression case

$$\mathbb{E}Y = B_1 x \oplus B_2. \tag{32}$$

Here $F^t = \binom{x_1,\dots,x_n}{1,\dots,1}$, $\Lambda = \binom{\lambda_{11}\cdots\lambda_{1n}}{\lambda_{21}\cdots\lambda_{2n}}$ and the second equation of (31) writes

$$\begin{aligned} &\sum_i |\lambda_{1i}| = 1 \qquad \sum_i |\lambda_{1i}||x_i| = 0 \\ &\sum_i |\lambda_{2i}| = 0 \qquad \sum_i |\lambda_{2i}||x_i| = 1, \end{aligned} \tag{33}$$

which is inconsistent, since e.g. from $\sum |\lambda_{2i}| = 0$ it follows $\lambda_{2i} = 0$ for all i. But then we cannot obtain $\sum |\lambda_{2i}||x_i| = 1$.

Therefore it holds:

Theorem 1. *For the model (24) with $m \geq 2$, there is in general no unbiased linear estimator for $\boldsymbol{B}$ of the form (26).*

For $m = 1$, i.e. $\mathbb{E}Y = f(x)B_1$, unbiasedness can be forced. Then (31) reduces to

$$\sum_i \lambda_i f(x_i) = 1, \quad \sum_i |\lambda_i||f(x_i)| = 1 \tag{34}$$

which is equivalent to $\sum_i \lambda_i f(x_i) = 1$ and sign $\lambda_i =$ sign $f(x_i)$. E.g. $\lambda_i = 1/nf(x_i)$ automatically leads to an unbiased linear estimator for B_1. In this one-dimensional special case it is easy to find the BLUE: Using (23) with $\operatorname{Var} Y_i =: \sigma^2$ we have

$$\operatorname{Var} \hat{B} = \sum_{i=1}^{n} \lambda_i^2 \sigma^2. \tag{35}$$

The BLUE coefficients

$$\lambda_i^* = f(x_i) \Big/ \sum_{j=1}^{n} \left(f(x_j)\right)^2 \tag{36}$$

are the solutions of $\sum_{i=1}^{n} \lambda_i^2 = \text{Min}$ with side condition (34). As a special case of (36), clearly, the arithmetic fuzzy mean

$$\overline{Y} := \frac{1}{n} Y_1 \oplus \cdots \oplus \frac{1}{n} Y_n \tag{37}$$

is BLUE for the expectation $\mathbb{E}Y = B$.

3.3 Weak BLUE

One way out of the situation described in Theorem 1 is to make setups and requirements only for the modal values of the data and not for the spreads (see Näther, 1997). Given data of the form (25), i.e. $Y_i = (y_i, \Delta_i)_L$; $i = 1, \dots, n$; instead of (24) we use a model $Y_i = Y(x_i)$ with

$$\mathbb{E}Y(x) = (f_1(x)\beta_1 + \dots + f_m(x)\beta_m,\ \Delta_0(x))_L = \left(f(x)^t\beta,\ \Delta_0(x)\right)_L. \quad (38)$$

For estimating the β_j we consider analogously to (26) for $j = 1, \dots, m$

$$\hat{\beta}_j = \lambda_{1j}Y_1 \oplus \dots \oplus \lambda_{nj}Y_n = \lambda_j^t \boldsymbol{Y} = \left(\lambda_j^t \boldsymbol{y}, |\lambda_j|^t \boldsymbol{\Delta}\right)_L ; \quad (39)$$

where $\boldsymbol{y} = (y_1, \dots, y_n)^t$, $\boldsymbol{\Delta} = (\Delta_1, \dots, \Delta_n)^t$. Now, with (39) and with the terminology from (29) to (31) we have for $j = 1, \dots, m$

$$\mathbb{E}\hat{\beta}_j = \left(\lambda_j^t F\beta, |\lambda_j|^t \boldsymbol{\Delta}_0\right)_L ; \qquad \boldsymbol{\Delta}_0 = (\Delta_0(x_1), \dots, \Delta_0(x_n))^t$$

or more condensed

$$\mathbb{E}\hat{\beta} = (\Lambda F\beta,\ |\Lambda|\boldsymbol{\Delta}_0)_L . \quad (40)$$

Much weaker as in Subsection 3.2, we only require unbiasedness of the modal value, i.e.

$$\Lambda F = I_m. \quad (41)$$

An estimator $\hat{\beta}$ with (41) is called weak unbiased. Analogously to classical linear inference a weak unbiased linear estimator $\hat{\beta}$ exists iff F is of full rank. To find the so called *weak BLUE* for β let us consider $\operatorname{Var}\hat{\beta}_j$ which we can compute by use of (23) as

$$\begin{aligned} \operatorname{Var}\hat{\beta}_j &= \operatorname{Var}(\lambda_j^t y) + 2L_2 \operatorname{Var}(|\lambda_j|^t \Delta) \\ &= \lambda_j^t \Sigma_y \lambda_j + 2L_2 |\lambda_j|^t \Sigma_\Delta |\lambda_j|;\ j = 1, \dots, m; \end{aligned} \quad (42)$$

where Σ_y, Σ_Δ are the covariance matrices of the observed modal values y and of the observed spreads Δ. Minimization of (42) w.r.t. λ_j under side condition (41) gives the coefficient vector λ_j^* of the weak BLUE $\hat{\beta}_j^*$. As an essential difference to the classical linear estimation theory, (42) can in general not be reduced to minimization of a quadratic form. However, reduction to a quadratic form is possible, if the spreads are uncorrelated, i.e. if

$$\Sigma_\Delta = \operatorname{diag}(\sigma_\Delta(x_i)). \quad (43)$$

Then (42) reduces to

$$\operatorname{Var}\hat{\beta}_j = \lambda_j^t \Sigma \lambda_j ; \qquad \Sigma = \Sigma_y + 2L_2\Sigma_\Delta. \quad (44)$$

Minimization of (44) under side condition (41) coincides with the classical BLUE-problem for linear regression with observations correlated by Σ. Thus, the solution is given by

Theorem 2. *If the matrix $F^t\Sigma^{-1}F$ is regular, the weak BLUE for β in the model (38) is given by*

$$\hat{\beta}^* = (F^t\Sigma^{-1}F)^{-1}F^t\Sigma^{-1}\mathbf{Y}\,. \tag{45}$$

Note that in this case the extended BLUE-estimator according to Subsection 2.3 looks like

$$\breve{B} = (F^t\Sigma_y^{-1}F)^{-1}F^t\Sigma_y^{-1}\mathbf{Y}, \tag{46}$$

which is different from (45). Coincidence, however, happens if $\Sigma_y = \sigma_Y^2 I_n$, $\Sigma_\Delta = \sigma_\Delta^2 I_n$. Then both, (45) and (46) reduce to the extended least squares estimator $\breve{\boldsymbol{B}} = (F^tF)^{-1}F^t\mathbf{Y}$.

The main disadvantage of weak BLUE is that the spreads are uncontrolled, i.e. examples show too large spreads of the estimated regression function, similar to Example 1.

3.4 Componentwise BLUE and further modifications

A more satisfactory approach which can be justified in terms of statistics seems to be the following: The idea is, for estimation of the modal value of the regression parameter only to use the observed modal values y_i and for estimation of the spread only to use the observed spreads Δ_i. I.e. we give up the requirement that the estimator should be a linear form of the "unsplitted" fuzzy data $Y_i = (y_i, \Delta_i)_L$. Somewhat more detailed: As model now is used

$$\mathbb{E}Y(x) = (\mathbb{E}y, \mathbb{E}\Delta)_L = \left(f(x)^t\beta,\ g(x)^t\gamma\right)_L;\ \beta \in \mathbb{R}^m, \gamma \in \mathbb{R}^q \tag{47}$$

i.e. for the modal value and for the spread different setups are used. Clearly, in the interesting region, say $H \subseteq \mathbb{R}^k$, positivity of spread must be ensured, i.e.

$$\forall x \in H : g(x)^t\gamma \geq 0.$$

Given observations $Y_i = (y_i, \Delta_i)_L$ we consider estimators of the form

$$\hat{\beta} = \Lambda\boldsymbol{y},\ \hat{\gamma} = \Gamma\boldsymbol{\Delta},\ \forall x \in H : g(x)^t\hat{\gamma} \geq 0$$

where $\boldsymbol{y} = (y_1, \ldots, y_n)^t$ and $\boldsymbol{\Delta} = (\Delta_1, \ldots, \Delta_n)^t$. $\hat{\beta}$ is a classical linear estimator of the centre-parameter β based only on the observed modal values y and $\hat{\gamma}$ is a classical linear estimator of the spread-parameter γ based on the observed spreads $\boldsymbol{\Delta}$ with a side condition for positivity. Unbiasedness of $\hat{\beta}, \hat{\gamma}$ is ensured if

$$\Lambda F = I_m,\ \Gamma G = I_q;\ G = (g(x_1), \ldots, g(x_n))^t.$$

For estimation of the regression function $\mathbb{E}Y(x)$ we will use

$$\hat{Y}(x) = \left(f(x)^t\hat{\beta},\ g(x)^t\hat{\gamma}\right)_L. \tag{48}$$

An unbiased $\hat{Y}^*$ of the form (48) with minimal variance is called *componentwise BLUE.* Clearly, $\hat{Y}$ is unbiased iff $\hat{\beta}$ and $\hat{\gamma}$ are unbiased. To find the componentwise BLUE, first $\operatorname{Var}\hat{Y}(x)$ is obtained from (23) as

$$\begin{aligned}\operatorname{Var}\hat{Y}(x) &= \operatorname{Var}(f(x)^t\hat{\beta}) + 2L_2\operatorname{Var}(g(x)^t\hat{\gamma})\\ &= f(x)^t\operatorname{Cov}\hat{\beta}f(x) + 2L_2 g(x)^t\operatorname{Cov}\hat{\gamma}g(x).\end{aligned}$$

From this the following is clear:

Theorem 3. *If $\hat{\beta}^*$ is BLUE for β in the linear model $Ey = F\beta$ and if $\hat{\gamma}^*$ is BLUE for γ in the linear model $E\Delta = G\gamma$ under the side condition $\forall x \in H : g(x)^t\hat{\gamma}^* \geq 0$, then*

$$\hat{Y}^*(x) = \left(f(x)^t\hat{\beta}^*,\ g(x)^t\ \hat{\gamma}^*\right)_L$$

is componentwise BLUE. Clearly, if regularity of the matrices is ensured, $\hat{\beta}^*$ is given by

$$\hat{\beta}^* = (F^t\Sigma_y^{-1}F)^{-1}F^t\Sigma_y^{-1}y$$

and $\hat{\gamma}^*$ by

$$\hat{\gamma}^* = (G^t\Sigma_\Delta^{-1}G)^{-1}G^t\Sigma_\Delta^{-1}\Delta \tag{49}$$

if

$$\forall x \in H : g(x)^t\hat{\gamma}^* \geq 0 \tag{50}$$

is fulfilled.

The crucial point is that $\hat{\gamma}^*$ from (49) fulfils (50) only in special cases. For example, if $\{H_j\}_{j=1,\dots,q}$, is a partition of H and if we model the spread on H_j by $\gamma_j \geq 0$, i.e. in terms of (47) we use $g(x) = (1_{H_1}(x),\dots,1_{H_q}(x))^t$, then $\hat{\gamma}^*$ from (49) is given by $\hat{\gamma}^* = (\overline{\Delta}_1,\dots,\overline{\Delta}_q)^t$ where $\overline{\Delta}_j \geq 0$ for $x \in H_j$ and the requirement (50) is automatically satisfied.

By straightforward considerations, the results of this section can be generalized for LR-type-data using the model

$$\mathbb{E}Y(x) = (\mathbb{E}y, \mathbb{E}l, \mathbb{E}r)_{LR} = (f(x)^t\beta, g_l(x)^t\gamma_l, g_r(x)^t\gamma_r)_{LR}.$$

Example 2. We use the data of Example 1 and the following setups: for the modal values: $\mathbb{E}y = \beta_1 x + \beta_2$

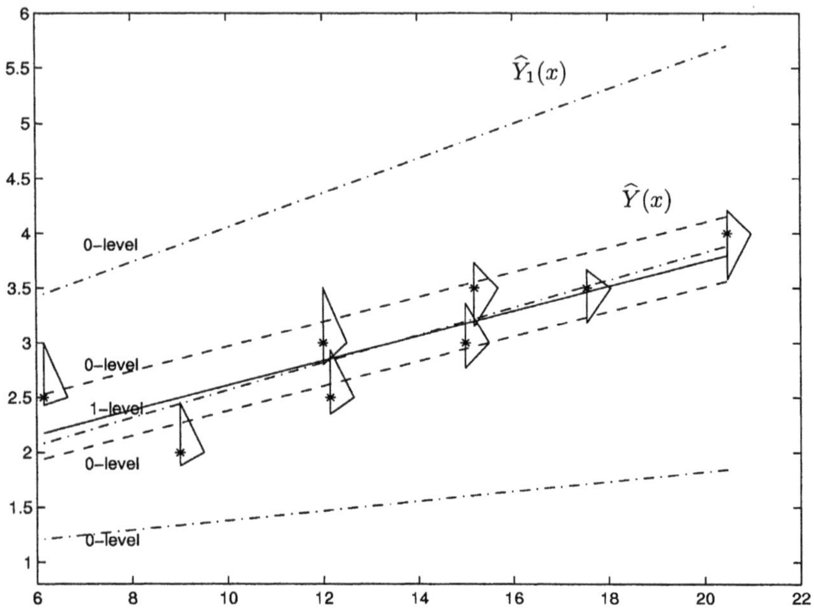

Fig. 2. Comparison of Examples 1 and 2

for the spreads: $\mathbb{E}l = \gamma_l$, $\mathbb{E}r = \gamma_r$.
The componentwise BLUE is given by

$$\hat{Y}(x) = (0.1133x + 1.4774, 0.2325, 0.3562)_L \ .$$

In Figure 2 the dashed-dotted lines represent the 0-level set of the fuzzy line given by $\widehat{Y}_1(x)$ from Example 1.

The solid line and the dashed lines denote the 1-level set and the 0-level set of $\widehat{Y}(x)$ from Example 2.

However, if we model $\mathbb{E}\Delta(x)$ in (47) by a straight line then it is easy to see that (50) cannot be satisfied for all possible observation results Δ_i. Consider e.g. $H = [-1, 1]$ and the three design points $x_1 = 1$, $x_2 = 0$, $x_3 = -1$ with the results $\Delta_1 = 1$, $\Delta_2 = 0$, $\Delta_3 = 0$. The setup $\mathbb{E}\Delta = \gamma_1 + \gamma_2 x$ leads to $\hat{\gamma}^* = (\overline{\Delta},\ (\Delta_1 - \Delta_3)/2)^t = (1/3,\ 1/2)^t$, but unfortunately we have $g(x)^t \hat{\gamma}^* = -1/6$ for $x = -1$.

This is the reason for looking for further modifications. One is discussed in some detail in Subsection 4.2. Here we will sketch briefly some further ideas. The positivity of the spread is automatically included if we model $\mathbb{E}\Delta$ e.g. by $\mathbb{E}\Delta(x) = \gamma^t\ G(x)\gamma$, $\gamma \in R^q$ unknown, $G(x)$ known and positive semidefinite for $x \in H$ or by $\mathbb{E}\Delta(x) = g(x)^t D g(x)$, $g(x)$ known, D positive semidefinite and unknown.

Another possibility is to give up the requirement that $g(x)^t\hat{\gamma}$ in (48) has to be a linear estimate. Denote $M(x) \in \mathbb{R}^n$ the set of possible observations $\boldsymbol{\Delta} = (\Delta_1, \ldots, \Delta_n)^t$ which ensure (50) for given $x \in H$. Then it seems to be reasonable to estimate $\mathbb{E}\Delta(x)$ by some kind of truncated BLUE, i.e.

$$\hat{\Delta}(x) = g(\underline{x})^t \; 1_{M(x)}(\boldsymbol{\Delta})\hat{\gamma}.$$

This means: Take $\hat{\gamma}^*$ from (49) as long as w.r.t. $x \in H$ and the observed Δ positivity of the spread estimator is guaranteed, take zero if positivity fails.

4 Least squares principle

In fuzzy regression, a vector of fuzzy parameters $\boldsymbol{B} = (B_1, \ldots, B_m)^t$ is searched, which give, in some sense, the best fit of the fuzzy output $Y_1, \ldots, Y_n$ to the fuzzy model

$$Y_j = F(x_j, \boldsymbol{B}) \qquad \text{for} \qquad j = 1, 2, \ldots, n. \tag{51}$$

Most commonly used are the *linear models*, where the relation F is a linear function, for example,

$$Y(x) = B_0 \oplus B_1 x_1 \oplus \ldots \oplus B_m x_m,$$

in the crisp input, fuzzy output case.

By the same way, fuzzy input, fuzzy output models such as

$$Y(\boldsymbol{X}) = B_0 \oplus B_1 X_1 \oplus \ldots \oplus B_m X_m,$$

for fuzzy inputs $\boldsymbol{X} = (X_1, \ldots, X_m)^t$ can be developed. A number of different approaches have been used for the notion of bestness of fit. Tanaka and his school (Tanaka, 1987, and references therein) have used possibilistic methods which reduce to optimization by linear programming (cf. Subsection 2.1).

If a suitable metric D is defined on appropriate spaces of fuzzy sets, *least squares methods* can be developed by this metric (cf. Diamond, 1987, 1988). The fuzzy parameter $\boldsymbol{B}$ is searched, which fit the fuzzy data $Y_1, \ldots, Y_n$ to the parametric fuzzy function $F(x_j, \boldsymbol{B})$ in such a way that sum of the quadratic distances

$$\sum_{j=1}^{n} D\left(F(x_j, \boldsymbol{B}), Y_j\right)^2$$

leads D to a minimum.

Celminš (1987) has considered a method which incorporates some elements of both these techniques, i.e. linear programming and least squares. Because the observations x may contain errors and the model equations are not satisfied at x, but in the vicinity of x, he introduced a correction term c_x and replaced the model (51) by the constraint equations

$$Y_j = F(x_j + c_{x_j}, \boldsymbol{B}) \qquad \text{for} \qquad j = 1, 2, \ldots, n.$$

The constraint model is fitted by minimizing of the correction terms (minimizing the sum of a squared distance of the membership values from 1).

Recently, Chang and Lee (1994) have interpreted conflicting trends between modal values and spreads in fuzzy data by allowing choice of some parameters with *negative spreads.* This interesting device certainly fits the data in a formal sense, but these parameters are no longer fuzzy numbers.

Here we reinterpret the problem of negative spreads in a broader context and obtain least squares estimates without negative spreads by introducing linear models incorporating a *generalized Hukuhara difference* $B \ominus_H A$. This is defined as the least squares solution of the equation $A \oplus X = B$. That is,

$$B \sim_H A := \arg \inf_{X \in \mathcal{F}_c(\mathbb{R})} D(A \oplus X, B) \tag{52}$$

in some L_2–type metric space. When the usual Hukuhara difference $B \ominus_H A$ exists (Hukuhara, 1967, Diamond and Kloeden, 1994), it coincides with the least squares solution $B \sim_H A$.

4.1 Least squares estimation

For given fuzzy data $Y_1, \ldots, Y_n \in \mathcal{F}_c(\mathbb{R})$ $(x_1, \ldots, x_n)$, the fuzzy parameter $\widehat{\boldsymbol{B}} = (\widehat{B}_1, \ldots, \widehat{B}_m)^t$ is called *least squares estimation* of the model $Y(x) = F(x, \boldsymbol{B})$ if the quadratic distance

$$\sum_{j=1}^{n} \delta_2 \left(F(x_j, \boldsymbol{B}), Y_j\right)^2$$

with the L_2-metric (cf. Körner and Näther in Part 2 in this book)

$$\delta_2(A, B) = \sqrt{d \int_0^1 \int_{S^{d-1}} |s_A(u, \alpha) - s_B(u, \alpha)|^2 \nu(du) d\alpha} \tag{53}$$

leads to a minimum.

As in classical statistical regression the *coefficient of determination*

$$r^2 := 1 - \frac{\sum_{j=1}^{n} \delta_2 \left(F(x_j, \boldsymbol{B}), Y_j\right)^2}{\sum_{j=1}^{n} \delta_2 \left(Y_j, \overline{Y}_n\right)^2}$$

is used as a measure of the goodness-of-fit of the fuzzy model to the fuzzy data (see, for example, [Mendenhall, 1983, p. 436]). The closer the coefficient r^2 to unity, the better the fit.

The existence and uniqueness of the least squares solution follows by the well-known projection theorem (cf. Luenberger, 1968) in the Hilbert space $L_2(S^{d-1} \times (0, 1])$:

Theorem 4. *Let be given $Y_1, \ldots, Y_n \in \mathcal{F}_c(\mathbb{R})$ at the design points $x_1, \ldots, x_n$ and $F(x, \boldsymbol{B}) = B_1 f_1(x) + \cdots + B_m f_m(x)$. Then the least squares solution $\widehat{\boldsymbol{B}} = (\widehat{B}_1, \ldots, \widehat{B}_m)^t \in (\mathcal{F}_c(\mathbb{R}))^m$ with respect to the metric δ_2 from (53), i.e.*

$$\inf_{\boldsymbol{B} \in (\mathcal{F}_c(\mathbb{R}))^m} \sum_{j=1}^{n} \delta_2 \left(f(x_j)^t \boldsymbol{B}, Y_j\right)^2 = \sum_{j=1}^{n} \delta_2 \left(f(x_j)^t \widehat{\boldsymbol{B}}, Y_j\right)^2 \qquad (54)$$

is uniquely determined.

If the included fuzzy sets are LR-fuzzy numbers, then the problem (54) can be transformed into a quadratic optimization problem and can be solved with the help of a Kuhn-Tucker-Theorem (cf. Rockafellar, 1970).

Theorem 5. *The least squares problem (54) with respect to LR-fuzzy numbers*

$$B_k = (b_k, l_{B_k}, r_{B_k})_{LR}\,, \qquad Y_j = (y_j, l_j, r_j)_{LR}\,,$$

is equivalent to the quadratic optimization problem

$$\inf_{\substack{\boldsymbol{l}_B, \boldsymbol{r}_B, \boldsymbol{b} \in \mathbb{R}^m \\ \boldsymbol{l}_B, \boldsymbol{r}_B \geq 0}} \begin{pmatrix} \boldsymbol{b} - \check{\boldsymbol{b}} \\ \boldsymbol{l}_B - \check{\boldsymbol{l}}_B \\ \boldsymbol{r}_B - \check{\boldsymbol{r}}_B \end{pmatrix}^t C \begin{pmatrix} \boldsymbol{b} - \check{\boldsymbol{b}} \\ \boldsymbol{l}_B - \check{\boldsymbol{l}}_B \\ \boldsymbol{r}_B - \check{\boldsymbol{r}}_B \end{pmatrix} + c\,, \qquad (55)$$

where the abbreviations

$$\boldsymbol{b} := (b_1, \ldots b_m)^t; \qquad \boldsymbol{y} := (y_1, \ldots, y_n)^t;\ \check{\boldsymbol{b}} = \left(F^t F\right)^{-1} F^t \boldsymbol{y};$$
$$\boldsymbol{l}_B := (l_{B_1}, \ldots, l_{B_m})^t;\ \ \boldsymbol{l} := (l_1, \ldots, l_n)^t;\ \ \check{\boldsymbol{l}}_B = \left(F^t F\right)^{-1} F^t \boldsymbol{l};$$
$$\boldsymbol{r}_B := (r_{B_1}, \ldots, r_{B_m})^t;\ \boldsymbol{r} := (r_1, \ldots, r_n)^t;\ \check{\boldsymbol{r}}_B = \left(F^t F\right)^{-1} F^t \boldsymbol{r}\,,$$

are used and

$$C := \begin{pmatrix} F^t F & -L_1 F^t F & R_1 F^t F \\ -L_1 F^t F & L_2 F^t F & 0 \\ R_1 F^t F & 0 & R_2 F^t F \end{pmatrix}$$

is a non-negative definite matrix and c is a constant with respect to the optimization.

Proof. Formula (55) is calculated by expansion of (54) and (53) to

$$\begin{aligned} D^2 &= \sum_{j=1}^{n} \delta_2 \left(f(x_j)^t \boldsymbol{B},\ Y_j\right)^2 \\ &= \parallel \boldsymbol{y} - F\boldsymbol{b} \parallel^2 + L_2 \parallel \boldsymbol{l} - F\boldsymbol{l}_B \parallel^2 + R_2 \parallel \boldsymbol{r} - F\boldsymbol{r}_B \parallel^2 \\ &\quad -2L_1(\boldsymbol{y} - F\boldsymbol{b})^t(\boldsymbol{l} - F\boldsymbol{l}_B) + 2R_1(\boldsymbol{y} - F\boldsymbol{b})^t(\boldsymbol{r} - F\boldsymbol{r}_B)\,, \end{aligned}$$

where $\parallel \boldsymbol{y} - F\boldsymbol{b} \parallel^2 = (\boldsymbol{y} - F\boldsymbol{b})^t(\boldsymbol{y} - F\boldsymbol{b})$. Since for vectors $\boldsymbol{b}_1, \boldsymbol{b}_2 \in \mathbb{R}^m$ and $\boldsymbol{y}_1, \boldsymbol{y}_2 \in \mathbb{R}^n$ with the abbreviation $\check{\boldsymbol{b}}_k = (F^t F)^{-1} F^t \boldsymbol{y}_k$ the formula

$$(\boldsymbol{y}_1 - F\boldsymbol{b}_1)^t(\boldsymbol{y}_2 - F\boldsymbol{b}_2) = (\boldsymbol{b}_1 - \check{\boldsymbol{b}}_1)^t F^t F(\boldsymbol{b}_2 - \check{\boldsymbol{b}}_2) + \boldsymbol{y}_1^t(I_n - F(F^t F)^{-1} F^t)\boldsymbol{y}_2$$

is valid, the distance D^2 is written as

$$\begin{aligned} D^2 =& (\boldsymbol{b}-\check{\boldsymbol{b}})^t F^t F(\boldsymbol{b}-\check{\boldsymbol{b}}) \\ &+L_2\,(\boldsymbol{l}_B-\check{\boldsymbol{l}}_B)^t F^t F(\boldsymbol{l}_B-\check{\boldsymbol{l}}_B) + R_2\,(\boldsymbol{r}_B-\check{\boldsymbol{r}}_B)^t F^t F(\boldsymbol{r}_B-\check{\boldsymbol{r}}_B) \\ &-2L_1(\boldsymbol{b}-\check{\boldsymbol{b}})^t F^t F(\boldsymbol{l}_B-\check{\boldsymbol{l}}_B) + 2R_1(\boldsymbol{b}-\check{\boldsymbol{b}})^t F^t F(\boldsymbol{r}_B-\check{\boldsymbol{r}}_B) + c\,, \end{aligned}$$

where c is constant with respect to optimization.

Now, write the D^2 in the matrix structure (55).

Obviously, the determinant of C is obtained by

$$\mathbf{det}(C) = \left(L_2R_1^2 + L_1^2R_2 - L_2R_2\right)^m \mathbf{det}(F^tF)^3\,.$$

The Cauchy-Schwartz inequality implies that $L_1^2 \le L_2$ and $R_1^2 \le R_2$. Therefore, the matrix C is positive definite if F^tF is positive definite and $L_2R_1^2 + L_1^2R_2 \ne L_2R_2$. ■

Note that the constraints $\boldsymbol{l}_B,\ \boldsymbol{r}_B \ge \mathbf{o}$ show that the trend of fuzziness within the model can not be decreasing. This is a consequence of the non-linearity of the space of fuzzy sets.

If the terms $\check{\boldsymbol{l}}_B$, $\check{\boldsymbol{r}}_B$ fulfill the non-negativity conditions $\check{\boldsymbol{l}}_B$, $\check{\boldsymbol{r}}_B \ge \mathbf{o}$, then the solution of (55) is given by the linear estimators

$$\begin{aligned} \widehat{\boldsymbol{b}} &= \left(F^tF\right)^{-1} F^t\boldsymbol{y} \\ \widehat{\boldsymbol{l}}_B &= \left(F^tF\right)^{-1} F^t\boldsymbol{l}_B \\ \widehat{\boldsymbol{r}}_B &= \left(F^tF\right)^{-1} F^t\boldsymbol{r}_B\,. \end{aligned} \tag{56}$$

Example 3. Reconsider the Example 1. Then the calculation for the least squares problem $Y(x) = Ax \oplus B$ yields $\check{\boldsymbol{b}} = (0.1259, 1.3079)^t$,

$$\check{\boldsymbol{l}}_B = (0.0250, -0.1040)^t \quad \text{and} \quad \check{\boldsymbol{r}}_B = (-0.0254, 0.6980)^t\,.$$

The estimators $\check{\boldsymbol{l}}_B, \check{\boldsymbol{r}}_B$ are not admissible (there are negative components) for $\widehat{A}$ and $\widehat{B}$. The quadratic optimization problem (55) is solved by

$$\widehat{A} = (0.1157, 0.0098, 0)_{LL}\,, \qquad \widehat{B} = (1.4446, 0.1010, 0.3562)_{LL}$$

and the model

$$\widehat{Y}(x) = (0.1157, 0.0098, 0)_{LL}\; x \oplus (1.4446, 0.1010, 0.3562)_{LL}$$

is fitted (see Figure 3) by a coefficient of determination of $r^2 = 0.7558$.

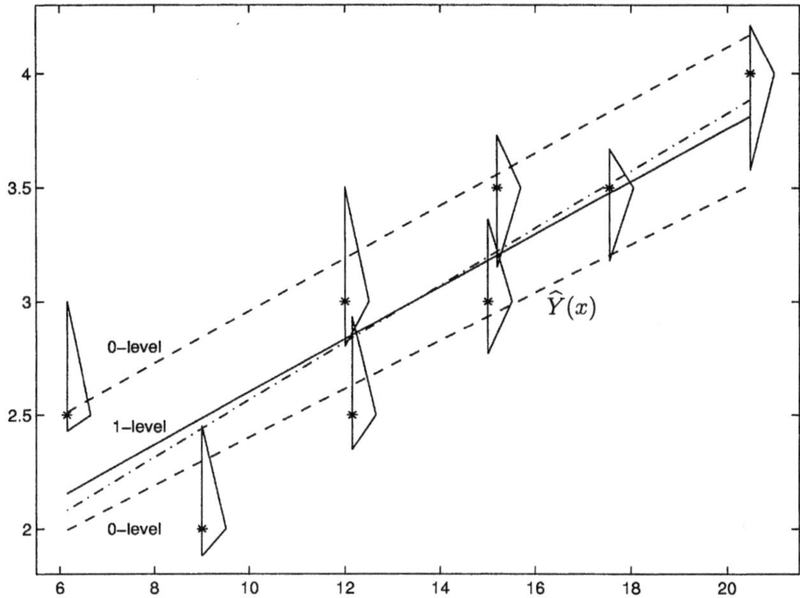

Fig. 3. Least squares regression $\widehat{Y}(x)$

4.2 Hukuhara models

As mentioned above, $B \oplus (-1)A$ is not really a difference and is rather unnatural with respect to a linear structure. For example, the models $Y = A \oplus (-b)X$ and $Y \oplus bX = A$ differ from each other. More importantly, $B \oplus (-1)A$ is not compatible with the difference in the function space $L_2(S^{d-1} \times [0,1])$, that is, it does not have the support function $s_B - s_A$. But the Hukuhara difference (see Hukuhara, 1967, Diamond and Kloeden, 1994), defined as the solution for X in the equation $A \oplus X = B$ if it exists, does coincide with the difference in $L_2(S^{d-1} \times [0,1])$. This property justifies the application of this difference instead of the fuzzy number $B \oplus (-1)A$.

For fuzzy numbers $A, B \in \mathcal{F}_c(\mathbb{R})$ the *Hukuhara difference* $B \ominus_H A$ (if it exists) is given by

$$(B \ominus_H A)_\alpha = \left\{a \in \mathbb{R}^d \,:\, A_\alpha \oplus \{a\} \subseteq B_\alpha\right\}\,, \qquad \alpha \in [0,1]\,. \tag{57}$$

In particular, the Hukuhara difference $B \ominus_H A$ of two symmetric triangular fuzzy numbers $A = (m_A, l_A)_\Delta$ and $B = (m_B, l_B)_\Delta$ is well-defined if $l_B \geq l_A$ and $B \ominus_H A = (m_B - m_A, l_B - l_A)_\Delta$. However, this breaks down if $l_B < l_A$ and $B \ominus_H A$ does not exist. Since an exact solution of the equation $A \oplus X = B$ is then impossible, we find $X \in \mathcal{F}_c(\mathbb{R})$ such that $A \oplus X$ is the L_2-approximant to B:

$$B \sim_H A = \mathbf{arg} \inf_{X \in \mathcal{F}_c(\mathbb{R})} \delta_2\,(A \oplus X, B)\ . \tag{58}$$

For symmetric triangular fuzzy numbers, $B \sim_H A = (m_B - m_A, 0)_\Delta$ if $l_B < l_A$, which is actually a crisp number.

When $B \ominus_H A$ exists, it coincides with $B \sim_H A$, so we use the latter notation for both in all what follows. This operation has some of the formal properties of a difference which make it useful in fuzzy least squares regression. Note that the existence and uniqueness of $B \sim_H A$ is given by the projection theorem in $L_2(S^{d-1} \times [0,1])$.

The calculation of the difference is more complicated. Only solutions in special cases can easily be described. Certainly, with LR-fuzzy numbers a radical simplification of the calculation is possible.

Theorem 6. *Let $A = (m_A, l_A, r_A)_{LR}$ and $B = (m_B, l_B, r_B)_{LR}$ be LR-fuzzy numbers.*
Then the solution $C = B \sim_H A$ in the sense of (58) is given by

$$l_C = \begin{cases} \max\{l_B - l_A, 0\} & \text{if } r_B - r_A \geq 0 \\ \max\{l_B - l_A + \frac{L_1 R_1}{L_2 - L_1^2}(r_B - r_A), 0\} & \text{if } r_B - r_A < 0 \end{cases}$$

$$r_C = \begin{cases} \max\{r_B - r_A, 0\} & \text{if } l_B - l_A \geq 0 \\ \max\{r_B - r_A + \frac{L_1 R_1}{R_2 - R_1^2}(l_B - l_A), 0\} & \text{if } l_B - l_A < 0 \end{cases}$$

and

$$m_C = m_B - m_A + R_1(r_B - r_A - r_C) - L_1(l_B - l_A - l_C)\,.$$

Proof. By the relation (see Körner and Näther, 1998)

$$\delta_2(A,B)^2 = \begin{pmatrix} m_A - m_B \\ l_A - l_B \\ r_A - r_B \end{pmatrix}^t \begin{pmatrix} 1 & -L_1 & R_1 \\ -L_1 & L_2 & 0 \\ R_1 & 0 & R_2 \end{pmatrix} \begin{pmatrix} m_A - m_B \\ l_A - l_B \\ r_A - r_B \end{pmatrix}$$

we obtain for $C = B \sim_H A = (m_C, l_C, r_C)_{LR}$

$$\inf_{\substack{m_C \in \mathbb{R}, \\ l_C, r_C \geq 0}} \begin{pmatrix} m_A + m_C - m_B \\ l_A + l_C - l_B \\ r_A + r_C - r_B \end{pmatrix}^t \begin{pmatrix} 1 & -L_1 & R_1 \\ -L_1 & L_2 & 0 \\ R_1 & 0 & R_2 \end{pmatrix} \begin{pmatrix} m_A + m_C - m_B \\ l_A + l_C - l_B \\ r_A + r_C - r_B \end{pmatrix}.$$

By standard manipulation this is simplified to

$$m_C = m_B - m_A + R_1(r_B - r_A - r_C) - L_1(l_B - l_A - l_C)$$

and

$$\inf_{l_C, r_C \geq 0} \begin{pmatrix} l_C - (l_B - l_A) \\ r_C - (r_B - r_A) \end{pmatrix}^t \begin{pmatrix} L_2 - L_1^2 & L_1 R_1 \\ L_1 R_1 & R_2 - R_1^2 \end{pmatrix} \begin{pmatrix} l_C - (l_B - l_A) \\ r_C - (r_B - r_A) \end{pmatrix}.$$

Furthermore, this last optimization problem is again solved by application of the Kuhn-Tucker theorem and gives the expressions for l_C, r_C. ■

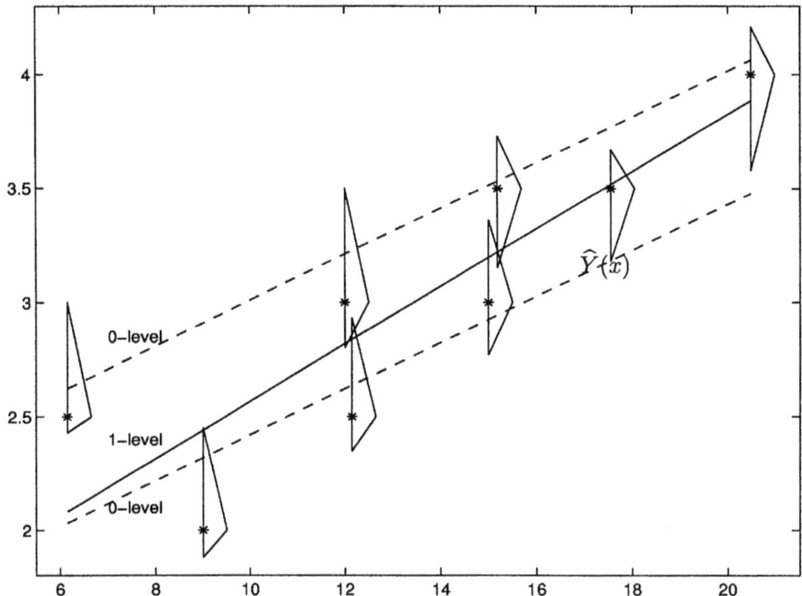

Fig. 4. Least squares regression for the Hukuhara model

To describe problems with "negative spreads", the linear model $Y = Ax \oplus B$ is extended with help of the generalized Hukuhara difference $B \sim_H A$ to

$$(x) = (A_1 x \oplus B_1) \sim_H (A_2 x \oplus B_2)\,,$$

where A_1, B_1 models the "positive spreads" and A_2, B_2 the "negative spreads".

For simplicity, any calculation like $B \sim_H A$ is done in the function space $L_2(S^{d-1} \times [0,1])$ and the results is isomorphically mapped to the space of fuzzy sets afterwards. This is done by $s_{B \sim_H A} = \mathcal{P}(s_B - s_A)$, where $\mathcal{P}$ denotes the projection on the cone of fuzzy sets.

Theorem 7. *Let $(Y_j, x_j)_{j=1}^n$ $(x_j \geq 0)$ be given data for the extended linear model*

$$Y(x) = (A_1 x \oplus B_1) \sim_H (A_2 x \oplus B_2)\,.$$

Then, the least squares approach is equivalent to the problem

$$\inf_{A_1, A_2, B_1, B_2 \in \mathcal{F}_c(\mathrm{IR})} \sum_{j=1}^{n} \delta_2(Y_j \oplus A_2 x_j \oplus B_2, A_1 x_j \oplus B_1)^2$$

and gives the estimators

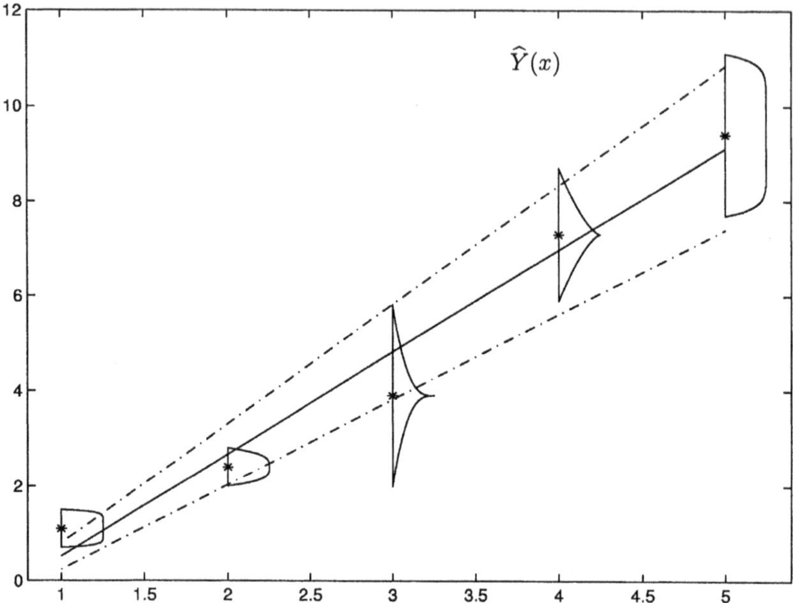

Fig. 5. Linear regression with *Lp*-fuzzy numbers

$$\widehat{A_1} = \frac{1}{nC_{xx}} \sum_{j=1}^{n} Y_j x_j, \quad \widehat{A_2} = \frac{1}{C_{xx}} \overline{Y}\,\overline{x}, \quad \widehat{B_1} = \overline{Y} \oplus \widehat{A_2}\,\overline{x} \quad \text{and} \quad \widehat{B_2} = \widehat{A_1}\,\overline{x},$$

where $C_{xx} := \frac{1}{n} \sum_{j=1}^{n} (x_j - \overline{x})^2$.

Proof. It is easy to prove that the solution of the Normal Equations in the function space gives the estimators. ■

If the Hukuhara difference $\widehat{A_1} \ominus_H \widehat{A_2}$ exists, the common estimator $\widehat{A}$ is built by $\widehat{A} := \widehat{A_1} \ominus_H \widehat{A_2}$ ($\widehat{B_1}$ and $\widehat{B_2}$ analogously).

Example 4. Reconsider the Example 1. The calculation for the Hukuhara model $Y(x) = (A_1 x \oplus B_1) \sim_H (A_2 x \oplus B_2)$ yields

$$\widehat{A_1} = (2.2984, 0.1934, 0.2326)_{LR}, \qquad \widehat{A_2} = (2.1725, 0.1684, 0.2580)_{LR},$$

$$\widehat{B_1} = (32.1959, 2.4952, 3.8233)_{LR}, \qquad \widehat{B_2} = (30.8881, 2.5992, 3.1253)_{LR}.$$

Note that the Hukuhara difference $(A_1 x \oplus B_1) \ominus_H (A_2 x \oplus B_2)$ exists for $x \in [4, 22]$ and the model can be simplified to

$$Y(x) = (0.1259x + 1.3079,\ 0.0250x - 0.1040,\ -0.0254x + 0.6980)_{LR}$$

(see Figure 4) with a coefficient of determination of $r^2 = 0.7583$.

Moreover, fuzzy linear models in $\mathbb{R}^d$, without the assumption of LR-fuzzy numbers, can be constructed in an appropriate linear space of functions.

Example 5. Consider the class of symmetric Lp-fuzzy numbers $Y = (y, l, p)_{Lp}$, given by the α-levels

$$Y_\alpha = [y - l\,(1-\alpha)^{1/p},\ y + l\,(1-\alpha)^{1/p}] \qquad \alpha \in [0,1]\,.$$

Fit the data set

$$Y(1) = (1.1, 0.4, 0.1)_{Lp}\ Y(2) = (2.4, 0.4, 0.3)_{Lp}\ Y(3) = (3.9, 1.9, 3.0)_{Lp}$$
$$Y(4) = (7.3, 1.4, 1.5)_{Lp}\ Y(5) = (9.4, 1.7, 0.1)_{Lp}$$

to the model $Y(x) = (A_1 x \oplus B_1) \sim_H (A_2 x \oplus B_2)$.

The estimated predictor is shown in Figure 5 and the estimators for the parameters in Figure 6. The constant term B is approximately $m_B = -1.63$, with a spread of $l_B = 0.08$, and the trend A is approximately $m_A = 2.15$, with a spread of $l_A = 0.36$. The coefficient of determination is $r^2 = 0.9592$.

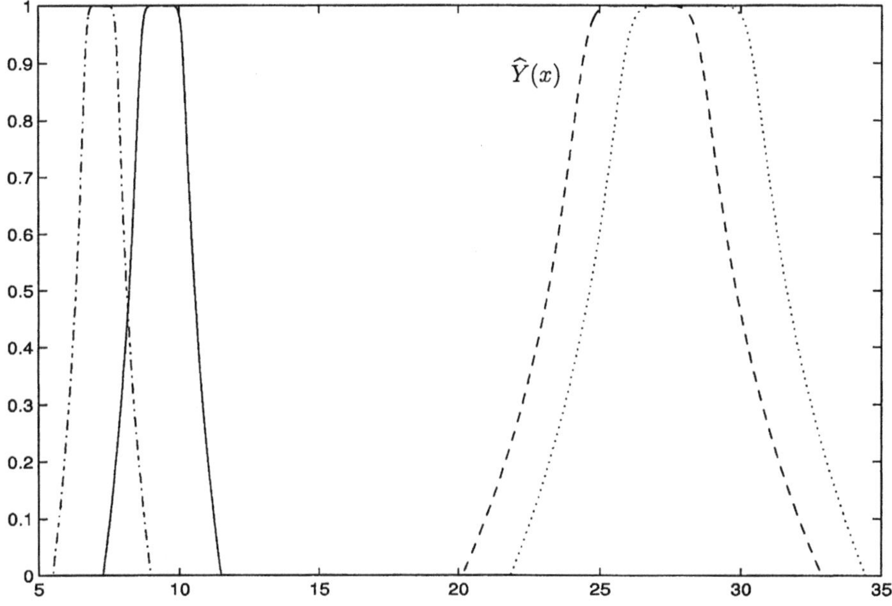

Fig. 6. Estimators for A_1, A_2, B_1 and B_2

Here, the Hukuhara difference $\widehat{A_1} \ominus_H \widehat{A_2}$ exists (not for $\widehat{B_1}$, $\widehat{B_2}$).
Hence set $\widehat{A} := \widehat{A_1} \ominus_H \widehat{A_2}$ and the linear model $Y(x) = (Ax \oplus B_1) \sim_H B_2$ is obtained.

References

1. Bandemer, H. (1985). Evaluating explicit functional relationships from fuzzy observations, *Fuzzy Sets and Systems* **16**, 41–52.
2. Bandemer, H. and Näther, W. (1992). *Fuzzy Data Analysis.* Kluwer Academic Publishers, Dordrecht-Boston-London.
3. Bardossy, A. (1990). Note on fuzzy regression, *Fuzzy Sets and Systems* **37**, 65–75.
4. Bardossy, A., Hagaman, R., Duckstein, L., Bogardi, I. (1992). Fuzzy least squares regression: Theory and application. In *Fuzzy Regression Analysis* (J. Kacprzyk and M. Fedrizzi, Eds.). Physica-Verlag, Heidelberg, 183–193.
5. Celminš, A. (1987). Least squares model fitting to fuzzy vector data, *Fuzzy Sets and Systems* **22**, 245–269.
6. Chang, P.T. and Lee, E.S. (1994). Fuzzy linear regression with spreads unrestricted in sign, *Computers Math. Applic.* **28**, 61–70.
7. Diamond, P. (1987). Least squares fitting of several fuzzy variables. *Proc. 2nd IFSA Congress*, Tokyo.
8. Diamond, P. (1988). Fuzzy least squares, *Inform. Sci.* **46**, 141–157.
9. Diamond, P. (1992). Least squares and maximum likelihood regression for fuzzy linear models. In *Fuzzy Regression Analysis* (J. Kacprzyk and M. Fedrizzi, Eds.). Physica-Verlag, Heidelberg, 137–151.
10. Diamond, P. and Kloeden, P. (1994). *Metric Space of Fuzzy Sets.* World Scientific, New Jersey.
11. Hukuhara, M. (1967). Integration des applications mesurables dont la valeur est un compact convexe, *Funkc. Ekvacioj.* 205–223.
12. Kacprzyk, J. and Fedrizzi, M. (1992). *Fuzzy Regression Analysis.* Omnitech Press, Warsaw, and Physica-Verlag, Heidelberg.
13. Körner, R. (1997a). *Linear Models with Random Fuzzy Variables.* PhD thesis, Faculty of Mathematics and Computer Sciences, Freiberg University of Mining and Technology.
14. Körner, R. (1997b). On the variance of fuzzy random variables, *Fuzzy Sets and Systems* **92**, 83–93.
15. Körner, R. (2000). An asymptotic α-test for the expectation of random fuzzy variables, *J. Statist. Plan. Infer.* **83**, 331-346.
16. Körner, R. and Näther, W. (1998). Linear regression with random fuzzy variables: extended classical estimates, best linear estimates, least squares estimates, *Inform. Sci.* **109**, 95-118.
17. Körner, R. and Näther, W. (2001). On the Variance of Random Fuzzy Variables. (In Part 2 in this volume).
18. Luenberger, K. (1968). *Optimization by Vector Space Methods.* J. Wiley & Sons, New York-London-Sydney-Toronto.
19. Mendenhall, W. (1983). *Introduction to Probability and Statistics.* 6th Edition. Duxbury Press, Boston.
20. Näther, W. (1997). Linear statistical inference for random fuzzy data , *Statistics* **29**, 221–240.
21. Rockafellar, R.T. (1970). *Convex Analysis.* Princeton Univ. Press, Princeton-New Jersey.
22. Sakawa, M. and Yano, M: (1992). Fuzzy linear regression and its applications. In *Fuzzy Regression Analysis* (J. Kacprzyk and M. Fedrizzi, Eds.). Omnitech Press, Warsaw, and Physica-Verlag, Heidelberg, 61–80.

23. Tanaka, H., Uejima, S. and Asai, K. (1980). Fuzzy linear regression model, *IEEE Trans. Syst. Man Cybern.* **10**, 2933–2938.
24. Tanaka, H. (1987). Fuzzy data analysis by possibilistic linear models, *Fuzzy Sets and Systems* **24**, 363–375.
25. Tanaka, H. and Watada, J. (1988). Possibilistic linear systems and their application to the linear regression model, *Fuzzy Sets and Systems* **27**, 275–289.

Index

GPSR Compliance
The European Union's (EU) General Product Safety Regulation (GPSR) is a set of rules that requires consumer products to be safe and our obligations to ensure this.

If you have any concerns about our products, you can contact us on

ProductSafety@springernature.com

In case Publisher is established outside the EU, the EU authorized representative is:

Springer Nature Customer Service Center GmbH
Europaplatz 3
69115 Heidelberg, Germany

www.ingramcontent.com/pod-product-compliance
Ingram Content Group UK Ltd.
Pitfield, Milton Keynes, MK11 3LW, UK
UKHW021834190726
13853UKWH00003B/1298

* 9 7 8 3 6 6 2 0 0 3 3 2 9 *